OXFORD MATHEMATICAL MONOGRAPHS

Editors

I. G. MACDONALD   H. MCKEAN   R. PENROSE

OXFORD MATHEMATICAL MONOGRAPHS

Semigroups of Linear Operators and Applications

JEROME A. GOLDSTEIN

Tulane University

OXFORD UNIVERSITY PRESS · *New York*
CLARENDON PRESS · *Oxford*
1985

OXFORD UNIVERSITY PRESS

Oxford   London   New York   Toronto
Delhi   Bombay   Calcutta   Madras   Karachi
Kuala Lumpur   Singapore   Hong Kong   Tokyo
Nairobi   Dar es Salaam   Cape Town
Melbourne   Auckland

and associated companies in
Beirut   Berlin   Ibadan   Mexico City   Nicosia

Library of Congress Cataloging in Publication Data
Goldstein, Jerome A., 1941–
Semigroups of linear operators and applications.
(Oxford mathematical monographs)
Bibliography: p.
Includes index.
1. Semigroups of operators.   2. Linear operators.
I. Title.   II. Series.
QA329.2.G64 1985   512′.2   84-27216
ISBN 0-19-503540-2

Printing (last digit): 9   8   7   6   5   4   3   2   1

Printed in the United States of America

To my parents,
Henrietta and Morris Goldstein

Preface

Like Monsieur Jordan in *Le Bourgeois Gentilhamme*, who found to his great surprise that he had spoken prose all his life, mathematicians are becoming aware of the fact that they have used semigroups extensively if not consciously.

EINAR HILLE

It is difficult to tell when semigroup theory began. The concept was formulated and named in 1904, but in an 1887 paper, Giuseppe Peano [1][†] wrote the system of linear ordinary differential equations

$$du_1/dt = a_{11} u_1 + \cdots + a_{1n} u_n + f_1(t)$$

$$\vdots$$

$$du_n/dt = a_{n1} u_1 + \cdots + a_{nn} u_n + f_n(t)$$

in matrix form as $du/dt = Au + f$ and solved it using the explicit formula

$$u(t) = e^{tA} u(0) + \int_0^t e^{(t-s)A} f(s)\, ds,$$

where $e^{tA} = \sum_{k=0}^{\infty} t^k A^k / k!$. That is, he transformed a complicated problem in one dimension to a formally simple one in higher dimensions and used the ideas of one-variable calculus to solve it. That is the essence of this book.

The spectral theory of self-adjoint and normal operators on Hilbert space is based on the same idea. The notion of a self-adjoint operator is very special, and spectral theory enables one to take more-or-less arbitrary functions of it. In semigroup theory one only wants to take the exponential function of an operator, so one can work in much greater generality. This allows for the possibility of many surprising applications and the extension to a setting of nonlinear operators.

Mathematicians started taking one-parameter semigroup theory seriously in the 1930s. Perhaps its development became inspired when it was realized that the theory had immediate applications to partial differential equations, Markov processes, and ergodic theory. In 1948, Einar Hille published his monograph *Functional Analysis and Semi-Groups* in the American Mathematical Society Colloquium Series. The theory continued to develop rapidly in the fifties, thanks largely to Ralph Phillips; Hille's monograph then evolved into the Hille-Phillips book the same title, which makes substantial additions and deletions to the material in Hille's original book. The Hille-Phillips book, together with Part I of the three-volume series by Dunford and Schwartz, served as a Bible for my generation, the students of the sixties.

[†] I am indebted to Eugenio Sinestrari for this reference.

In 1970 I wrote a short set of lecture notes for a course on semigroups of linear operators at Tulane University. The most striking feature of those notes was how disjoint they were from Hille-Phillips. Post-1957 results (such as the Neveu-Trotter-Kato approximation theorem and the Chernoff product formula) played a prominent role, and simple transparent proofs had been found for many of the older results. The main flaw in those notes was that they did not begin to indicate the wide scope of applications of the simple theory. In 1972, I wrote another set of notes for an analogous course on nonlinear semigroups. Some years later I accepted Gian-Carlo Rota's suggestion to expand the lecture notes, linear and nonlinear, into a book, with a great emphasis on the applications.

This project turned out to be bigger and much more complicated and time consuming than I anticipated. The result is two volumes: the present one on linear theory and applications and a forthcoming nonlinear one.

The emphasis is on motivation, heuristics, and applications. It is hoped (and planned) that this work will be of use to graduate students and professionals in science and engineering as well as mathematics. All the main results are well-known, but several of the proofs are new. An effort was made to solve some nontrivial initial value problems for parabolic and hyperbolic differential equations without doing the hard work associated with elliptic theory. The reason is pedagogical; we wish to get across some of the main ideas involved in Cauchy problems for partial differential equations as an easy consequence of semigroup theory. Besides partial differential equations, other areas of application include mathematical physics (Feynman integrals, scattering theory, etc.), approximation theory, ergodic theory, potential theory, classical inequalities, fluid motion, and so on.

The exercises marked with an asterisk range from difficult to very difficult indeed. Some of them are research results which are incidental to the text but of sufficient interest to deserve to be stated. The bibliography requires some explanation. As a graduate student, I was very impressed with the large list of references in Dunford-Schwartz. It led to many enjoyable evenings of browsing in the library. I thought it would be useful to compile a complete list of references on the theory and applications of operator semigroups. I tried, but I have not succeeded because, as I painfully discovered, the literature is simply too vast for me to keep up with. Nevertheless, a large list was compiled. This list, which covers more than three hundred single-spaced type pages, is cited in the References at the end of the book as Goldstein [24]. To include all of the relevant references here would have made the book unnecessarily long and expensive. Thus many important articles have not been included. Nevertheless, the bibliography presented here contains a fairly substantial and representative sampling of the literature. This should help those readers interested in learning more about the theory and applications than the text presents.

We use the Halmos symbols iff for "if and only if" and ■ to signify the end of a proof.

Over the years, I have had various opportunities to lecture on the material in this book and in the forthcoming one. For their kind invitations or helpful comments or encouragement (or usually all three), I thank Geraldo Avila and

Djairo de Figueiredo (Brasília), Luiz Adauto Medeiros and Gustavo Perla-Menzala (Rio de Janeiro), Dick Duffin and Vic Mizel (Carnegie-Mellon), M. M. Rao (California-Riverside), David Edmunds and Eduard Fraenkel (Sussex), John Erdos (Kings-London), Rosanna Villella-Bressan (Padova), and my Tulane colleagues, Tom Beale, Ed Conway, Karl Hofmann, and Steve Rosencrans. I thank Brian Davies, Frank Neubrander, Simeon Reich, and Eric Schechter for correcting errors in the typescript. I thank Gian-Carlo Rota for his suggestion to write this book and for his encouragement. I record my admiration of Haim Brezis, the late Einar Hille, Tosio Kato, Peter Lax, Ralph Phillips, and Kôsaku Yosida for publishing such beautiful articles and for being constant sources of inspiration. I thank Susan Lam who typed the manuscript beautifully and efficiently. I gratefully acknowledge the partial support of the National Science Foundation. Finally, I thank my wife, Liz, and my children, David and Devra, for putting up with me and this project for all these years.

New Orleans   J. A. G.
December, 1983

Contents

Semigroups of Linear
Operators and Applications

Chapter 0

A Heuristic Survey of the Theory and Applications of Semigroups of Operators

The evolution of a physical system in time is usually described by an initial value problem for a differential equation. (The differential equations can be ordinary or partial, and mixed initial value-boundary value problems are included.) The general setup is as follows. Let $u(t)$ describe the state of some physical system at time t. Suppose that the time rate of change of $u(t)$ is given by some function A of the state of the system $u(t)$. The initial data $u(0) = f$ is also given. Thus

$$\left.\begin{aligned}\frac{d}{dt}u(t) &= A[u(t)] \qquad (t \geq 0),\\[2mm] u(0) &= f.\end{aligned}\right\}\tag{0.1}$$

(For short $du/dt = Au$, $u(0) = f$.)

First of all we must make sense out of

$$du(t)/dt = \lim_{h \to 0} h^{-1}[u(t+h) - u(t)].$$

The function u takes values in a set $\mathcal{X}$. In order for $u(t+h) - u(t)$ to make sense, $\mathcal{X}$ is taken to be a vector space. In order that limits make sense in $\mathcal{X}$, $\mathcal{X}$ is taken to be a Banach space. (More generally, $\mathcal{X}$ could be a topological vector space or a differentiable manifold. But the desire to present a clean and complete theory with lots of applications in a reasonable number of pages led us to omit any setting more general than a Banach space.)

A is an operator (i.e. a function) from its domain $\mathcal{D}(A)$ in $\mathcal{X}$ to $\mathcal{X}$. The equation $du/dt = Au$ is interpreted to mean that $u(t)$ belongs to $\mathcal{D}(A)$ and that

$$\lim_{h \to 0} \|h^{-1}[u(t+h) - u(t)] - A[u(t)]\| = 0,$$

where $\|\cdot\|$ denotes the norm in $\mathcal{X}$.

Here are three examples.

Example 1.   Let Ω be a bounded domain in n-dimensional Eudidean space $\mathbb{R}^n$ and let $\partial\Omega$ denote the (nice) boundary of Ω. Let $\Delta = \sum_{i=1}^{n} \partial^2/\partial x_i^2$ denote the Laplacian. Consider the following classical mixed initial-boundary value problem for the heat equation. We seek a function $w = w(t,x)$, defined for $0 \leq t < \infty$, $x \in \bar{\Omega} = \Omega \cup \partial\Omega$, such that

$$\left.\begin{aligned}\frac{\partial w}{\partial t} &= \Delta w \qquad && \text{for } (t,x) \in [0,\infty[\,\times\,\Omega,\\[2mm] w(0,x) &= f(x) && \text{for } x \in \Omega,\\[2mm] w(t,x) &= 0 && \text{for } x \in \partial\Omega, t \geq 0.\end{aligned}\right\}\tag{0.2}$$

3

(For consistency we should have $f(x) = 0$ for $x \in \partial\Omega$.) Write $u(t) = w(t,\cdot)$, regarded as a function of x, and take $\mathscr{X}$ to be a space of functions on Ω, e.g. $L^p(\Omega)$ for some $p \geq 1$ or $C(\bar{\Omega})$, the continuous functions on the closure of Ω. The derivatives du/dt and $\partial w/\partial t$ are both limits of the difference quotient $h^{-1}[w(t+h,x) - w(t,x)]$, the first limit being in the sense of the norm of $\mathscr{X}$ and the second limit being a pointwise one. Even so, we can formally identify $\partial w/\partial t$ with du/dt. Clearly the functions denoted by f in both (0.1) and (0.2) can be identified with each other. To define A we take $\mathscr{X} = C(\bar{\Omega})$ for definiteness. Let $\mathscr{D}(A) = \{v \in C(\bar{\Omega}) : v$ is twice differentiable, $\Delta v \in C(\bar{\Omega})$, and $v(x) = 0$ for each $x \in \partial\Omega\}$. Define $Av = \Delta v$ for $v \in \mathscr{D}(A)$. Equations (0.2) are thus written in the form (0.1). Note that the boundary condition of (0.2) is absorbed into the domain of definition of the operator A and into the requirement that $u(t) \in \mathscr{D}(A)$ for all $t \geq 0$.

Example 2.   Consider the initial value problem for the wave equation

$$\frac{\partial^2 w}{\partial t^2} = \Delta w \qquad \text{for } (t,x) \in [0,\infty[\times \mathbb{R}^n,$$

$$w(0,x) = f_1(x) \quad \text{for } x \in \mathbb{R}^n, \tag{0.3}$$

$$\frac{\partial w}{\partial t}(0,x) = f_2(x) \quad \text{for } x \in \mathbb{R}^n.$$

For $\mathscr{X}$ we take a space of pairs of functions on $\mathbb{R}^n$. We set

$$u(t) = \begin{pmatrix} w(t,\cdot) \\ \dfrac{\partial w}{\partial t}(t,\cdot) \end{pmatrix}, \ f = \begin{pmatrix} f_1 \\ f_2 \end{pmatrix}, \text{ and } A = \begin{pmatrix} 0 & 1 \\ \Delta & 0 \end{pmatrix}$$

i.e.

$$A\begin{pmatrix} v_1 \\ v_2 \end{pmatrix} = \begin{pmatrix} v_2 \\ \Delta v_1 \end{pmatrix}.$$

Then formally (0.3) becomes (0.1).

Example 3.   Consider the initial value problem for the one-dimensional Hamilton-Jacobi equation

$$\left.\begin{array}{ll} \dfrac{\partial w}{\partial t} + F\left(\dfrac{\partial w}{\partial x}\right) = 0 & (t \geq 0, x \in \mathbb{R}), \\[2mm] w(0,x) = f(x) & (x \in \mathbb{R}). \end{array}\right\} \tag{0.4}$$

We take $\mathscr{X}$ to be a space of functions on $\mathbb{R}$ and set $u(t) = w(t,\cdot)$, $Av = -F(dv/dx)$. Then (0.4) formally becomes (0.1). Note that the operator A of this example is nonlinear, in contrast to the two preceding linear examples.

We return to the notion of a physical system which we imagine being housed in our (imaginary) experimental laboratory. In a *well-posed* physical experiment something happens, only one thing happens, and repeating the experiment with only small changes in the initial conditions or physical parameters produces only small changes in the outcome of the experiment. This suggests that if the initial value problem (0.1) is to correspond to a well-posed physical experiment, then we must establish an existence theorem, a uniqueness theorem, and a (stability) theorem which says that the solution depends continuously on the ingredients of the problem, namely the initial condition f and the operator A.

Suppose (0.1) is well-posed in the above (informal) sense. Let $T(t)$ map the solution $u(s)$ at time s to the solution $u(t + s)$ at time $t + s$. The assumption that A does not depend on time implies that $T(t)$ is independent of s; the physical meaning of this is that the underlying physical mechanism does not depend on time.

The solution $u(t + \tau)$ at time $t + \tau$ can be computed as $T(t + \tau)f$ or, alternatively, we can solve for $u(\tau) = T(\tau)f$, take this as initial data, and t units of time later the solution becomes $u(t + \tau) = T(t)(T(\tau)f)$. The uniqueness of the solution implies the semigroup property

$$T(t + \tau) = T(t)T(\tau) \qquad t,\tau > 0.$$

Also, $T(0) = I =$ the identity operator (this means that the initial condition is assumed), $t \to T(t)f$ is differentiable on $[0,\infty[$ [and $(d/dt)T(t)f = AT(t)f$ so that $u(t) = T(t)f$ solves (0.1)], and each $T(t)$ is a continuous operator on $\mathscr{X}$. (This reflects the continuous dependence of $u(t)$ on f.) The initial data f should belong to the domain of A, which is assumed to be dense in $\mathscr{X}$. Finally, each $T(t)$ is linear if A is linear.

We are thus led to the notion of a strongly continuous one-parameter semigroup of bounded linear operators on a Banach space $\mathscr{X}$. Such a semigroup is called a (C_0) semigroup; this terminology, introduced by Hille, has become standard. The definition is as follows. A family $T = \{T(t) : 0 \le t < \infty\}$ of linear operators from $\mathscr{X}$ to $\mathscr{X}$ is called a (C_0) semigroup if

(i) $\|T(t)\| < \infty$ (i.e. $\sup\{\|T(t)f\| : f \in \mathscr{X}, \|f\| \le 1\} < \infty$) for each $t \ge 0$,
(ii) $T(t + s)f = T(t)T(s)f$ for all $f \in \mathscr{X}$ and all $t,s \ge 0$,
(iii) $T(0)f = f$ for all $f \in \mathscr{X}$,
(iv) $t \to T(t)f$ is continuous for $t \ge 0$ for each $f \in \mathscr{X}$.

T is called a (C_0) contraction semigroup if, in addition,

(v) $\|T(t)f\| \le \|f\|$ for all $t \ge 0$ and all $f \in \mathscr{X}$,
       i.e. $\|T(t)\| \le 1$ for each $t \ge 0$.

Roughly speaking, for most purposes it is enough to consider only (C_0) contraction semigroups. (This will be fully explained in Section 2 of Chapter I.)

Let T be a (C_0) semigroup. Define its *generator* (or *infinitesimal generator*) A by the equation

$$Af = \lim_{t \to 0} \frac{T(t)f - f}{t}.$$

where f is in the domain of A iff this limit exists. Formally, the semigroup property suggests that $T(t) = $ "e^{tA}" where $A = (d/dt)T(t)|_{t=0}$. This also suggests that the solution of (0.1) is given by $u(t) = T(t)f$, where T is the semigroup generated by A. The following result is basic.

THEOREM I (well-posedness theorem).   *The initial value problem* (0.1) (*with A linear*) *is "well-posed" iff A is the generator of a* (C_0) *semigroup T. In this case the unique solution of* (0.1) *is given by* $u(t) = T(t)f$ *for f in the domain of A.*

See Chapter II, Theorem 1.2 and Exercise 1.5.4 for precise versions of this.

The obvious question that arises at this point is: which operators A generate (C_0) semigroups? For simplicity we work with a (C_0) contraction semigroup T. If A is the generator of T, think of $T(t)$ as e^{tA}. The formula

$$\frac{1}{\lambda - A} = \int_0^\infty e^{-\lambda t} e^{tA}\, dt,$$

which is valid when A is a number and $\lambda > \text{Re}(A)$, suggests the operator version

$$(\lambda I - A)^{-1}f = \int_0^\infty e^{-\lambda t} T(t)f\, dt, \tag{0.5}$$

which turns out to be valid for all $\lambda > 0$ and all $f \in \mathcal{X}$; here I is the identity operator on $\mathcal{X}$. The f is there to make the integrand nice, namely, continuous and bounded in norm by the integrable function $\|f\|e^{-\lambda t}$. The estimate

$$\left\| \int_0^\infty e^{-\lambda t} T(t)f\, dt \right\| \le \int_0^\infty e^{-\lambda t} \|T(t)f\|\, dt$$

$$\le \int_0^\infty e^{-\lambda t}\|f\|\, dt = \|f\|/\lambda$$

suggests that:

$$\left.\begin{array}{l} \text{for each } \lambda > 0, \\[1.5ex] \lambda I - A \text{ maps the domain of } A \text{ onto } \mathcal{X} \\[1.5ex] \text{and } \|(\lambda I - A)^{-1}f\| \le \dfrac{1}{\lambda}\|f\| \text{ for all } f \in \mathcal{X}. \end{array}\right\} \tag{0.6}$$

THEOREM II (Hille-Yosida generation theorem).   *A linear operator A generates a* (C_0) *contraction semigroup iff the domain of A is dense in* $\mathcal{X}$ *and* (0.6) *holds.*

One can recover the semigroup T from the generator A by inverting the Laplace transform (0.5) or by other methods, such as the formula

$$T(t)f = \lim_{n \to \infty} \left(I - \frac{t}{n} A \right)^{-n} f;$$

note that $(I - \alpha A)^{-1} = \lambda(\lambda I - A)^{-1}$ where $\lambda = 1/\alpha$.

The important implications in Theorems I and II are: (i) A densely defined operator A satisfying (0.6) generates a (C_0) semigroup. (ii) If A generates a (C_0) semigroup T, then the initial value problem (0.1) is well-posed and is governed by T. In other words, we solve (0.1) by solving equations of the form $\lambda h - Ah = g$ and getting a solution h satisfying the estimate $\|h\| \le \|g\|/\lambda$; this should be true for all $g \in \mathcal{X}$ and $\lambda > 0$.

The semigroup method has certain limitations. To illustrate this, consider the heat equation initial value problem

$$\frac{\partial u}{\partial t} = \frac{\partial^2 u}{\partial x^2} \qquad (x \in \mathbb{R}, t \geq 0) \qquad u(0,x) = f(x)(x \in \mathbb{R}).$$

One can show that this is governed by a (C_0) contraction semigroup on $L^P(\mathbb{R})$, $1 \leq p < \infty$ or on the space of bounded uniformly continuous functions on $\mathbb{R}$. It is well-known that another solution of this problem is given by

$$v(t,x) = u(t,x) + \sum_{k=0}^{\infty} g^{(k)}(t) x^{2k}/(2k)!$$

where u is the semigroup solution and $g(t) = e^{-1/t^2}$. Here $v(t,x)$ does not grow more slowly then e^{x^2} as $x \to \pm\infty$ for fixed $t > 0$. Thus, while the semigroup approach gives the unique solution in a certain Banach space context, there may be other solutions (if one widens one's notion of solution).

The next two results are samples from the theory of (C_0) semigroups.

THEOREM III (perturbation theorem). *If A generates a (C_0) semigroup and if B is a bounded linear operator on $\mathcal{X}$, then $A + B$ generates a (C_0) semigroup.*

THEOREM IV (approximation theorem). *For $n = 0, 1, 2, \ldots$ let A_n generate a (C_0) contraction semigroup T_n. Then*

$$\lim_{n \to \infty} (\lambda I - A_n)^{-1} f = (\lambda I - A_0)^{-1} f$$

holds for all $\lambda > 0$ and all $f \in \mathcal{X}$ iff

$$\lim_{n \to \infty} T_n(t)f = T_0(t)f$$

holds for all $t > 0$ and all $f \in \mathcal{X}$. Sufficient for this is that $\mathscr{D}(A_0) \subset \mathscr{D}(A_n)^\dagger$ and

$$\lim_{n \to \infty} A_n f = A_0 f$$

holds for all f in the domain of A_0.

The perturbation theorem says, roughly, that once we know how to solve (0.1), we can alter A in certain way and solve the resulting problem. The approximation theorem says that the solution of (0.1) depends continuously on A.

Let A be an operator like the one in Example 1, e.g. $Au(x) = \sum_{i,j=1}^{n} a_{ij}(x) \partial^2 u/\partial x_i \partial x_j$ and $u(x) = 0$ for $x \in \partial\Omega$ whenever u is in the domain of A. After one shows that A generates a (C_0) semigroup, a variant of Theorem III can be applied with $Bu(x) = a_0(x)u + \sum_{i=1}^{n} a_i(x) \partial u/\partial x_i$. Thus when we add certain variable coefficient differential operators to constant coefficient operators which are generators, the resulting sums are generators. Theorem IV says that for A, B as above, the unique solution of $du/dt = (A + B)u, u(0) = f$ depends continuously on the coefficients a_i, a_{jk} ($0 \leq i \leq n, 1 \leq j,k \leq n$).

---

† $\mathscr{D}(A_n)$ denotes the domain of A_n.

We emphasize that solving (0.1) by showing that A generates a (C_0) semigroup produces a number of bonuses besides existence, uniqueness, and continuous dependence on the data f — namely continuous dependence on A. Moreover, we can perturb A by a large class of allowable perturbations and still get a semigroup generator; thus (0.1) is automatically imbedded in a large class of well-posed problems. Many problems of the form (0.1) can be solved by a variety of techniques, but few other methods have the advantages just described.

The theory of (C_0) semigroups has many applications to problems that are not concerned with the solution of differential equations. A consequence of the Hille-Yosida theorem (Theorem II) is Stone's theorem: *Let A be a densely defined operator on a complex Hilbert space. Then A and $-A$ both generate (C_0) contraction semigroups iff A generates a (C_0) group of unitary operators iff iA is self-adjoint.*

As a consequence of this we deduce the spectral theorem, i.e. the structure theorem for self-adjoint operators. One version of this theorem concludes that *every self-adjoint operator on a complex Hilbert space is* (unitarily equivalent to) *an operator on an L^2 space given by multiplication by a measurable real-valued function.* Other results follow. For instance, perturbation theorems for self-adjoint operators follow from corresponding results for generators.

Concerning applications to quantum physics, we use semigroup theory to explain and justify the Feynman path integral formula for the wave function of nonrelativistic quantum mechanics, and we show that the Klein-Gordon equation has the correct nonrelativistic limit as the speed of light becomes infinite. We also discuss scattering theory (in some detail), asymptotic equipartition of energy, and the ergodic theorem.

Semigroups of operators are intimately connected with probability theory, particularly with Markov processes. We sketch a number of aspects of this relationship, including potential theory. There are other connections with probability theory as well. We illustrate this by proving the central limit theorem as a consequence of the approximation theorem.

Other applications include connections with analyticity, various kinds of approximation processes (e.g. difference schemes), and classical inequalities.

The solution of the inhomogeneous problem

$$du/dt = Au + g(t) \qquad u(0) = f$$

is given by

$$u(t) = T(t)f + \int_0^t T(t-s)g(s)\,ds,$$

where T is the (C_0) semigroup generated by A. This formula can be "derived" by pretending that A and $g(t)$ are real numbers and $T(t) = e^{tA}$. This "variation of parameters" formula is also familiar from the theory of systems of first-order linear ordinary differential equations. It follows that certain nonlinear equations can be solved by the above device coupled with successive approximations. More precisely, to solve

$$du/dt = Au + g[u(t)] \qquad u(0) = f, \tag{0.7}$$

convert it to the integral equation

$$u(t) = T(t)f + \int_0^t T(t - s)g[u(s)]\, ds,$$

set $u_0(t) \equiv f$,

$$u_{n+1}(t) = T(t)f + \int_0^t T(t - s)g[u_n(s)]\, ds \qquad n \geq 0,$$

and hope that $u_n(t)$ converges to the desired solution $u(t)$, at least for t in some interval containing the origin. This technique enables us to treat the Navier-Stokes system of equations (of fluid dynamics) for small values of t.

Problems of the form (0.7) are often called semilinear; they are nonlinear perturbations of linear problems and are not "fully" nonlinear. We next turn our attention to fully nonlinear problems.

At this point we make the startling observation, discovered in 1967, that in the theory of (C_0) semigroups of linear operators, *linearity is irrelevant*. This statement, which is outrageous, is (to a large extent) true. To justify it, we begin as follows. For B, an operator, linear or not, with domain and range contained in a Banach space $\mathcal{X}$, let the Lipschitz seminorm of B be

$$\|B\|_{\text{Lip}} = \sup\{\|Bf - Bg\|/\|f - g\|\}$$

the supremum being over all f, g in the domain of B with $f \neq g$. When B is linear, $\|B\|_{\text{Lip}}$ is just the usual operator norm of B. Return now to the initial value problem (0.1) and suppose A is nonlinear. Replace (0.1) by the difference equation

$$\frac{1}{\epsilon}[u_\epsilon(t) - u_\epsilon(t - \epsilon)] = A[u_\epsilon(t)] \qquad t \geq 0$$

with initial condition $u_\epsilon(s) = f$ for $-\epsilon \leq s \leq 0$. The solution is given by

$$u_\epsilon(t) = (I - \epsilon A)^{-1}[u_\epsilon(t - \epsilon)],$$

or $u_\epsilon(t) = (I - t/n(A))^{-n}f$ if $\epsilon = t/n$. This leads us to the assumption:

$$\left.\begin{array}{l}\text{for each } \alpha > 0, \\[4pt] \text{the range of } (I - \alpha A) \text{ is all of } \mathcal{X} \\[4pt] \text{and } \|(I - \alpha A)^{-1}\|_{\text{Lip}} \leq 1.\end{array}\right\} \qquad (0.8)$$

When A is linear, the identity $(I - \alpha A)^{-1} = \lambda(\lambda I - A)^{-1}$ with $\alpha = 1/\lambda$ shows that (0.8) is equivalent to (0.6). Before stating the nonlinear version of Theorem II, we need one additional notion. Let A be a possibly multivalued function, which we identify with its graph in $\mathcal{X} \times \mathcal{X}$. Then (0.8) means that for $\alpha > 0$,

$$(I - \alpha A)^{-1} = \left\{(g,f) \in \mathcal{X} \times \mathcal{X} : \left(f, \frac{1}{\alpha}(f - g)\right) \in A\right\}$$

is the graph of a contraction, i.e. a single-valued function defined on $\mathcal{X}$ of Lipschitz seminorm at most one.

As an example, let $A : \mathbb{R} \to \mathbb{R}$ be a single-valued real function of a real variable. Then $\|(I - \alpha A)^{-1}\|_{\mathrm{Lip}} \leq 1$ for all $\alpha > 0$ iff A is monotone nonincreasing. This is easy to verify; we assume it. Then the range of $I - \alpha A$ is all of $\mathbb{R}$ iff A is continuous. To make (0.8) hold when A has discontinuities, i.e. jumps, let A be multivalued at the jumps so that if A jumps at x, $A(x)$ can be redefined as the closed interval $[A(x - 0), A(x + 0)]$. Then (0.8) holds. Thus the multivaluedness of A has to do with the range condition in (0.8). Also, if A is a single-valued function which satisfies the second part of (0.8) and the range of $I - \alpha A$ is dense in $\mathscr{X}$, let $\bar{A}$ denote the function whose graph is the closure of the graph of A. Then $\bar{A}$ satisfies (0.8), and $\bar{A}$ may be multivalued.

We are now ready to state a generation-type theorem for nonlinear contraction semigroups. A *strongly continuous contraction semigroup* on $\mathscr{X}$ is family $T = \{T(t) : t \geq 0\}$ of operators on $\mathscr{X}$ to $\mathscr{X}$ such that for $t, \tau \geq 0$, and $f \in \mathscr{X}$, $T(t + \tau)f = T(t)T(\tau)f$, $T(0)f = f$, $s \to T(s)f$ is continuous on $[0,\infty[$, and $\|T(t)\|_{\mathrm{Lip}} \leq 1$.

THEOREM II' (Crandall-Liggett theorem).   *Let A satisfy (0.8) and be densely defined. Then for each $f \in \mathscr{X}$ and $t \geq 0$,*

$$T(t)f = \lim_{n \to \infty} \left(I - \frac{t}{n}A \right)^{-n} f$$

exists and defines a strongly continuous contraction semigroup.

Here "contraction" means $\|T(t)\|_{\mathrm{Lip}} \leq 1$ for each $t \geq 0$. Theorem II' thus generalizes the important half of Theorem II. There is also an analogue of Theorem I.

THEOREM I'.   *Let A satisfy (0.8). Then the initial value problem (0.1) (with f in the domain of A) is "well-posed."*

This is the analogue of the important half of Theorem I. If T is the semigroup of Theorem II', the solution of (0.1) is given by $u(t) = T(t)f$. When $\mathscr{X}$ is reflexive (and more generally when $\mathscr{X}$ has the Radon-Nikodym property), $du/dt \in A(u)$ holds almost everywhere, for f in the domain of A. However, u may be nowhere differentiable when $\mathscr{X}$ is nonreflexive. In this case, the sense in which u is a solution is a very general sense, which we do not explain here, except to say that it involves a family of inequalities. The notion of well-posedness in Theorem I' differs somewhat from the corresponding notion in Theorem I. But in each case the solution, which exists in a certain sense, is unique and depends continuously on A and f in a suitable sense.

The converse of Theorem II' is false in general. However, when $\mathscr{X}$ is a Hilbert space, the converse of Theorem II' holds. Moreover, in this case, the infinitesimal generator [i.e. $T'(0)$] is a restriction of A and determines both A and T uniquely.

A variant of Theorem II' holds when A is not densely defined. Versions of Theorem III and IV also hold in the nonlinear case. Thus the reasons for wanting to solve (0.1) by semigroup methods in the linear case apply equally well in the nonlinear case.

We shall solve a variety of nonlinear parabolic and hyperbolic partial differential equations by semigroup methods. We shall also indicate some applications to probability theory. However, the nonlinear theory, which is still young, is not yet as rich in applications as the linear theory.

We close this section with a brief discussion of the one-dimensional *conservation law*

$$\frac{\partial u}{\partial t} + \frac{\partial}{\partial x}(\phi(u)) = 0 \qquad (x \in \mathbb{R}, t \geq 0) \tag{0.9}$$

where $\phi: \mathbb{R} \to \mathbb{R}$ is a smooth function. In general, the term *conservation law* describes a situation in which a change in the total amount of a physical quantity in a region is due to the flux of that quantity across the boundary of the region. Mathematically, this is described by an equation which says that the divergence of a certain vector field is zero; (0.9) is of this form. Equations such as (0.9) occur in gas dynamics and other areas. For a specific example involving traffic flow on a long road (identified with $\mathbb{R}$), let $u(t,x)$ be the traffic density at the point x on the road at time t. The rate at which cars flow past x at time t is assumed to be a function ϕ of the traffic density; then one can "derive" (0.9).

Let u be a classical solution of (0.9). Then on the (characteristic) curves $x = x(t)$ defined by the ordinary differential equation

$$dx(t)/dt = \phi'\{u[t,x(t)]\}, \tag{0.10}$$

we quickly calculate $(d/dt)\{u[t,x(t)]\} = 0$, i.e. u is constant along these curves. By (0.10), then, these curves have constant slope and so are straight lines. One therefore attempts to solve the initial value problem (0.9) together with the initial condition

$$u(0,x) = f(x) \tag{0.11}$$

as follows. Pick (t_0,x_0). Find the characteristic line through (t_0,x_0); say it hits the x-axis at $(0,x_1)$. Then $u(t_0,x_0) = f(x_1)$. Unfortunately, it can (and usually does) happen that the characteristic lines intersect, and so the solution is constrained to take on distinct values at the same point. The conclusion is that we cannot expect to have classical solutions of (0.9) defined for all $t \geq 0$; we must allow for discontinuous solutions if we want solutions to exist globally in time (and we do). We can define solutions by multiplying (0.9) by a smooth function g and integrating by parts. We call a locally bounded measurable function u a *weak solution* of (0.9), (0.11) if

$$\int_0^\infty \int_{-\infty}^\infty \left(u \frac{\partial g}{\partial t} + \phi(u) \frac{\partial g}{\partial x} \right) dx\, dt + \int_{-\infty}^\infty f(x)g(0,x)\, dx = 0$$

holds for all smooth g which vanish outside of a bounded subset of $\{(t,x) : t \geq 0, x \in \mathbb{R}\}$. Unfortunately, while an existence theorem is valid in this context, uniqueness fails. The "physically correct" solution must be singled out. This solution turns out to be $\lim_{\epsilon \to 0} u_\epsilon(t,x)$ where

$$\frac{\partial u_\epsilon}{\partial t} + \frac{\partial}{\partial x}[\phi(u_\epsilon)] = \epsilon \frac{\partial^2 u_\epsilon}{\partial x^2} \qquad u_\epsilon(0,x) = f(x).$$

This must be incorporated into the notion of solution for (0.9), (0.11), and the most convenient way of doing this is to follow Kružkov and define a notion of solution of (0.9), (0.11) involving a family of inequalities.

Since we require the abstract theory of nonlinear semigroups to govern (0.9), (0.11) as a special case, the notion of solution of (0.1) must be general enough to cover the correct notion of solution of (0.9), (0.11). The space in which (0.9), (0.11) is governed by a strongly continuous contraction semigroup turns out to be the nonreflexive space $L^1(\mathbb{R})$. The correct space for the Hamilton-Jacobi equation involves the supremum norm. Thus, despite the fact that nonlinear semigroup theory is more difficult in nonreflexive spaces, we must develop it in this context to include many of the significant applications.

Chapter I

Semigroups of Linear Operators

1. Notation; Closed Operators

1.1. Let $\mathcal{X}, \mathcal{Y}$ be real or complex Banach spaces. Let $\|\cdot\|$ (or sometimes $\|\cdot\|_{\mathcal{X}}$) denote the norm in $\mathcal{X}$. $\mathbb{K}$ will denote the underlying scalar field. Thus $\mathbb{K} = \mathbb{R}$, the real numbers, or $\mathbb{K} = \mathbb{C}$, the complex numbers. Let $\mathbb{R}^+ = [0, \infty[$ be the non-negative real numbers, let $\mathbb{N} = \{1, 2, \ldots\}$ be the positive integers, and let $\mathbb{N}_0 = \{0, 1, 2, \ldots\}$ be the nonnegative integers. $\mathcal{B}(\mathcal{X}, \mathcal{Y})$ is the space of all bounded linear operators from $\mathcal{X}$ to $\mathcal{Y}$. $\mathcal{B}(\mathcal{X}) = \mathcal{B}(\mathcal{X}, \mathcal{X})$. "$A$ is an operator on $\mathcal{X}$ to $\mathcal{Y}$" means A is a linear operator from its domain $\mathcal{D}(A) \subset \mathcal{X}$ to $\mathcal{Y}$. "A is an operator on $\mathcal{X}$" means A is an operator on $\mathcal{X}$ to $\mathcal{X}$. An operator A on $\mathcal{X}$ to $\mathcal{Y}$ is *closed* if its graph $\mathcal{G}(A) = \{(f, Af) : f \in \mathcal{D}(A)\}$ is a closed subspace of $\mathcal{X} \times \mathcal{Y}$, or equivalently, if $f_n \in \mathcal{D}(A)$, $f_n \to f$, and $Af_n \to g$ imply $f \in \mathcal{D}(A)$ and $Af = g$.

Recall the following basic facts about closed operators. Let A on $\mathcal{X}$ to $\mathcal{Y}$ be closed. Then

- (i) $A + B$ [with domain $\mathcal{D}(A)$] is closed for each $B \in \mathcal{B}(\mathcal{X}, \mathcal{Y})$,
- (ii) A^{-1} is closed if A is injective (i.e. one-to-one),
- (iii) (closed graph theorem) A is bounded if $\mathcal{D}(A)$ is closed in $\mathcal{X}$,
- (iv) a bounded operator is closed iff its domain if closed.

An operator A on $\mathcal{X}$ to $\mathcal{Y}$ is *closable* if the closure of its graph $\overline{\mathcal{G}(A)}$ is a graph, i.e. $(0, y) \in \overline{\mathcal{G}(A)}$ implies $y = 0$. Then $\overline{\mathcal{G}(A)}$ is the graph of a closed operator, which is called the *closure* of A and is denoted by $\bar{A}$.

1.2. Let A be an operator on $\mathcal{X}$. The *resolvent set* of A is $\rho(A) = \{\lambda \in \mathbb{K} : \lambda I - A : \mathcal{D}(A) \to \mathcal{X}$ is bijective and

$$(\lambda I - A)^{-1} \in \mathcal{B}(\mathcal{X})\}.$$

$\rho(A) \neq \emptyset$ implies A is closed. If A is closed, then

$$\rho(A) = \{\lambda \in \mathbb{K} : \lambda I - A : \mathcal{D}(A) \to \mathcal{X} \text{ is bijective}\}$$

by the closed-graph theorem. Here I is the identity operator on $\mathcal{X}$. We shall usually write $\lambda - A$ for $\lambda I - A$. $(\lambda - A)^{-1}$ is called the *resolvent (operator)* of A. $\sigma(A) = \mathbb{K} \backslash \rho(A)$ is the *spectrum* of A.

1.3. LEMMA (the resolvent identity). *Let A be closed on $\mathcal{X}$. Then for all $\lambda, \mu \in \rho(A)$,*

$$(\lambda - A)^{-1} - (\mu - A)^{-1} = (\mu - \lambda)(\lambda - A)^{-1}(\mu - A)^{-1}; \qquad (1.1)$$

hence $(\lambda - A)^{-1}$ *and* $(\mu - A)^{-1}$ *commute.*

Proof. For $\lambda, \mu \in \rho(A)$, $f \in \mathscr{D}(A)$,

$$(\lambda - A)[(\lambda - A)^{-1} - (\mu - A)^{-1}](\mu - A)f = (\mu - A)f - (\lambda - A)f$$
$$= (\mu - \lambda)f;$$

and (1.1) follows. The last part of the lemma follows by interchanging λ and μ in (1.1). ∎

1.4. For functions $h: J \to \mathscr{X}$ where J is an interval in $\mathbb{R}$, notions such as dh/dt, $\int_a^b h(t)\, dt$ are defined in the same way as in the case $\mathscr{X} = \mathbb{R}$ or $\mathbb{C}$; the limits of the difference quotients or of the Riemann sums are to be taken in the norm topology of $\mathscr{X}$. All the usual theorems hold; for instance, if $h \in C([a,b],\mathscr{X})$ (i.e. if $h: [a,b] \to \mathscr{X}$ is continuous), then $\int_a^b h(t)\, dt$ exists.

1.5. LEMMA. *Let A be a closed operator on $\mathscr{X}$, $h \in C([a,b],\mathscr{X})$, Range $(h) \subset \mathscr{D}(A)$, and $Ah \in C([a,b],\mathscr{X})$. Then $\int_a^b h(t)\, dt \in \mathscr{D}(A)$ and*

$$A \int_a^b h(t)\, dt = \int_a^b Ah(t)\, dt.$$

Proof. If $\sum_{i=1}^n h(t_i')(t_i - t_{i-1})$ is a typical Riemann sum for $\int_a^b h$, then $\sum_{i=1}^n Ah(t_i')(t_i - t_{i-1})$ is a typical Riemann sum for $\int_a^b Ah$; moreover

$$A \sum_{i=1}^n h(t_i')(t_i - t_{i-1}) = \sum_{i=1}^n Ah(t_i')(t_i - t_{i-1}).$$

The result follows since A is closed. ∎

2.   The Hille-Yosida Generation Theorem

2.1. DEFINITION.   A (C_0) *semigroup* T on $\mathscr{X}$ is a family of operators $T = \{T(t): t \in \mathbb{R}^+\} \subset \mathscr{B}(\mathscr{X})$ satisfying

   (i) $T(t)T(s) = T(t + s)$ for each $t, s \in \mathbb{R}^+$,
   (ii) $T(0) = I$,
   (iii) $T(\cdot)f \in C(\mathbb{R}^+,\mathscr{X})$ (i.e. $T(\cdot)f: \mathbb{R}^+ \to \mathscr{X}$ is continuous) for each $f \in \mathscr{X}$.

2.2. DEFINITION.   A (C_0) *contraction semigroup* T on $\mathscr{X}$ is a (C_0) semigroup T on $\mathscr{X}$ such that for each $t \in \mathbb{R}^+$, $T(t)$ is a contraction, i.e. $\|T(t)\| \le 1$.

2.3. DEFINITION.   Let T be a (C_0) semigroup on $\mathscr{X}$. The *(infinitesimal) generator* A of T is defined by the formula

$$Af = \lim_{t \to 0} \frac{T(t)f - f}{t} = \frac{d}{dt}\, T(t)f \Big|_{t=0},$$

the domain $\mathscr{D}(A)$ of A being the set of all $f \in \mathscr{X}$ for which the limit defined above exists.

2.4. Formally, $u(\cdot) = T(\cdot)f$ solves the initial value problem

$$\frac{du(t)}{dt} = Au(t) \qquad (t \in \mathbb{R}^+), u(0) = f,$$

where A is the generator of T. Thus, from the point of view of solving initial value problems (or abstract Cauchy problems), it is natural to ask: which operators A generate (C_0) semigroups? This section is devoted to answering this question fully.

2.5. PROPOSITION.   *Let $A \in \mathcal{B}(\mathcal{X})$. Then*

$$T = \left\{ T(t) = e^{tA} = \sum_{n=0}^{\infty} \frac{(tA)^n}{n!} : t \in \mathbb{R}^+ \right\}$$

is a (C_0) semigroup satisfying

(iii') $\|T(t) - I\| \to 0$ as $t \to 0$.

Moreover, A is the generator of T. Conversely, if T is a (C_0) semigroup satisfying (iii'), then the generator A of T belongs to $\mathcal{B}(\mathcal{X})$ and $T(t) = e^{tA}$.

Proof.   The direct part is straightforward. For N and M positive integers

$$\left\| \sum_{n=N}^{N+M} \frac{(tA)^n}{n!} \right\| \leq \sum_{n=N}^{N+M} \frac{t^n \|A\|^n}{n!}.$$

Thus the series $\sum_{n=0}^{\infty} (tA)^n/n!$ converges in the uniform operator topology [i.e. in the norm topology of $\mathcal{B}(\mathcal{X})$] to an operator $T(t)$. The semigroup property (i) is an immediate consequence of the (formal power series) fact that

$$\left(\sum_{n=0}^{\infty} \frac{x^n}{n!} \right) \left(\sum_{m=0}^{\infty} \frac{y^m}{m!} \right) = \sum_{p=0}^{\infty} \frac{(x+y)^p}{p!}.$$

$T(0) = I$ is clear, and

$$\|T(t) - I\| = \left\| \sum_{n=1}^{\infty} \frac{t^n A^n}{n!} \right\| \leq \sum_{n=1}^{\infty} \frac{t^n \|A\|^n}{n!} = e^{t\|A\|} - 1 \to 0$$

as $t \to 0$. Similarly,

$$\left\| \frac{T(t) - I}{t} - A \right\| \leq \left\| \sum_{n=2}^{\infty} \frac{t^{n-1} A^n}{n!} \right\| \leq t\|A\|^2 e^{t\|A\|} \to 0$$

as $t \to 0$. Thus A is the generator of T.

A simple proof of the converse part can be based on some simple constructions we shall establish in the proof of the Hille-Yosida theorem. The details will be given in Section 2.11.

The above proposition implies that the (C_0) semigroups satisfying the strong continuity condition 2.1 (iii) but not the uniform continuity condition (iii') are precisely the ones having unbounded generators; these are the interesting (C_0) semigroups.

2.6. HILLE-YOSIDA THEOREM.   *A is the generator of a (C_0) contraction semigroup iff A is closed, densely defined, and for each $\lambda > 0$, $\lambda \in \rho(A)$ and*

$$\|\lambda(\lambda - A)^{-1}\| \leq 1.$$

Proof (*Necessity*).   For each $f \in \mathscr{D}(A)$,

$$\frac{d^+}{dt} T(t)f = \lim_{h \to 0^+} h^{-1}[T(t+h) - T(t)]f = T(t)Af$$

$$= \lim_{h \to 0^+} h^{-1}(T(h) - I)T(t)f;$$

Thus $T(t)[\mathscr{D}(A)] \subset \mathscr{D}(A)$ and

$$\frac{d^+}{dt} T(t)f = AT(t)f = T(t)Af \qquad f \in \mathscr{D}(A).$$

Also, if $t > 0$,

$$\frac{d^-}{dt} T(t)f = \lim_{h \to 0^+} h^{-1}[T(t) - T(t-h)]f$$

$$= \lim_{h \to 0^+} T(t-h)h^{-1}[T(h) - I]f = T(t)Af = AT(t)f.$$

Thus for each $f \in \mathscr{D}(A)$, $T(\cdot)f \in C^1(\mathbb{R}^+, \mathscr{X})^\dagger$ and

$$T(t)f - f = \int_0^t \frac{d}{ds} T(s)f \, ds$$

$$= \int_0^t AT(s)f \, ds = \int_0^t T(s)Af \, ds. \qquad (2.1)$$

To see that $\overline{\mathscr{D}(A)} = \mathscr{X}$, let $f \in \mathscr{X}$ and set $f_t = \int_0^t T(s)f \, ds$. Clearly $\lim_{t \to 0^+} t^{-1}f_t = f$, and

$$h^{-1}[T(h) - I]f_t = h^{-1} \int_t^{t+h} T(s)f \, ds - h^{-1} \int_0^h T(s)f \, ds$$

$$\to T(t)f - f(= Af_t) \text{ as } h \to 0^+;$$

thus $f_t \in \mathscr{D}(A)$ and so $\overline{\mathscr{D}(A)} = \mathscr{X}$. Moreover, we have shown

$$T(t)f - f = A \int_0^t T(s)f \, ds \text{ for all } f \in \mathscr{X}. \qquad (2.2)$$

A is closed, for if $f_n \in \mathscr{D}(A)$, $f_n \to f$, $Af_n \to g$, then

$$t^{-1}[T(t) - I]f_n = t^{-1} \int_0^t T(s)Af_n \, ds \qquad \text{by (2.1)}$$

$$\downarrow n \to \infty \qquad\qquad\qquad \downarrow n \to \infty$$

$$t^{-1}[T(t) - I]f \qquad t^{-1} \int_0^t T(s)g \, ds \to g \qquad \text{as } t \to 0^+.$$

Thus $f \in \mathscr{D}(A)$ and $Af = g$.

Note that for each $\lambda > 0$, $\{e^{-\lambda t}T(t) : t \in \mathbb{R}^+\}$ is a (C_0) contraction semigroup with generator $A - \lambda I$ [with domain $\mathscr{D}(A)$]. Applying (2.2) to this semigroup

---

$\dagger$ $C^n(J, \mathscr{X})$ denotes the n times continuously differentiable functions from J to $\mathscr{X}$.

gives

$$-e^{-\lambda t}T(t)f + f = (\lambda - A)\int_0^t e^{-\lambda s}T(s)f\,ds \qquad f \in \mathcal{X},$$

$$-e^{-\lambda t}T(t)f + f = \int_0^t e^{-\lambda s}T(s)(\lambda - A)f\,ds \qquad f \in \mathcal{D}(A).$$

Now let $t \to \infty$; the closedness of A and the dominated convergence theorem imply $\int_0^\infty e^{-\lambda s}T(s)f\,ds \in \mathcal{D}(A)$ and

$$f = (\lambda - A)\int_0^\infty e^{-\lambda s}T(s)f\,ds \qquad f \in \mathcal{X},$$

$$f = \int_0^\infty e^{-\lambda s}T(s)(\lambda - A)f\,ds \qquad f \in \mathcal{D}(A).$$

Thus $\lambda \in \rho(A)$ and

$$(\lambda - A)^{-1}g = \int_0^\infty e^{-\lambda s}T(s)g\,ds \qquad g \in \mathcal{X}, \lambda > 0. \qquad (2.3)$$

Moreover,

$$\|(\lambda - A)^{-1}g\| \le \int_0^\infty e^{-\lambda s}\|T(s)\|\,\|g\|\,ds \le \|g\|/\lambda, g \in \mathcal{X}, \lambda > 0.$$

This completes the proof of the necessity.

2.7. REMARK. Equation (2.3) shows that $(\lambda - A)^{-1}$, the resolvent of A, is given by the Laplace transform of the semigroup T. Hence the sufficiency part of the Hille-Yosida theorem can be regarded as an inversion theorem for the Laplace transform in an infinite dimensional setting.

2.8. Sections 2.4 and 2.5 (together with some optimism) suggest that if A generates a (C_0) semigroup T, then the formula $T(t) = e^{tA}$ should admit some interpretation that makes sense. Thus it seems reasonable to try to base a proof of the sufficiency part of the Hille-Yosida theorem on one of the classical formulas for the exponential function. For instance, if $A \in \mathbb{C}$,

(i) $e^{tA} = \lim_{n \to \infty} \sum_{k=0}^n (tA)^k/k!$
(ii) $e^{tA} = \lim_{n \to \infty} (1 + tA/n)^n$
(iii) $e^{tA} = \lim_{n \to \infty} (1 - tA/n)^{-n}$
(iv) $e^{tA} = \lim_{\lambda \to \infty} e^{tA_\lambda}$, where $\lim_{\lambda \to \infty} A_\lambda = A$.

Equations (i) and (ii) involve limits of unbounded operators; on the other hand, (iii) or (iv) can be used as a basis for the sufficiency proof. We shall use (iv), with $A_\lambda \in \mathcal{B}(\mathcal{X})$ (so that e^{tA_λ} is easy to construct). First we prepare a lemma.

2.9. LEMMA. *Let C,D generate (C_0) contraction semigroups U, V such that $U(t)V(s) = V(s)U(t)$ for all $s,t \in \mathbb{R}^+$. Then for each $f \in \mathcal{D}(C) \cap \mathcal{D}(D)$,*

$$\|U(t)f - V(t)f\| \le t\|Cf - Df\|.$$

Proof.

$$U(t)f - V(t)f = \int_0^1 \frac{d}{ds}\{U(ts)V[t(1-s)]f\}\,ds$$

$$= \int_0^1 U(ts)V(t(1-s))t(Cf - Df)\,ds,$$

whence the result follows.   ∎

2.10. **Sufficiency Proof of the Hille-Yosida Theorem.** Set

$$A_\lambda = \lambda A(\lambda - A)^{-1} = \lambda^2(\lambda - A)^{-1} - \lambda I \in \mathcal{B}(\mathcal{X}).$$

Then $\{e^{tA_\lambda} : t \in \mathbb{R}^+\}$ is a (C_0) contraction semigroup by Proposition 2.5 and since

$$\|e^{tA_\lambda}\| = e^{-t\lambda}\|\exp[t\lambda^2(\lambda - A)^{-1}]\| \le e^{-t\lambda}\exp[t\lambda^2\|(\lambda - A)^{-1}\|] \le 1$$

for all $t \ge 0$, $\lambda > 0$. Next,

$$\lim_{\lambda \to \infty} A_\lambda f = Af \text{ for all } f \in \mathcal{D}(A),$$

for, if $g \in \mathcal{D}(A)$, then

$$\lambda(\lambda - A)^{-1}g - (\lambda - A)^{-1}Ag = g, \qquad (\lambda - A)^{-1}Ag \to 0,$$

and so $\lambda(\lambda - A)^{-1}g \to g$ as $\lambda \to \infty$, for all $g \in \mathcal{D}(A)$ and hence for all $g \in \overline{\mathcal{D}(A)} = \mathcal{X}$. Thus if $f \in \mathcal{D}(A)$,

$$A_\lambda f = \lambda(\lambda - A)^{-1}Af \to Af \text{ as } \lambda \to \infty.$$

We apply Lemma 2.9 with $C = A_\lambda$, $D = A_\mu$. We have

$$\|e^{tA_\lambda}f - e^{tA_\mu}f\| \le t\|A_\lambda f - A_\mu f\| \to 0 \text{ as } \lambda,\mu \to \infty$$

for each $f \in \mathcal{D}(A)$ (t fixed). Define

$$T(t)f = \lim_{\lambda \to \infty} e^{tA_\lambda}f \qquad f \in \mathcal{D}(A).$$

Clearly $\|T(t)\| \le 1$, and the equation above holds for all $f \in \mathcal{X}$; moreover $T(t)T(s) = T(t + s)$, $T(0) = I$. Next, for $f \in \mathcal{D}(A)$,

$$T(t)f - f = \lim_{\lambda \to \infty} e^{tA_\lambda}f - f = \lim_{\lambda \to \infty}\int_0^t e^{sA_\lambda}A_\lambda f\,ds$$

$$= \int_0^t T(s)Af\,ds \tag{2.4}$$

by the bounded convergence theorem. Thus $T(\cdot)f$ is continuous on $\mathbb{R}^+$ for each $f \in \mathcal{D}(A)$ and hence for each $f \in \mathcal{X}$. Hence T is a (C_0) contraction semigroup. Let B denote its generator. Then (2.4) implies $B \supset A$ i.e. $\mathcal{D}(B) \supset \mathcal{D}(A)$ and $B|_{\mathcal{D}(A)} = A$. By the necessity part of the Hille-Yosida theorem, $1 \in \rho(B)$; also $1 \in \rho(A)$. Hence $(I - B)^{-1} = (I - A)^{-1}$ since $(I - B)^{-1} \supset (I - A)^{-1}$ and both are in $\mathcal{B}(\mathcal{X})$. It follows that $B = A$.   ∎

2.11. Proof of the Converse Part of Proposition 2.5.
Let

$$B(t)f = \frac{1}{t}\int_0^t T(s)f\,ds$$

for $f \in \mathcal{X}$ and $t > 0$. (iii) implies

$$\|B(t)f - f\| = \left\|\frac{1}{t}\int_0^t [T(s) - I]f\,ds\right\|$$

$$\leq \frac{1}{t}\int_0^t \|T(s) - I\|ds\|f\| \to 0$$

as $t \to 0$; in particular, $\|B(t) - I\| < 1/2$ for $0 < t < \delta$ for some $\delta > 0$. Since

$$B(t) = I + [B(t) - I],$$

the usual geometric series argument shows that $B(t)^{-1} \in \mathcal{B}(\mathcal{X})$ for $0 < t < \delta$ and

$$B(t)^{-1} = \sum_{n=0}^{\infty} (-1)^n[B(t) - I]^n.$$

Let A be the generator of T and let $\mathcal{Y}$ be $\mathcal{D}(A)$ with its graph norm (i.e. $\|f\|_{\mathcal{Y}} = \|Af\| + \|f\|$). Then (2.2) implies $B(t) \in \mathcal{B}(\mathcal{X},\mathcal{Y})$. Since $A \in \mathcal{B}(\mathcal{Y},\mathcal{X})$ it follows that $AB(t) \in \mathcal{B}(\mathcal{X})$ for $0 < t < \delta$. [Note that the contraction assumption of Theorem 2.6 was not used in deriving (2.2).] Fix t with $0 < t < \delta$. Then for all $f \in \mathcal{X}$,

$$\|Af\| \leq \|AB(t)\|\|B(t)^{-1}\|\|f\|,$$

whence $A \in \mathcal{B}(\mathcal{X})$. The result now follows from the fact that the semigroup is uniquely determined by its generator (see Remark 2.14 below). ∎

2.12. We now turn to the problem of characterizing the generators of arbitrary (C_0) semigroups. The general generation theorem will be derived as a consequence of the Hille-Yosida theorem (i.e. the general case follows from the contraction case).

LEMMA. *Let T be a (C_0) semigroup. Then constants $M \geq 1, \omega \geq 0$ exist such that*

$$\|T(t)\| \leq Me^{\omega t} \text{ for each } t \in \mathbb{R}^+.$$

Proof. Since $T(\cdot)f \in C([0,1],\mathcal{X})$, $\sup_{0 \leq t \leq 1}\|T(t)f\| < \infty$ for each $f \in \mathcal{X}$. The Banach-Steinhaus theorem (or uniform boundedness principle) implies the existence of a constant M such that

$$\|T(t)\| \leq M \qquad 0 \leq t \leq 1.$$

($M \geq 1$ since $T(0) = I$.) Let $\omega = \log M$. Then for each $t > 0$, if n is the least integer $\geq t$,

$$\|T(t)\| = \|T(t/n)^n\| \leq M^n \leq M^{t+1} = Me^{\omega t}.$$ ∎

2.13. Generation Theorem. *A is the generator of a (C_0) semigroup T iff A is closed, densely defined, and constants $M \geq 1$, $\omega \in \mathbb{R}$ exist such that $\lambda \in \rho(A)$ for each $\lambda > \omega$ and*

$$\|(\lambda - \omega)^n(\lambda - A)^{-n}\| \leq M \tag{2.5}$$

whenever $\lambda > \omega$ and $n = 1, 2, 3, \ldots$. In this case $\|T(t)\| \leq Me^{\omega t}$, $t \in \mathbb{R}^+$. Moreover, an equivalent norm $\|\|\cdot\|\|$ exists on $\mathscr{X}$ such that $S = \{S(t) = e^{-\omega t}T(t): t \in \mathbb{R}^+\}$ is a (C_0) contraction semigroup on $(\mathscr{X}, \|\|\cdot\|\|)$ with generator $A - \omega I$.

Proof (Necessity). By Lemma 2.12, there are constants M, ω such that $\|T(t)\| \leq Me^{\omega t}$, $t \in \mathbb{R}^+$. Moreover, $A - \omega I$ generates the (C_0) semigroup $S = \{S(t) = e^{-\omega t}T(t): t \in \mathbb{R}^+\}$ which is uniformly bounded. Let

$$\|\|f\|\| = \sup_{t \geq 0} \|S(t)f\|.$$

Then $\|\|\cdot\|\|$ and $\|\cdot\|$ are equivalent norms on $\mathscr{X}$ ($\|f\| \leq \|\|f\|\| \leq M\|f\|$) and S is a (C_0) contraction semigroup on $(\mathscr{X}, \|\|\cdot\|\|)$ since

$$\|\|S(t)f\|\| = \sup_{\tau \geq 0} \|S(t + \tau)f\| \leq \|\|f\|\|.$$

By the Hille-Yosida theorem, $A - \omega I$ is closed, densely defined, and for all $\mu > 0$, $\mu \in \rho(A - \omega I)$ and

$$\|\|\mu[\mu - (A - \omega)]^{-1}f\|\| \leq \|\|f\|\|$$

for each $f \in \mathscr{X}$; thus

$$\|\|\mu^n(\mu + \omega - A)^{-n}f\|\| \leq \|\|f\|\|$$

whenever $n = 1, 2, \ldots$, $\mu > 0$, $f \in \mathscr{X}$. This implies (2.5) (with $\lambda = \mu + \omega$), thereby proving the necessity.

For the sufficiency we replace A by $A - \omega I$ and thereby assume $\omega = 0$ (without loss of generality). For $\lambda > 0$ let $R(\lambda) = (\lambda - A)^{-1}$ denote the resolvent of A. From the resolvent identity (Lemma 1.3) it follows that $(d/d\lambda)R(\lambda) = -R(\lambda)^2$, and, more generally,

$$\frac{d^n}{d\lambda^n} R(\lambda) = (-1)^n n! R(\lambda)^{n+1} \tag{2.6}$$

for $\lambda > 0$. Thus for $0 < \lambda < \mu$, the Taylor series for the analytic function $R(\cdot)$ is the convergent geometric series

$$R(\lambda) = \sum_{k=0}^{\infty} (\mu - \lambda)^k R(\mu)^{k+1}.$$

Differentiating this series term-by-term $n - 1$ times with respect to λ and using (2.6) yields

$$R(\lambda)^n = \sum_{k=n-1}^{\infty} \binom{k}{n-1}(\mu - \lambda)^{k-n+1} R(\mu)^{k+1}$$

where

$$\binom{k}{n-1} = \frac{k!}{(n-1)!(k-n+1)!}$$

is the familiar binomial coefficient. Consequently

$$\|\lambda^n R(\lambda)^n f\| \le (\lambda/\mu)^n \sum_{k=n-1}^{\infty} \binom{k}{n-1}\left(\frac{\mu-\lambda}{\mu}\right)^{k-n+1} \|\mu^{k+1}R(\mu)^{k+1}f\|. \quad (2.7)$$

Set

$$|f|_\mu = \sup\{\|\mu^n R(\mu)^n f\| : n = 0, 1, 2, \ldots\}$$

for $f \in \mathscr{X}$. We claim that

$$|f|_\lambda \le |f|_\mu \text{ for } 0 < \lambda \le \mu. \quad (2.8)$$

To see this, note that for $0 \le x \le 1$, $(1-x)^{-1} = \sum_{k=0}^{\infty} x^k$, $x^{-n} = \sum_{k=n-1}^{\infty} \binom{k}{n-1}(1-x)^{k-n+1}$. Let $x = \lambda/\mu$. Then (2.7) implies for $n \ge 1$,

$$\|\lambda^n R(\lambda)^n f\| \le x^n \sum_{k=n-1}^{\infty} \binom{k}{n-1}(1-x)^{k-n+1}|f|_\mu = |f|_\mu.$$

Taking the supremum over n yields (2.8). Now let

$$|f| = \lim_{\lambda \to \infty} |f|_\lambda = \sup_{\lambda > 0} |f|_\lambda.$$

Hypothesis (2.5) implies that

$$\|f\| \le |f| \le M\|f\|,$$

and it follows easily that $|\cdot|$ is a norm on $\mathscr{X}$ equivalent to $\|\cdot\|$.

For $0 < \lambda \le \mu$ and n a nonnegative integer, (2.8) implies

$$\|\mu^n R(\mu)^n \lambda R(\lambda)f\| = \|\lambda R(\lambda)\mu^n R(\mu)^n f\|$$

$$\le |\mu^n R(\mu)^n f|_\lambda \le |\mu^n R(\mu^n)f|_\mu \le |f|_\mu \le |f|.$$

Taking the supremum over n yields

$$|\lambda R(\lambda)f|_\mu \le |f|;$$

then letting $\mu \to \infty$ gives

$$|\lambda R(\lambda)f| \le |f|$$

for all $\lambda > 0$ and all $f \in \mathscr{X}$. Thus by the Hille-Yosida theorem, A generates a (C_0) contraction semigroup on $(\mathscr{X}, |\cdot|)$. It follows that A generates a (C_0) semigroup T on $(\mathscr{X}, \|\cdot\|)$ satisfying $\|T(t)\| \le M$ for all $t \in \mathbb{R}^+$. ∎

2.14. REMARK.   *A semigroup is uniquely determined by its generator.*

Proof.   Let T and S be two (C_0) semigroups having the same generator A. Let $f \in \mathscr{D}(A)$ and let $t > 0$. Define $u : [0, t] \to \mathscr{X}$ by $u(s) = T(s)S(t-s)f$. Then

$$\frac{du(s)}{ds} = T(s)(-A)S(t-s)f + (T(s)A)S(t-s)f = 0,$$

whence u is constant on $[0,t]$. Therefore

$$T(t)f = u(t) = u(0) = S(t)f.$$

This holds for each $t > 0$ and each $f \in \mathcal{D}(A)$, which is dense in $\mathcal{X}$. $T = S$ follows.
∎

2.15. **DEFINITIONS.** A (C_0) *group* on $\mathcal{X}$ is a family of operators $T = \{T(t): t \in \mathbb{R}\} \subset \mathcal{B}(\mathcal{X})$ satisfying the conditions of Definition 2.1 but with $\mathbb{R}^+$ replaced by $\mathbb{R}$. The *generator* A of a (C_0) group T on $\mathcal{X}$ is defined by

$$Af = \lim_{t \to 0} t^{-1}[T(t)f - f],$$

the domain $\mathcal{D}(A)$ of A being the set of all $f \in \mathcal{X}$ for which the limit given above exists. Note that this limit is a two-sided one (i.e. $t \to 0$, not $t \to 0^+$).

2.16. **REMARK.** A is the generator of a (C_0) group T iff $\pm A$ generates a (C_0) semigroup $T_\pm$. In this case,

$$T(t) = \begin{cases} T_+(t), t \geq 0 \\ T_-(-t), t \leq 0. \end{cases}$$

This is left as an exercise, as is the following consequence of this remark and Theorem 2.13.

2.17. **GENERATION THEOREM FOR GROUPS.** *A generates a (C_0) group T on $\mathcal{X}$ iff A is closed, densely defined, and there exist constants M, ω such that $\lambda \in \rho(A)$ whenever λ is real, $|\lambda| > \omega$ and*

$$\|(|\lambda| - \omega)^n(\lambda - A)^{-n}\| \leq M,$$

$|\lambda| > \omega$, $n = 1, 2, \ldots$. *In this case, $\|T(t)\| \leq Me^{\omega|t|}$, $t \in \mathbb{R}$.*

2.18. **EXERCISES**

1. Let $T = \{T(t): t \in \mathbb{R}^+\} \subset \mathcal{B}(\mathcal{X})$ satisfy (i) and (ii) of Definition 2.1 and (iii″) $\lim_{t \to 0} \|T(t)f - f\| = 0$ for each $f \in \mathcal{X}$. Show that T is a (C_0) semigroup.
2. Let $\mathbb{K} = \mathbb{C}$. If A generates a (C_0) contraction semigroup, then $\lambda \in \rho(A)$ for all λ with $\mathrm{Re}(\lambda) > 0$ and $\|(\lambda - A)^{-1}\| \leq 1/\mathrm{Re}(\lambda)$ holds in this case. Show this and also state and prove the analogous result in the context of Theorem 2.13.
3. Let $\{S_\lambda : \lambda \in \Lambda\} \subset \mathcal{B}(\mathcal{X})$ be a family of commuting operators. There exists an equivalent norm $|\cdot|$ on $\mathcal{X}$ such that each S_λ is a contraction on $(\mathcal{X}, |\cdot|)$ iff

$$\sup\{\|S_{\lambda_1} \cdots S_{\lambda_n}\| : \lambda_i \in \Lambda, n \in \mathbb{N}\} < \infty.$$

This condition implies

$$\sup\{\|S_\lambda^n\| : \lambda \in \Lambda, n \in \mathbb{N}\} < \infty,$$

and the converse holds whenever $S_\lambda = (\lambda - A)^{-1}$ where A generates a (C_0) semigroup and λ is sufficiently large. To see that the converse fails in general, let $\{e_n : n \in \mathbb{N}_0\}$ be an orthonormal basis for a Hilbert space $\mathcal{X}$ and define $S_m e_n = \delta_{mn}e_0 + (1 - \delta_{mn})e_n$, δ_{mn} being the Kronecker delta (i.e. $\delta_{mn} = 1$ or 0 according as $m = n$ or $m \neq n$). Check the details.

4. Let $\mathscr{X} = BUC(\mathbb{R})$ or $C_0(\mathbb{R})$ or $C[-\infty,\infty]$. Here $BUC(\mathbb{R})$ is the bounded uniformly continuous functions on $\mathbb{R}$, and $C_0(\mathbb{R})$ consists of those functions in $BUC(\mathbb{R})$ which vanish at $\pm\infty$. $\mathscr{X}$ is a Banach space under the supremum norm. Define T by

$$[T(t)f](x) = f(x+t)$$

for $t \in \mathbb{R}^+, f \in \mathscr{X}, x \in \mathbb{R}$. Then T is a (C_0) contraction semigroup on $\mathscr{X}$. The generator of T is $A = d/dx$ with domain

$$\mathscr{D}(A) = \{f \in \mathscr{X}: f \text{ absolutely continous}, f' \in \mathscr{X}\}.$$

In fact, $\{T(t): t \in \mathbb{R}\}$ is a (C_0) group of isometries on $\mathscr{X}$.
5. Do Exercise 4 in the context of $\mathscr{X} = L^p(\mathbb{R})$, $1 \le p < \infty$.
6. Let $f: \mathbb{R} \to \mathbb{R}$ be bounded. If $\sup|f(x+t) - f(x)| \to 0$ as $t \to 0$, then $f \in BUC(\mathbb{R})$. Thus $BUC(\mathbb{R})$ is the largest subspace of $L^\infty(\mathbb{R})$ on which the translation semigroup is strongly continuous (cf. Exercise 4).
*7. Let $\mathscr{X} = BUC(\mathbb{R})$ or $L^p(\mathbb{R})$, $1 \le p < \infty$. Let $A = \frac{1}{2}d^2/dx^2$ with $\mathscr{D}(A) = \{f \in \mathscr{X}: f', f'' \in \mathscr{X}\}$. Then A generates a (C_0) contraction semigroup on $\mathscr{X}$ given by

$$[T(t)f](x) = \frac{1}{\sqrt{2\pi t}} \int_{-\infty}^{\infty} \exp(-y^2/2t)f(x-y)\, dy.$$

Hints: Its easiest to check first that T is a (C_0) contraction semigroup, using $1/(\sqrt{2\pi t}) \int_{-\infty}^{\infty} \exp(-y^2/2t)\, dy = 1$. Use

$$g_\lambda(x) = \lambda(\lambda - A)^{-1}f(x) = \lambda \int_0^\infty e^{-\lambda t}T(t)f(x)\, dt$$

and

$$\int_0^\infty \exp-\left(y^2 + \frac{c^2}{y^2}\right) dy = \frac{\sqrt{\pi}}{2}e^{-2c}(c>0)$$

to conclude

$$g_\lambda(x) = \int_{-\infty}^{\infty} \sqrt{\frac{\lambda}{2}} f(y) \exp(-\sqrt{2\lambda}|x-y|)\, dy$$

and $g_\lambda''(x) = -2\lambda f(x) + 2\lambda g_\lambda(x)$. Finally get $Ag_\lambda = \frac{1}{2}g_\lambda''$. Do not expect to carry out all the details in five minutes!
8. Let $\mathscr{X} = BUC(\mathbb{R})$ or $L^p(\mathbb{R})$, $1 \le p < \infty$. The formula

$$T(t)f(x) = \frac{t}{\pi} \int_{-\infty}^{\infty} \frac{f(x-y)}{t^2 + y^2}\, dy$$

defines a (C_0) contraction semigroup on $\mathscr{X}$.
{Actually, the easiest way to do Exercises 7 and 8 is in the context of the Fourier transform; see Section 3 of Chapter 2 for an introduction to this notion. Hint for Exercise 8: What is the Fourier transform of $\phi(x) = e^{-t|x|}$?}
9. T defined by

$$T(t)f(x) = e^{-\lambda t} \sum_{k=0}^{\infty} \frac{(\lambda t)^k}{k!} f(x - k\mu) \qquad (\lambda,\mu > 0)$$

is a (C_0) contraction semigroup on $BUC(\mathbb{R})$ whose generator is the difference operator given by

$$Af(x) = \lambda\{f(x-\mu) - f(x)\}.$$

10. Let A generate a (C_0) semigroup. Then for each $f \in \mathcal{X}$,

$$\lim_{\lambda \to \infty} \lambda(\lambda - A)^{-1}f = f.$$

*11. Let $T = \{T(t): t \in \mathbb{R}^+\} \subset \mathcal{B}(\mathcal{X})$ satisfy (i) and (ii) of Definition 2.1 and also

(iii″) $\phi[T(\cdot)f] \in C(\mathbb{R}^+, \mathbb{K})$ for each $f \in \mathcal{X}$ and each $\phi \in \mathcal{X}^*$,

where $\mathcal{X}^*$ is the dual space of $\mathcal{X}$. Then T is a (C_0) semigroup on $\mathcal{X}$. In other words, weak continuity implies strong continuity for semigroups.

2.19.   REMARKS AND EXERCISES ON EQUIVALENT NORMS. Let T be a uniformly bounded (C_0) semigroup on $\mathcal{X}$, i.e. $\|T(t)\| \leq M$ for some real M and all $t \in \mathbb{R}^+$. Then by Feller's trick (cf. the necessity proof in 2.13) $\mathcal{X}$ can be renormed with an equivalent norm $\|\|\cdot\|\|$ such that

$$\|\|T(t)\|\| = \sup\{\|\|T(t)f\|\| : \|\|f\|\| \leq 1\} \leq 1$$

for each $t \in \mathbb{R}^+$, i.e. T is a (C_0) contraction semigroup on $(\mathcal{X}, \|\|\cdot\|\|)$.

(1) Given two uniformly bounded (C_0) semigroups T, S in $\mathcal{B}(\mathcal{X})$, can $\mathcal{X}$ be given an equivalent norm $\|\|\cdot\|\|$ so that on $(\mathcal{X}, \|\|\cdot\|\|)$, both T and S are contraction semigroups? The answer is no. A counterexample is given by $\mathcal{X} = L^1(\mathbb{R})$, $T(t)f(x) = f(x + t)$, $S(t)f(x) = f(\psi[\psi^{-1}(x) - t])$ where $\psi(x) = x$ for $x \leq 0$ and $\psi(x) = 2x$ for $x \geq 0$. Show that for $t > 0$, $\|[T(t)S(t)]^k\| \geq k$ by looking at $[T(t)S(t)]^k f$ where f is the characteristic function (or indicator function) $\chi_{[0,r]}$ of the interval $[0,r]$ with r suitably chosen.

*(2) If $\mathcal{X}$ is a Hilbert space, can $\|\|\cdot\|\|$ be taken to be an inner product norm? The answer is again no. For a counterexample, let $\mathcal{Y} = L^2([0,\infty[)$ and let $\mathcal{X} = \mathcal{Y} \oplus \mathcal{Y}$. For $f \in \mathcal{Y}, t > 0, x \in \mathbb{R}^+$ let

$$V(t)f(x) = \begin{cases} f(x - t) & \text{for } x \geq t \\ 0 & \text{for } x < t, \end{cases}$$

$$V^*(t)f(x) = f(x + t),$$

$$P(t)f(x) = f(2 \cdot 4^k - t - x),$$

whenever $x \in [0, 2 \cdot 4^l - t]$ or $x \in \,]4^k - t, 4^k]$ for $k \geq l$, and $P(t)f(x) = 0$ otherwise; here $k \geq l$ and l is the unique integer such that $4^l < t \leq 4^{l+1}$. Then

$$T(t) = \begin{pmatrix} V^*(t) & P(t) \\ 0 & V(t) \end{pmatrix}$$

is a uniformly bounded (C_0) semigroup on $\mathcal{X}$. If it is a contraction semigroup with respect to an equivalent inner product norm on $\mathcal{H}$, then it is similar to a (C_0) contraction semigroup, i.e. there is an $S \in \mathcal{B}(\mathcal{X})$ such that $S^{-1} \in \mathcal{B}(\mathcal{X})$ and $\|ST(t)S^{-1}\| \leq 1$ for each $t \in \mathbb{R}^+$. The generator A of such a semigroup necessarily satisfies

$$W(A) \cap [W(A^*)]^\perp = \{0\},$$

where $W(A) = \{ f \in \mathscr{X} : \text{weak } \lim_{t \to \infty} T(t)f = 0 \}$ and similarly for $W(A^*)$. But

$$0 \neq \chi_{]0,1[} \oplus 0 \in W(A) \cap [W(A^*)]^\perp,$$

$\chi_{]0,1[}$ being the characteristic (or indicator) function of the open interval $]0,1[$.

(3) If T is a uniformly bounded *group*, then

$$||| f ||| = \sup \{ \| T(t)f \| : t \in \mathbb{R} \}$$

defines an equivalent norm on $\mathscr{X}$ so that T is a (C_0) group of isometries on $(\mathscr{X}, ||| \cdot |||)$. If $\mathscr{X}$ is a Hilbert space, then there is an equivalent inner product norm $|\cdot|$ on $\mathscr{X}$ (different from $||| \cdot |||$) such that T is a (C_0) group of isometries on $(\mathscr{X}, |\cdot|)$. To see this let LIM be any fixed Banach limit (as $t \to \infty$) and let $[f,g] = \text{LIM} \langle T(t)f, T(t)g \rangle$, where $\langle \cdot, \cdot \rangle$ is the inner product of $\mathscr{X}$. Then $[\cdot, \cdot]$ is an inner product, $|f| = [f,f]^{1/2}$ is equivalent to $\| f \|$, and $|T(t)f| = |f|$ for all $t \in \mathbb{R}$ and all $f \in \mathscr{X}$. {LIM is a norm one-linear functional on the bounded functions on $\mathbb{R}$ such that $\text{LIM} f(\cdot) = \lim_{t \to \infty} t^{-1} \int_0^t f(s)\, ds$ provided this limit exists. The key property is $\text{LIM} f(\cdot) = \text{LIM} f(t + \cdot)$.}

3. Dissipative Operators: The Hille-Yosida Theorem Again

In this section we give Lumer and Phillips' alternative formulation of the Hille-Yosida theorem.

3.1.   DEFINITIONS.   Let $\langle \cdot, \cdot \rangle$ denote the pairing between $\mathscr{X}$ and $\mathscr{X}^*$ ($=$ the dual space of $\mathscr{X}$) ; that is, $\langle f, \phi \rangle = \phi(f)$ for $f \in \mathscr{X}$, $\phi \in \mathscr{X}^*$. For each $f \in \mathscr{X}$ define

$$\mathscr{J}(f) = \{ \phi \in \mathscr{X}^* : \| \phi \|^2 = \| f \|^2 = \langle f, \phi \rangle \}.$$

For each $f \in \mathscr{X}$, $\mathscr{J}(f)$ is nonempty by the Hahn-Banach theorem. $\mathscr{J}$ is a (multivalued) function called the *duality map* of $\mathscr{X}$. Let J be a section of $\mathscr{J}$, i.e. $J : \mathscr{X} \to \mathscr{X}^*$ and $J(f) \in \mathscr{J}(f)$ for each $f \in \mathscr{X}$. Such a function J will be called a *duality section*. An operator A on $\mathscr{X}$ is called *dissipative with respect to a duality section* J if for each $f \in \mathscr{D}(A)$, $\text{Re} \langle Af, J(f) \rangle \leq 0$. A on $\mathscr{X}$ is *dissipative* means that A is dissipative with respect to some duality section. A dissipative operator A is *m-dissipative* provided $\rho(A) \cap]0, \infty[\neq \emptyset$. An operator B is *accretive, accretive with respect to J, or m-accretive* iff $A = -B$ is dissipative, dissipative with respect to J, or m-dissipative.

3.2.   REMARKS.   When $\mathscr{X}$ is a Hilbert space which we identify with its dual, $\mathscr{J}$ becomes the identity, (i.e. $\mathscr{J}(f) = \{ f \}$) and $\langle \cdot, \cdot \rangle$ the inner product. $\mathscr{J}$ is single-valued in many cases, but not always, cf. Exercises 3.10.2,3,4. For (Ω, Σ, μ) a measure space and $\mathscr{X} = L^p(\Omega, \Sigma, \mu)$, $1 \leq p < \infty$, $\mathscr{J}(0) = \{ 0 \}$; and for $0 \neq f \in \mathscr{X}$ and $\phi \in \mathscr{J}(f)$,

$$\phi(\omega) = \| f \|_p^{2-p} \overline{f(\omega)} | f(\omega) |^{p-2}$$

whenever $f(\omega) \neq 0$, $\phi(\omega) = 0$ whenever $f(\omega) = 0$ and $p > 1$, and for $f(\omega) = 0$ and $p = 1$, $\phi(\omega)$ is arbitrary, subject to the conditions that ϕ is Σ-measurable and $|\phi(\omega)| \leq \|f\|_1$. In particular, $\mathscr{J}$ is single valued for L^p with $1 < p < \infty$ but not for L^1 (nor for $C[0,1]$).

3.3. HILLE-YOSIDA THEOREM: LUMER-PHILLIPS FORM. *Suppose A generates a* (C_0) *contraction semigroup. Then*

(i) $\overline{\mathscr{D}(A)} = \mathscr{X}$,
(ii) *A is dissipative with respect to any duality section,*
(iii) $(0,\infty) \subset \rho(A)$.

Conversely, if A satisfies

(i') $\overline{\mathscr{D}(A)} = \mathscr{X}$
(ii') *A is dissipative with respect to some duality section,*
(iii') $(0,\infty) \cap \rho(A) \neq \emptyset$,

then A generates a (C_0) *contraction semigroup on* $\mathscr{X}$. *Thus A generates a* (C_0) *contraction semigroup iff A is densely defined and m-dissipative.*

For the proof we prepare two lemmas.

3.4. LEMMA. *Let* $f,g \in \mathscr{X}$. *Then* $\|f\| \leq \|f - \alpha g\|$ *for all* $\alpha > 0$ *iff there is a* $\phi \in \mathscr{J}(f)$ *such that* $\mathrm{Re}\langle g,\phi \rangle \leq 0$.

Proof. This is trivial for $f = 0$, so suppose $f \neq 0$. If $\mathrm{Re}\langle g,\phi \rangle \leq 0$ for some $\phi \in \mathscr{J}(f)$, then for $\alpha > 0$,

$$\|f\|^2 = \langle f,\phi \rangle \leq \mathrm{Re}\langle f - \alpha g,\phi \rangle \leq \|f - \alpha g\|\|\phi\|$$

whence $\|f\| \leq \|f - \alpha g\|$. Conversely, if $\|f\| \leq \|f - \alpha g\|$ for all $\alpha > 0$, let $\phi_\alpha \in \mathscr{J}(f - \alpha g)$ and let $\psi_\alpha = \phi_\alpha \|\phi_\alpha\|^{-1}$. Then

$$\|f\| \leq \|f - \alpha g\| = \langle f - \alpha g,\psi_\alpha \rangle = \mathrm{Re}\langle f,\psi_\alpha \rangle - \alpha \mathrm{Re}\langle g,\psi_\alpha \rangle$$

$$\leq \|f\| - \alpha \mathrm{Re}\langle g,\psi_\alpha \rangle.$$

Therefore

$$\limsup_{\alpha \to 0^+} \mathrm{Re}\langle g,\psi_\alpha \rangle \leq 0, \quad \liminf_{\alpha \to 0^+} \mathrm{Re}\langle f,\psi_\alpha \rangle \geq \|f\|.$$

But since the closed unit ball in $\mathscr{X}^*$ is weak* compact, we have for a subnet $\alpha_\nu \to 0^+$, weak* $\lim \psi_\alpha = \psi_0$, $\mathrm{Re}\langle g,\psi_0 \rangle \leq 0$, $\mathrm{Re}\langle f,\psi_0 \rangle \geq \|f\|$. Hence $\phi = \psi_0\|f\| \in \mathscr{J}(f)$ and $\mathrm{Re}\langle g,\phi \rangle \leq 0$.   ∎

3.5. COROLLARY. *A is dissipative iff for each* $\alpha > 0$, $I - \alpha A$ *is injective and* $\|(I - \alpha A)^{-1}\| \leq 1$.

Proof. Immediate.   ∎

3.6. LEMMA. *Let S on* $\mathscr{X}$ *be closed and densely defined, and let* $\mu \in \rho(S)$. *Then* $\lambda \in \rho(S)$ *iff* $I - (\mu - \lambda)(\mu - S)^{-1}$ *has an inverse in* $\mathscr{B}(\mathscr{X})$; *in this case*

$$(\lambda - S)^{-1} = (\mu - S)^{-1}[I - (\mu - \lambda)(\mu - S)^{-1}]^{-1}. \tag{3.1}$$

Proof. Suppose $U = I - (\mu - \lambda)(\mu - S)^{-1}$ has an inverse in $\mathcal{B}(\mathcal{X})$. Then

$$(\lambda - S)(\mu - S)^{-1}U^{-1} = [(\lambda - \mu)I + (\mu - S)](\mu - S)^{-1}U^{-1}$$
$$= [(\lambda - \mu)(\mu - S)^{-1} + I]U^{-1} = I,$$

and it is also easy to show

$$(\mu - S)^{-1}U^{-1}(\lambda - S) = I|_{\mathcal{D}(S)}.$$

Thus $\lambda \in \rho(A)$ and (3.1) holds.

The proof of the converse is equally trivial and is omitted. ∎

3.7. Proof of Theorem 3.3. Note first that $(\lambda I - A)^{-1} = \alpha(I - \alpha A)^{-1}$ with $\alpha = 1/\lambda$. Thus by the Hille-Yosida theorem and Corollary 3.5, A generates a (C_0) contraction semigroup iff (i), (ii′), and (iii) of Theorem 3.3 hold. Thus it only remains to show that (i), (ii′), and (iii′) imply (ii) and (iii). We do this now.

By (iii′), there is some $\mu \in (0,\infty) \cap \rho(A)$. If $|\alpha| < \mu$, then

$$\|\alpha(\mu - A)^{-1}\| \le \frac{|\alpha|}{\mu} < 1,$$

so that $I - \alpha(\mu - A)^{-1}$ is invertible in $\mathcal{B}(\mathcal{X})$ (and its inverse is given by the convergent geometric series $\sum_{n=0}^{\infty} \alpha^n(\mu - A)^{-n}$). By Lemma 3.6, $\lambda \in \rho(A)$ if $|\lambda - \mu| < \mu$, i.e. if $0 < \lambda < 2\mu$. Applying Lemma 3.6 again (with $3\mu/2$ in place of μ) we get $(0,3\mu) \subset \rho(A)$. A simple induction argument completes the proof of (iii).

For (ii) let T be the (C_0) contraction semigroup generated by A. Let J be any duality section. Then for $f \in \mathcal{D}(A)$,

$$|\langle T(t)f, J(f)\rangle| \le \|T(t)f\|\|J(f)\| \le \|f\|^2,$$

$$\mathrm{Re}\langle T(t)f - f, J(f)\rangle = \mathrm{Re}\langle T(t)f, J(f)\rangle - \|f\|^2 \le 0.$$

Dividing by t and letting $t \to 0$ gives (ii). ∎

3.8. Remarks. A dissipative operator A is called *maximal dissipative* if A has no proper dissipative extension. If A is *m*-dissipative, then A is maximal dissipative. If $\mathcal{X}$ is a Hilbert space, then the converse is valid, but this is not so in a general Banach space situation, as the following example shows. Let

$$\mathcal{X} = \{f \in C[0,1] : f(0) = f(1) = 0\}.$$

$\mathcal{X}$ is equipped with the supremum norm. Let $A = d/dx$,

$$\mathcal{D}(A) = \{f : f, f' \in \mathcal{X}\}.$$

Then A is densely defined and maximal dissipative, but

$$\bigcap \{\mathrm{Range}\,(I - \alpha A) : \alpha > 0\} = \{0\}.$$

Now let A be a maximal dissipative operator on a Hilbert space $\mathcal{X}$. A is densely defined, for otherwise we could define a dissipative extension of A which vanishes on $\mathcal{D}(A)^{\perp}$.

We must show that the range of $I - A$ is all of $\mathscr{X}$. Let

$$C = (I + A)(I - A)^{-1}$$

be the *Cayley transform* of A. We can check that for each

$$f \in \mathscr{D}(C) = \text{Range}(I - A)$$

there is a $g \in \mathscr{D}(A)$ such that $f = g - Ag$, $Cf = g + Ag$. By the law of cosines,

$$\|f\|^2 = \|g\|^2 + \|Ag\|^2 - 2\,\text{Re}\langle g, Ag \rangle \geq \|Cf\|^2.$$

Solving the above equations gives

$$g = \frac{1}{2}(Cf + f) \qquad Ag = \frac{1}{2}(Cf - f).$$

So C is a contraction with $C + I$ injective. If $\tilde{C}$ is a contraction extension of A such that $\tilde{C} + I$ is injective, then $\tilde{C}$ is the Cayley transform of a dissipative extension $\tilde{A}$ of A. The maximality of A implies $\mathscr{D}(C) = X$, and the result follows. (See Chapter II, Proposition 15.3 for more on the Cayley transform.)

3.9. It is often the case that (ii′) and (iii′) (Section 3.3) imply (i′) in the Lumer-Phillips theorem, i.e. the density of the domain of a dissipative operator often follows from the surjectivity of $I - \alpha A$ for some $\alpha > 0$.

PROPOSITION. *Let A be dissipative with respect to some duality section of $\mathscr{X}$. Suppose Range $(I - \alpha A) = \mathscr{X}$ for some $\alpha > 0$. Let $\mathscr{X}_0 = \overline{\mathscr{D}(A)}$.*
(α) $\mathscr{X}_0 = \mathscr{X}$ if $\mathscr{X}$ is reflexive.
(β) $\mathscr{X}_0 \subsetneqq \mathscr{X}$ can happen if $\mathscr{X}$ is not reflexive.

Proof. The proof given in 3.7 shows that the range of $I - \alpha A$ is all of $\mathscr{X}$ for each $\alpha > 0$. Let $f \in \mathscr{X}$ and let $g_\alpha = (I - \alpha A)^{-1}f \in \mathscr{D}(A) \subset \mathscr{X}_0$. Let $\{\alpha_n\}$ be a sequence of positive numbers converging to zero. If we write g_n for g_{α_n}, then dissipativity of A implies $\|g_n\| = \|(I - \alpha_n A)^{-1}f\| \leq \|f\|$ for each n. If $\mathscr{X}$ is reflexive, then by the weak sequential compactness of closed balls in $\mathscr{X}$, $\{g_n\}$ has a weakly convergent subsequence, which we take to be the whole sequence $\{g_n\}$ (by redefining the α_ns). Then $g_0 = \text{weak } \lim_{n\to\infty} g_n$ belongs to the weak closure of $\mathscr{D}(A)$, which is $\mathscr{X}_0$. Moreover, $g_n - \alpha_n Ag_n = f$ implies weak $\lim_{n\to\infty} A(\alpha_n g_n) = g_0 - f$. Thus weak $\lim_{n\to\infty} (\alpha_n g_n, g_n - f) = (0, g_0 - f)$ belongs to the weak closure of the graph $\mathscr{G}(A)$ of A, which is $\mathscr{G}(A)$ itself. (According to the Hahn-Banach theorem, a subspace of a Banach space is closed iff it is weakly closed.) Therefore $g_0 = f$ and so $f \in \mathscr{X}_0$. This proves (α).

The above argument actually shows that $\mathscr{X}_0 = \mathscr{X}$ whenever $\mathscr{X}$ is a dual space and the graph of A is closed in the weak* topology.

Here are two counterexamples which establish (β). Let

$$\mathscr{X} = L^\infty(\mathbb{R}), \; Af(x) = -|x|f(x) \text{ for } f \in \mathscr{D}(A) = \left\{ g \in L^\infty(\mathbb{R}) : \text{ess sup}_{x\in\mathbb{R}} |xg(x)| < \infty \right\}.$$

Then

$$\mathscr{X}_0 = \mathscr{D}(A) = \left\{ f \in L^\infty(\mathbb{R}) : \operatorname*{ess\,lim}_{|x| \to \infty} f(x) = 0 \right\}.$$

For the second example let $\mathscr{X} = BC(\mathbb{R})$ be the bounded continuous functions on $\mathbb{R}$, and let $A = d/dx$ on $\mathscr{D}(A) = \{ f \in \mathscr{X} : f$ absolutely continuous, $f' \in \mathscr{X} \}$. Then $\mathscr{X}_0 = BUC(\mathbb{R}) \ne \mathscr{X}$. ∎

3.10. EXERCISES

1. Verify the details of Remarks 3.8.
2. A Banach space $\mathscr{X}$ is called *strictly convex* whenever $\| f + g \| = \| f \| + \| g \|$ only holds for vectors $f,g \in \mathscr{X}$ that are linearly dependent, i.e. $\| f + g \| = \| f \| + \| g \|$ implies $f = tg$ or $g = tf$ for some $t \ge 0$. If $\mathscr{X}^*$ is strictly convex, then the duality map $\mathscr{J}$ of $\mathscr{X}$ is single-valued. For a proof let ϕ_1 and ϕ_2 belong to $\mathscr{J}(f)$ where $f \ne 0$. Then $\phi_3 = \tfrac{1}{2}(\phi_1 + \phi_2) \in \mathscr{J}(f)$, and $\| 2\phi_3 \| = \| \phi_1 + \phi_2 \| = \| \phi_1 \| + \| \phi_2 \|$, whence $\phi_1 = \phi_2$ by strict convexity. Fill in the missing details.
*3. A Banach space $\mathscr{X}$ is called *uniformly convex* provided that for every $\epsilon > 0$ there is a $\delta > 0$ such that $f,g \in \mathscr{X}$, $\| f \| = \| g \| = 1$, and $\| f - g \| \ge \epsilon$ imply $\| \tfrac{1}{2}(f + g) \| \le 1 - \delta$.
   (a) A uniformly convex space is strictly convex.
   (b) $L^p(\Omega,\Sigma,\mu)$ is uniformly convex for $1 < p < \infty$. A proof can be based on the Clarkson inequalities: for $f,g \in L^p$ with $1 < p < \infty$, $1/p + 1/q = 1$, and $\|\cdot\|$ the L^p norm,

$$\| f + g \|^p + \| f - g \|^p \le 2^{p-1}(\| f \|^p + \| g \|^p) \qquad (p \ge 2),$$

$$\| f + g \|^q + \| f - g \|^q \ge 2(\| f \|^p + \| g \|^p)^{q-1} \qquad (p \ge 2),$$

$$\| f + g \|^q + \| f - g \|^q \le 2(\| f \|^p + \| g \|^p)^{q-1} \qquad (p \le 2),$$

$$\| f + g \|^p + \| f - g \|^p \ge 2^{p-1}(\| f \|^p + \| g \|^p) \qquad (p \le 2).$$

Notice that when $p = 2$, equality holds in these inequalities according to the parallelogram law.
   (c) A uniformly convex space is reflexive.
4. If $\mathscr{X}^*$ is uniformly convex, the duality map $\mathscr{J}$ is continuous, uniformly so on bounded sets.
5. Let $l^1(2) = \{ (x_1,x_2) \in \mathbb{R}^2 : \| x \|_1 = |x_1| + |x_2| \}$,

$$l^\infty(2) = \{ (x_1,x_2) \in \mathbb{R}^2 : \| x \|_\infty = \max\{ |x_1|, |x_2| \} \}.$$

Find the duality map $\mathscr{J}$ for each of these spaces. Conclude that neither space is strictly convex nor smooth (see Exercise 10 below for the definition of smooth).
6. A *semi-inner product* on a vector space $\mathscr{X}$ is a mapping $[\cdot,\cdot] : \mathscr{X} \times \mathscr{X} \to \mathbb{K}$ satisfying, for all $f,g,h \in \mathscr{X}$ and all $\lambda \in \mathbb{K}$, $[f+g,h] = [f,h] + [g,h]$, $[\lambda f,h] = \lambda [f,h]$, $[f,f] > 0$ for $f \ne 0$, and $|[f,g]|^2 \le [f,f][g,g]$. Show that on a Banach space $\mathscr{X}$, the semi-inner products are in one-to-one correspondence with the duality sections. Formulate the notion of dissipativeness in terms of semi-inner products.
7. Find a duality section for $\mathscr{X} = C[0,1]$.
8. Calculate Range$(I - \alpha A)$ for the example in Remarks 3.8.

9. Let A be dissipative and densely defined. Then A is closeable and $\bar{A}$ is dissipative. Hints: Assume that $(0,g) \in \overline{\mathcal{G}(A)}$ with $\|g\| = 1$ and seek a contradiction. Choose $f_n \in \mathcal{D}(A)$ such that $f_n \to 0$ and $Af_n \to g$. Choose $f \in \mathcal{D}(A)$ with $\|f\| = 1$ and $\|f - g\| < 1/2$. Let J be a duality section with respect to which A is dissipative, let $c > 0$, and let ψ_c be a weak* limit point of $\{J(f + cf_n)\}$. Deduce $\|\psi_c\| = 1$, $\mathrm{Re}\langle Af + cg, \psi_c \rangle \leq 0$, and $1/2 \leq \mathrm{Re}\langle g, \psi_c \rangle \leq c^{-1} \, \mathrm{Re}\langle Af, \psi_c \rangle \leq \|Af\|/c$, and let $c \to \infty$.

10. A Banach space $\mathcal{X}$ is called *smooth iff* for each unit vector f in $\mathcal{X}$ there is a unique vector ϕ in $\mathcal{X}^*$ such that $\langle f, \phi \rangle = 1$. Thus at each point in the unit sphere of $\mathcal{X}$ there is a unique supporting hyperplane. Show that $\mathcal{J}$ is single-valued iff $\mathcal{X}$ is smooth. Moreover, in this case, $\mathcal{J} : (\mathcal{X}, \|\cdot\|) \to (\mathcal{X}^*, \text{weak*-topology})$ is continuous.

11. The first inequality in Exercise 3 above is easy to prove and yields uniform convexity of L^p for $2 \leq p < \infty$. The third is much harder. Give a direct proof that L^p is uniformly convex for $1 < p < 2$ using the following hints. For $1 < p < 2$ and $0 < \epsilon < 1$, $g : \mathbb{R}^+ \to \mathbb{R}^+$ defined by $g(r) = |1 + re^{i\theta}|^p + |1 - re^{i\theta}|^p$ satisfies

$$g(r) \geq 2 - \epsilon + 4^{-1}p(p-1)\epsilon^{1-p/2}r^p. \tag{3.2}$$

Show this using Taylor's formula and noting $h(r) = |w + re^{i\theta}|^p$ satisfies $h''(0) \geq p(p-1)|w|^{p-2}$. Put $re^{i\theta} = z/w$ in (3.2) to get a lower bound for $|w + z|^p + |w - z|^p$. Etc.

4. Adjoint Semigroups; Stone's Theorem

At first we shall only discuss adjoint semigroups in a Hilbert space context; later we indicate that some of the results (for instance Theorem 4.3) are valid in reflexive Banach spaces. $\mathcal{H}$ will always denote a Hilbert space with inner product $\langle \cdot, \cdot \rangle$.

4.1. Definition. Let A be densely defined on $\mathcal{H}$. The (hermitean) *adjoint* A^* of A is defined by the equation

$$\langle Af, g \rangle = \langle f, A^*g \rangle, \ f \in \mathcal{D}(A), \ g \in \mathcal{D}(A^*).$$

The precise meaning of this is: $g \in \mathcal{D}(A^*)$ iff $f \to \langle Af, g \rangle$ is a bounded linear functional on $\mathcal{D}(A)$ (which we extend by continuity to $\overline{\mathcal{D}(A)} = \mathcal{H}$). Thus there is a unique $h \in \mathcal{H}$ such that $\langle Af, g \rangle = \langle f, h \rangle$ for all $f \in \mathcal{D}(A)$. In this case we define $A^*g = h$.

4.2. Here are some elementary properties of adjoints.

  (i) $A \in \mathcal{B}(\mathcal{H})$ implies $A^* \in \mathcal{B}(\mathcal{H})$ and $\|A^*\| = \|A\|$.
  (ii) $A, B \in \mathcal{B}(\mathcal{H})$ implies $(AB)^* = B^*A^*$.
  (iii) $(cA)^* = \bar{c}A^*$ if $c \in \mathbb{K}\backslash\{0\}$ and $\overline{\mathcal{D}(A)} = \mathcal{H}$.
  (iv) $(A + B)^* \supset A^* + B^*$ if $\overline{\mathcal{D}(A) \cap \mathcal{D}(B)} = \mathcal{H}$.
  (v)   If A is closed and densely defined, so is A^*. Furthermore, $A^{**} = A$.

4.3. Theorem. *Let A generate a (C_0) semigroup T on $\mathcal{H}$. Then $T^* = \{T(t)^* : t \in \mathbb{R}^+\}$ is a (C_0) semigroup on $\mathcal{H}$ whose generator is A^*.*

Proof. $T(t + s)^* = [T(t)T(s)]^* = T(s)^*T(t)^*$ for all $t,s \in \mathbb{R}^+$. Also $T(0)^* = I^* = I$. For the strong continuity, assume for the moment that for some real ω,

$$(\|T(t)^*\| =)\|T(t)\| \leq e^{\omega t}, t \in \mathbb{R}^+. \tag{4.1}$$

Then for each $f \in \mathcal{H}$,

$$\|T(t)^*f - f\|^2 = \langle T(t)^*f - f, T(t)^*f - f \rangle$$
$$= \|T(t)^*f\|^2 + \|f\|^2 - \langle f,T(t)f \rangle - \langle T(t)f,f \rangle.$$

By (4.1) and the strong continuity of T, it follows that

$$\limsup_{t \to 0} \|T(t)^*f - f\| = 0.$$

Thus $T(\cdot)^*f$ is continuous at the origin. This together with the semigroup property and (4.1) implies $T(\cdot)^*f \in C(\mathbb{R}^+,\mathcal{H})$ for each $f \in \mathcal{H}$. If (4.1) does not hold, then the strong continuity of T^* follows from Exercise 2.18.11, which is proven in Section 10. Thus T^* is a (C_0) semigroup on $\mathcal{H}$. Let B denote its generator. For $f \in \mathcal{D}(A)$, $g \in \mathcal{D}(B)$,

$$\langle Af,g \rangle = \lim_{t \to 0} \langle t^{-1}[T(t) - I]f,g \rangle$$

$$= \lim_{t \to 0} \langle f,t^{-1}[T(t)^* - I]g \rangle = \langle f,Bg \rangle.$$

Therefore $B \subset A^*$. Recall that if $g \in \mathcal{D}(A^*)$, then for all $f \in \mathcal{D}(A)$,

$$\langle f,T(t)^*g - g \rangle = \langle T(t)f - f,g \rangle = \int_0^t \langle AT(s)f,g \rangle \, ds$$

$$= \int_0^t \langle T(s)f,A^*g \rangle \, ds = \int_0^t \langle f,T(s)^*A^*g \rangle \, ds.$$

Thus

$$T(t)^*g - g = \int_0^t T(s)^*A^*g \, ds.$$

Divide by t and let $t \to 0$. It follows that $A^* \subset B$. Hence $A^* = B$. ∎

Note that the above proof includes a proof of 4.2(v) for generators A. .

4.4. DEFINITIONS.   An operator B on $\mathcal{H}$ is *symmetric* if $B \subset B^*$, i.e. B is densely defined and $\langle Bf,g \rangle = \langle f,Bg \rangle$ for all $f,g \in \mathcal{D}(B)$. B is *skew-symmetric* if $B \subset -B^*$. B is *self-adjoint* if $B^* = B$. B is *skew-adjoint* if $B^* = -B$.

When $\mathcal{H}$ is complex, B is skew-adjoint iff iB is self-adjoint.

4.5. COROLLARY.   *T is a self-adjoint (C_0) semigroup on $\mathcal{H}$ (i.e. $T(t)$ is self-adjoint for each $t \in \mathbb{R}^+$) iff its generator is self-adjoint.*

Proof.   Apply Theorem 4.3. ∎

4.6.   DEFINITIONS. $U \in \mathscr{B}(\mathscr{H})$ is *unitary* if $U^* = U^{-1}$. It is easy to check that U is unitary iff U is isometric and surjective. A (C_0) *unitary group* is a (C_0) group of unitary operators.

4.7.   STONE'S THEOREM.   *A is the generator of a (C_0) unitary group U on $\mathscr{H}$ iff A is skew-adjoint.*

Proof (Necessity).   Let A generate a (C_0) unitary group U. Then

$$t^{-1}[U(-t)f - f] = t^{-1}(U(t)^*f - f);$$

thus using Remark 2.16 and Theorem 4.3, and letting $t \to 0$, we see that $f \in \mathscr{D}(A)$ iff $f \in \mathscr{D}(A^*)$, and $A^*f = -Af$ for such f. Thus A is skew-adjoint.
    For the converse, assume $A^* = -A$. Then for all $f \in \mathscr{D}(A)$,

$$\langle Af,f \rangle = \langle f,A^*f \rangle = -\langle f,Af \rangle = -\overline{\langle Af,f \rangle}.$$

Thus $\mathrm{Re}\langle Af,f \rangle = 0$ for all $f \in \mathscr{D}(A)$. Therefore $\pm A$ are dissipative. $\mathscr{D}(A) = \mathscr{H}$ (since we assume A^* exists) and A is closed since every adjoint operator is closed (for, if $g_n \in \mathscr{D}(A^*)$, $g_n \to g$, $A^*g_n \to h$, then for all $f \in \mathscr{D}(A)$,

$$\langle Af,g \rangle = \lim_{n \to \infty} \langle Af,g_n \rangle = \lim_{n \to \infty} \langle f,A^*g_n \rangle = \langle f,h \rangle;$$

thus $g \in \mathscr{D}(A^*)$ and $A^*g = h$).
    We next show that $\pm 1 \in \rho(A)$; it then follows by Theorem 3.3 that $\pm A$ generate (C_0) contraction semigroups $U_\pm$. Suppose $(I \pm A)f = g$. Then, by dissipativity,

$$\|f\|^2 = \mathrm{Re}\langle f,g \rangle \leq \|f\| \|g\|.$$

Therefore $I \pm A$ is injective and $\|(I \pm A)^{-1}\| \leq 1$. Range $(I \pm A) = \mathscr{D}[(I \pm A)^{-1}]$ is closed since $(I \pm A)^{-1}$ is closed and bounded. To show $\pm 1 \in \rho(A)$ it remains to show that Range $(I \pm A)$ is dense in $\mathscr{H}$. To that end, let $g \perp$ Range $(I \pm A)$. Then for all $f \in \mathscr{D}(A)$, $\langle (I \pm A)f,g \rangle = 0$. Hence $g \in \mathscr{D}(A^*) = \mathscr{D}(A)$ and $(I \pm A^*)g = 0$, i.e. $A^*g = -Ag = \pm g$. Hence $\mathrm{Re}\langle Ag,g \rangle = \pm\|g\|^2$. But $\mathrm{Re}\langle Ag,g \rangle = 0$. Therefore $g = 0$.
    By Remark 2.16, A generates a (C_0) contraction group U given by

$$U(t) = \begin{cases} U_+(t) & \text{if } t \in \mathbb{R}^+ \\ U_-(-t) & \text{if } -t \in \mathbb{R}^+ \end{cases}.$$

But $U(t)^{-1} = U(-t)$ and $\|U(t)\|$, $\|U(-t)\| \leq 1$ imply that $U(t)$ is a surjective isometry. Hence U is a (C_0) unitary group on $\mathscr{H}$.   ∎

4.8.   DEFINITION.   Let $\mathscr{X}$ be a Banach space, and let A be a densely defined operator on $\mathscr{X}$. The *adjoint* A^* of A is the operator on $\mathscr{X}^*$ defined by the equation

$$\langle Af,\phi \rangle = \langle f,A^*\phi \rangle \qquad f \in \mathscr{D}(A), \phi \in \mathscr{D}(A^*).$$

More precisely, $\phi \in \mathscr{D}(A^*)$ iff there is a $\psi \in \mathscr{X}^*$ such that $\langle Af,\phi \rangle = \langle f,\psi \rangle$ holds for all $f \in \mathscr{D}(A)$; in this case $A^*\phi = \psi$.

4.9. THEOREM. *Let A generate a (C_0) semigroup T on a reflexive Banach space $\mathscr{X}$. Then $T^* = \{T(t)^*: t \in \mathbb{R}^+\}$ is a (C_0) semigroup on $\mathscr{X}^*$ whose generator is A^*.*

4.10. EXERCISES

1. Prove 4.2 in both the Hilbert and Banach space versions.
2. Prove Theorem 4.9.
3. Construct a (C_0) semigroup T on a (nonreflexive) space $\mathscr{X}$ such that T^* is not strongly continuous on $\mathscr{X}^*$, i.e. there is a $\phi \in \mathscr{X}^*$ such that

$$\limsup_{t \to 0} \| T(t)^*\phi - \phi \| > 0.$$

4. Calculate the adjoint semigroups for the examples of Exercises 2.18.4,7,8.

5. Analytic Semigroups

Throughout this section, $\mathscr{X}$ is a complex Banach space.

5.1. DEFINITIONS. For $0 < \theta \le \pi$, let Σ_θ denote the sector

$$\Sigma_\theta = \{z \in \mathbb{C} : z \ne 0, |\arg z| < \theta\}.$$

If A is an operator on $\mathscr{X}$, we say that $A \in \mathscr{GA}_b(\theta, M)$ (where $\pi/2 < \theta \le \pi$, $M \ge 1$) if A is closed, densely defined, and for all $\lambda \in \Sigma_\theta$, $\lambda \in \rho(A)$ and

$$\|(\lambda - A)^{-1}\| \le M/|\lambda|.$$

We say that $A \in \mathscr{GA}_b(\theta)$ (where $\pi/2 < \theta \le \pi$) if for each $\epsilon > 0$ (with $\epsilon < \theta - \pi/2$), there exists an $M_\epsilon \ge 1$ such that $A \in \mathscr{GA}_b(\theta - \epsilon, M_\epsilon)$. (The notation $A \in \mathscr{GA}_b$ means that A generates a (C_0) uniformly bounded semigroup which has an analytic extension into a sector of the complex plane. This will be made precise in Theorem 5.3. Note that $\bigcup_{M \ge 1} \mathscr{GA}_b(\theta, M) \subset \mathscr{GA}_b(\theta)$.)

5.2. DEFINITIONS. *An analytic semigroup of type (α, M) (where $0 < \alpha \le \pi/2$, $M \ge 1$) is a family of operators $T = \{T(t): t \in \Sigma_\alpha \cup \{0\}\}$ satisfying*

(i) $T(t)T(s) = T(t + s)$ for all $t, s \in \Sigma_\theta$, $T(0) = I$;
(ii) for each $f \in \mathscr{X}$ and each $\phi \in \mathscr{X}^*$, the complex-valued function $\langle T(\cdot)f, \phi \rangle$ is analytic on Σ_α;
(iii) $\lim_{\substack{t \to 0 \\ t \in \Sigma_{\alpha - \epsilon}}} T(t)f = f$ for each $f \in \mathscr{X}$ and each $\epsilon \in (0, \alpha)$;
(iv) it follows from (i)–(iii) that $\{T(t): t \in \mathbb{R}^+\}$ is a (C_0) semigroup on $\mathscr{X}$; let A denote its generator; then for all $t \in \Sigma_\alpha$, $T(t)(\mathscr{X}) \subset \mathscr{D}(A)$ and

$$\|T(t)\| \le M, \|tAT(t)\| \le M.$$

We call A the *generator* of T.
   $T = \{T(t): t \in \Sigma_\alpha \cup \{0\}\}$ is an *analytic semigroup of type* (α) (where $0 < \alpha \le \pi/2$) if for each $\epsilon > 0$ (with $\epsilon < \alpha$) there exists an $M_\epsilon \ge 1$ such that $\{T(t): t \in \Sigma_{\alpha - \epsilon} \cup \{0\}\}$ is an analytic semigroup of type $(\alpha - \epsilon, M_\epsilon)$.

5.3. THEOREM. *Let $A \in \mathscr{GA}_b(\theta, M)$ for some $\theta \in]\pi/2, \pi]$. Then A is the generator of an analytic semigroup of type $(\theta - \pi/2)$.*

5.4. **Theorem.** $A \in \mathscr{GA}_b(\theta)$ (where $\pi/2 < \theta \leq \pi$) iff A generates an analytic semigroup of type $(\theta - \pi/2)$.

5.5. The necessity of Theorem 5.4 follows immediately from Theorem 5.3. We omit the proof of the sufficiency; for the details see Hille-Phillips [1, pp. 384–386], Yosida [10, pp. 255–259], or A. Friedman [1, pp.106–108]. (See also Exercise 5.10.4.)

5.6. *Proof of Theorem 5.3.* Let $\epsilon > 0$ (with $2\epsilon < \theta - \pi/2$). Define $T(t)$ by the Dunford-Taylor integral

$$T(t) = \frac{1}{2\pi i} \int_C e^{\lambda t}(\lambda - A)^{-1} \, d\lambda \qquad t \in \Sigma_\alpha; \tag{5.1}$$

here $\alpha = \theta - \pi/2 - 2\epsilon$, and C is a piecewise smooth curve in Σ_θ consisting of three pieces: a segment $\{re^{-i(\theta - \epsilon)}: 1 \leq r < \infty\}$, a segment $\{re^{i(\theta - \epsilon)}: 1 \leq r < \infty\}$, and a smooth curve in Σ_θ connecting $e^{-i(\theta - \epsilon)}$ and $e^{i(\theta - \epsilon)}$, e.g. the arc $\{e^{i\beta}: -(\theta - \epsilon) \leq \beta \leq \theta - \epsilon\}$. C is oriented so that it runs from $\infty e^{-i(\theta - \epsilon)}$ to $\infty e^{i(\theta - \epsilon)}$ (see Figure 1).

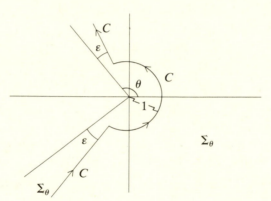

Figure 1

The integrand is continuous on $C \cap \{z: |z| \leq R\}$ for each $R > 0$, and the estimate

$$\|e^{\lambda t}(\lambda - A)^{-1}\| \leq e^{\mathrm{Re}\,\lambda t} M/|\lambda| \leq e^{-(\sin \epsilon)|\lambda t|} M/|\lambda|$$

holds for $|\lambda| \geq 1$; hence the improper integral appearing in (5.1) exists, converges absolutely, and defines a bounded operator $T(t)$ on $\mathscr{X}$.

Let C' be obtained from C by shifting each point to the right by a fixed (positive) amount. Then also

$$T(t) = \frac{1}{2\pi i} \int_{C'} e^{\lambda t}(\lambda - A)^{-1} \, d\lambda, \, t \in \Sigma_\alpha. \tag{5.2}$$

To see this, let $S(t)$ denote the bounded operator defined by the right side of (5.2), let $f \in \mathscr{X}$, and let $\phi \in \mathscr{X}^*$. Then

$$\langle T(t)f,\phi \rangle - \langle S(t)f,\phi \rangle = \frac{1}{2\pi i} \left\{ \int_C - \int_{C'} \right\} e^{\lambda t} \langle (\lambda - A)^{-1}f,\phi \rangle \, d\lambda = 0$$

by Cauchy's theorem. Since f and ϕ are arbitrary, it follows that $T(t) = S(t)$. Also 5.2.(ii) follows.

We next establish 5.2.(i). Using (5.1) and (5.2), if $t,s \in \Sigma_\alpha$,

$$T(t)T(s) = (2\pi i)^{-2} \int_C \int_{C'} e^{\lambda t} e^{\mu s} (\lambda - A)^{-1} (\mu - A)^{-1} \, d\mu \, d\lambda$$

$$= (2\pi i)^{-2} \int_C \int_{C'} e^{\lambda t + \mu s} (\mu - \lambda)^{-1} [(\lambda - A)^{-1} - (\mu - A)^{-1}] \, d\mu \, d\lambda \quad (5.3)$$

by the resolvent identity (Lemma 1.3). But

$$\int_C e^{\lambda t} (\mu - \lambda)^{-1} d\lambda = 0$$

since C lies to the left of C' and $\mu \in C'$; and

$$\int_{C'} e^{\mu s} (\mu - \lambda)^{-1} (\lambda - A)^{-1} \, d\mu = 2\pi i \, e^{\lambda s} (\lambda - A)^{-1}$$

by Cauchy's theorem, for $\lambda \in C$. Using these facts, Fubini's theorem, and (5.3) we see that

$$T(t)T(s) = (2\pi i)^{-1} \int_C e^{\lambda t} e^{\lambda s} (\lambda - A)^{-1} \, d\lambda = T(t + s) \qquad t,s \in \Sigma_\alpha.$$

It is clear that for $f \in \mathcal{X}$ and $\phi \in \mathcal{X}^*$,

$$\langle T(t)f, \phi \rangle = (2\pi i)^{-1} \int_C e^{\lambda t} \langle (\lambda - A)^{-1} f, \phi \rangle \, d\lambda$$

is an analytic function t in Σ_α, and its derivative can be computed by differentiating under the integral sign.

We next attack 5.2.(iv). In formula (5.1), change variables from λ to $\mu = |t|\lambda$, $t \in \Sigma_\alpha$. Then C is transformed to a curve $|t|C$ in Σ_θ running from $\infty e^{-i(\theta - \epsilon)}$ to $\infty e^{i(\theta - \epsilon)}$. Thus

$$T(t) = (2\pi i)^{-1} \int_{|t|C} e^{\mu t/|t|} \left(\frac{\mu}{|t|} - A \right)^{-1} \frac{d\mu}{|t|}$$

$$= (2\pi i)^{-1} \int_C e^{\mu t/|t|} \left(\frac{\mu}{|t|} - A \right)^{-1} \frac{d\mu}{|t|} \quad (5.4)$$

by Cauchy's theorem. Also, $\|(\mu/|t| - A)^{-1}\| \leq M|t|/|\mu|$, and so

$$\|T(t)\| \leq \frac{M}{2\pi} \int_C |e^{\mu t/|t|}| \, |d\mu|/|\mu| \leq M'_\epsilon < \infty.$$

Similarly, since A is closed and

$$\left\| A \left(\frac{\mu}{|t|} - A \right)^{-1} \right\| = \left\| -I + \frac{\mu}{|t|} \left(\frac{\mu}{|t|} - A \right)^{-1} \right\| \leq 1 + M,$$

we get [using Lemma 1.5 and (5.4)] $T(t)(\mathscr{X}) \subset \mathscr{D}(A)$ and

$$AT(t) = (2\pi i)^{-1} \int_C e^{\mu t/|t|} A\left(\frac{\mu}{|t|} - A\right)^{-1} \frac{d\mu}{|t|}.$$

It easily follows that

$$\|tAT(t)\| \le \frac{1+M}{2\pi} \int_C |e^{\mu t/|t|}| \, |d\mu| \le M''_\epsilon < \infty.$$

Thus, letting $M_\epsilon = \text{Max}(M'_\epsilon, M''_\epsilon)$, we have

$$\|T(t)\|, \|tAT(t)\| \le M_\epsilon$$

whenever $t \in \Sigma_\alpha = \Sigma_{\theta - \pi/2 - 2\epsilon}$.

Next we establish 5.2(iii). In the following calculation $t \in \Sigma_\alpha$. For $f \in \mathscr{D}(A)$,

$$
\begin{aligned}
T(t)f - f &= (2\pi i)^{-1} \int_C e^{\lambda t}[(\lambda - A)^{-1} - \lambda^{-1}]f \, d\lambda \\
&= (2\pi i)^{-1} \int_C e^{\lambda t}(\lambda - A)^{-1} Af \frac{d\lambda}{\lambda} \\
&\to (2\pi i)^{-1} \int_C (\lambda - A)^{-1} Af \frac{d\lambda}{\lambda} = 0
\end{aligned}
\tag{5.5}
$$

by the dominated convergence theorem (since the integrand is bounded by $M\|Af\|/|\lambda|^2$) and by Cauchy's theorem. (5.5) also holds for all $f \in \mathscr{X}$ since $\overline{\mathscr{D}(A)} = \mathscr{X}$ and since $\|T(t)\|$ is uniformly bounded in Σ_α.

It only remains to show that if B denotes the generator of the (C_0) semigroup $\{T(t): t \in \mathbb{R}^+\}$, then $B = A$. Let $f \in \mathscr{D}(A)$. Then

$$
\begin{aligned}
AT(t)f = T(t)Af &= (2\pi i)^{-1} \int_C e^{\lambda t}(\lambda - A)^{-1} Af \, d\lambda \\
&= (2\pi i)^{-1} \int_C e^{\lambda t}[-I + \lambda(\lambda - A)^{-1})f \, d\lambda \\
&= (2\pi i)^{-1} \int_C e^{\lambda t}\lambda(\lambda - A)^{-1}f \, d\lambda = \frac{d}{dt} T(t)f.
\end{aligned}
$$

Thus $T(\cdot)f \in C(\mathbb{R}^+, \mathscr{X}) \cap C^1[(0, \infty), \mathscr{X}]$ and

$$
\begin{aligned}
&\|t^{-1}[T(t)f - f] - Af\| \\
&\quad \le \|t^{-1}[T(t)f - f] - AT(t)f\| + \|AT(t)f - Af\| \\
&\quad = \|t^{-1} \int_0^t [T(s)Af - T(t)Af] \, ds\| + \|T(t)Af - Af\| \\
&\quad \to 0 \text{ as } t \to 0^+.
\end{aligned}
$$

Hence $B \supset A$. But $\rho(B) \cap \rho(A) \ne \emptyset$, hence $B = A$ (cf. 2.10). ∎

5.7. REMARK   It follows from Theorem 5.4 and from a calculation similar to the ones made in 5.6 that if T is an analytic semigroup of type (α) with generator

A, then

$$\lim_{\substack{t\to 0\\ t\in\Sigma_{\alpha-\epsilon}}} t^{-1}[T(t)f - f] = Af$$

for each $f \in \mathscr{D}(A)$ and each $\epsilon \in (0,\alpha)$.

The Hille-Yosida theorem (Theorem 2.6) and Hille's analytic semigroup theorem (Theorem 5.4) can be united into a single theorem. To do this we must first make some definitions.

5.8. DEFINITIONS. Let $0 \le a \le \infty$. Define the *sector* $S(a)$ by

$$S(a) = \begin{cases} [0,\infty[\text{ if } a = 0 \\ \{z \in \mathbb{C}: |\operatorname{Im} z| \le a\operatorname{Re} z\} \text{ if } 0 < a < \infty, \\ \{z \in \mathbb{C}: \operatorname{Re} z \ge 0\} \text{ if } a = \infty; \end{cases}$$

define further

$$-S(a) = \{z \in \mathbb{C}: -z \in S(a)\},$$
$$\tilde{S}(a) = S(1/a).$$

An *analytic contraction semigroup of type* $\tilde{S}(a)$ on $\mathscr{X}$ is a function $T\colon \tilde{S}(a) \to \mathscr{B}(\mathscr{X})$ satisfying

(i) $T(t)\,T(s) = T(t+s)$ for all $t,s \in \tilde{S}(a)$, $T(0) = I$;
(ii) for each $f \in \mathscr{X}$ and each $\phi \in \mathscr{X}^*$, $\langle T(\cdot)f,\phi\rangle$ is analytic in the interior (in $\mathbb{C}$) of $\tilde{S}(a)$;
(iii) $\lim_{\substack{t\to 0\\ t\in\tilde{S}(a+\epsilon)}} T(t)f = f$ for each $f \in \mathscr{X}$ and each $\epsilon \in (0,1/a]$ if $a < \infty$;

otherwise $\lim_{\substack{t\to 0\\ t>0}} T(t)f = f$ if $a = \infty$;
(iv) $\|T(t)\| \le 1$ for each $t \in \tilde{S}(a)$.

An operator A on $\mathscr{X}$ is *sectorial of type* $S(a)$ if, for all $f \in \mathscr{D}(A)$ and all $\phi \in \mathscr{J}(f)$ (see Definition 3.1 for the duality map $\mathscr{J}$),

$$\langle Af,\phi\rangle \in S(a).$$

If, in addition, $1 \in \rho(A)$, then A is *m-sectorial of type* $S(a)$. A is *dissipative of type* $\tilde{S}(a)$ if

$$\|(\lambda - A)f\| \ge \operatorname{dist}[-\lambda,S(a)]\,\|f\|$$

whenever $f \in \mathscr{D}(A)$ and $-\lambda \notin S(a)$; here $\operatorname{dist}[-\lambda,S(a)]$ is the distance (in $\mathbb{C}$) from $-\lambda$ to $S(a)$. A is *m-dissipative of type* $\tilde{S}(a)$ provided that, in addition, $1 \in \rho(A)$.

5.9. THEOREM. *The following are equivalent for $0 \le a \le \infty$.*
(1) *A generates an analytic contraction semigroup of type $\tilde{S}(a)$.*
(2) *$-A$ is m-sectorial of type $S(a)$ and densely defined on $\mathscr{X}$.*
(3) *A is m-dissipative of type $\tilde{S}(a)$ and densely defined in $\mathscr{X}$.*
  The proof is omitted.

5.10.   EXERCISES

1. Let B on $\mathcal{X}$ be closed and densely defined; suppose $\sigma(B)$ is bounded and there is an $M > 0$ such that

$$\|(\lambda - B)^{-1}\| \le M/|\lambda| \text{ whenever } |\lambda| > M.$$

   Then $B \in \mathcal{B}(\mathcal{X})$.

2. Let $A \in \mathcal{GA}_b(\theta,M)$ where $\pi/2 < \theta \le \pi$. Let $n \in \mathbb{N}$ and let $\epsilon > 0$. Then if T denotes the analytic semigroup generated by A, $T(t)(\mathcal{X}) \subset \mathcal{D}(A^n)$ and there is a constant $M_{n,\epsilon}$ such that $\|A^n T(t)\| \le M_{n,\epsilon}|t|^{-n}$ for $t \in \Sigma_{\theta - \pi/2 - \epsilon}$.

3. The above exercise shows that $T(t)(\mathcal{X}) \subset \bigcap_{n=1}^{\infty} \mathcal{D}(A^n)$ if T is an analytic semigroup. Now let T be any (C_0) semigroup with generator A. Then $\bigcap_{n=1}^{\infty} \mathcal{D}(A^n)$ is dense in $\mathcal{X}$. (Hint: for each $f \in \mathcal{X}$, and each $\phi \in C^{\infty}(0,\infty)$ having compact support, $\int_{\mathbb{R}^+} \phi(t)\, T(t)f\, dt \in \mathcal{D}(A^n)$ for $n = 1, 2, \ldots$.)

*4. Prove the sufficiency of Theorem 5.4. Here are some hints. Let A generate a (C_0) semigroup T satisfying $T(t)\mathcal{X} \subset \mathcal{D}(A)$ for $t > 0$ and $\sup_{t>0}\{\|T(t)\|, \|tAT(t)\|\} = C < \infty$. It suffices to show that for any $\epsilon > 0$, $A - \epsilon I \in \mathcal{GA}_b(\theta,M)$ for some $\theta \in (\pi/2,\pi]$ and some $M \ge 1$. Use

$$\frac{dT(t)}{dt} = AT(t) = T(t - s)AT(s)$$

   to obtain

$$d^n T(t)/dt^n = [AT(t/n)]^n \text{ and } \left(\frac{t}{n}\right)^n \left\|\frac{d^n T(t)}{dt^n}\right\| \le C^n.$$

   Then for fixed $t > 0$, $\Sigma_{n=0}^{\infty} 1/n!\, (\lambda - t)^n\, d^n T(t)/dt^n$ converges absolutely for $\{\lambda \in \mathbb{C}: |\lambda - z| < t/Ce\}$. This gives an analytic continuation of T into a sector $\{\lambda \in \mathbb{C}: |\lambda - t| < t/Ce\}$. Use

$$(1 + i\lambda - A)^{-1} = \int_0^{\infty} e^{-i\lambda t} e^{-t} T(t)\, dt$$

   to obtain $\lim_{\lambda \to \infty} \sup\|\lambda(1 + i\lambda - A)^{-1}\| < \infty$. This estimate together with $\sup_{n \ge 1, Re\,\lambda > 0}\|(Re\,\lambda)^n(\lambda - A)^{-n}\| < \infty$ implies the desired conclusion.

*5. Let A generate a (C_0) semigroup T satisfying $\lim_{t \to 0}\|T(t) - I\| < 2$. Then there exist constants $k \in \mathbb{R}^+$, $\theta \in (\pi/2,\pi)$ such that $A - kI \in \mathcal{GA}_b(\theta)$.

6. Let $A \in \mathcal{GA}_b(\theta)$ where $\theta > \pi/4$. Then $-A^2 \in \mathcal{GA}_b(2\theta - \pi/2)$. Thus if $A \in \mathcal{GA}_b(\pi/2)$, then $(-1)^{n+1}A^{2^n} \in \mathcal{GA}_b(\pi/2)$ for all $n \in \mathbb{N}$.

7. Which of the semigroups of Exercises 2.18.4,7,8 are analytic?

*8. Prove Remark 5.7 directly by noting that if $|\gamma| < \alpha$, $\{S_\gamma(t) = T(te^{i\gamma}): t \in \mathbb{R}^+\}$ is a (C_0) semigroup on $\mathcal{X}$, relating the generator A_γ of S_γ to A, and using the analyticity of $\langle T(\cdot)f,\phi\rangle$ on Σ_α for each $(f,\phi) \in \mathcal{X} \times \mathcal{X}^*$.

6.   Perturbation Theory

6.1.   THEOREM.   *Let A generate a (C_0) contraction semigroup. Let B be dissipative with $\mathcal{D}(B) \supset \mathcal{D}(A)$. Assume there are constants $0 \le a < 1$, $b \ge 0$ such that*

$$\|Bf\| \le a\|Af\| + b\|f\| \text{ for all } f \in \mathcal{D}(A). \tag{6.1}$$

Then $A + B$ [with domain $\mathcal{D}(A)$] generates a (C_0) contraction semigroup.

Proof. Before proving the theorem, we remark that since B is closable and since $D(B) \supset \mathcal{D}(A)$, the closed graph theorem implies the existence of $a,b \in \mathbb{R}^+$ such that (6.1) holds. What we are assuming is that $a < 1$.

Case (I): $a < 1/2$. Since B is dissipative with respect to some duality section J and A is dissipative with respect to each duality section (by Theorem 3.3), it follows that $A + B$ is dissipative (with respect to J). According to the Lumer-Phillips theorem (Theorem 3.3) it remains to show that Range $[\lambda - (A + B)] = \mathcal{X}$ for some $\lambda > 0$. But $\mathcal{D}(B) \supset \mathcal{D}(A)$ implies

$$\text{Range } [\lambda - (A + B)] \supset \text{Range } \{[\lambda - (A + B)](\lambda - A)^{-1}\}$$

$$= \text{Range } [I - B(\lambda - A)^{-1}].$$

Therefore it suffices to show that $\|B(\lambda - A)^{-1}\| < 1$ for some $\lambda > 0$ (since if $C \in \mathcal{B}(\mathcal{X})$ satisfies $\|C\| < 1$, then $I - C$ is invertible in $\mathcal{B}(\mathcal{X})$ and $(I - C)^{-1} = \sum_{n=0}^{\infty} C^n$). But

$$\|B(\lambda - A)^{-1}f\| \leq a\|A(\lambda - A)^{-1}f\| + b\|(\lambda - A)^{-1}f\|$$

$$= a\|\{\lambda(\lambda - A)^{-1} - I\}f\| + b\|(\lambda - A)^{-1}f\|$$

$$\leq 2a\|f\| + \frac{b}{\lambda}\|f\| = \left(2a + \frac{b}{\lambda}\right)\|f\|$$

for all $f \in \mathcal{X}$. Choosing λ so large that $2a + b/\lambda < 1$ completes the proof.

Case (II): $a < 1$. Let $0 \leq \alpha \leq 1$, $f \in \mathcal{D}(A)$. Then

$$\|(A + \alpha B)f\| \geq \|Af\| - \alpha\|Bf\| \geq \|Af\| - \|Bf\|$$

$$\geq (1 - a)\|Af\| - b\|f\|.$$

Choose and fix an integer n such that $a/n < (1 - a)/4$. Then

$$\left\|\frac{1}{n}Bf\right\| \leq \frac{a}{n}\|Af\| + \frac{b}{n}\|f\| \leq \frac{1 - a}{4}\|Af\| + \frac{b}{n}\|f\|$$

$$\leq \frac{1}{4}\|(A + \alpha B)f\| + \left(\frac{b}{n} + \frac{b}{4}\right)\|f\|.$$

Hence by Case (I), if $A + \alpha B$ generates a (C_0) contraction semigroup, so does $A + \alpha B + (1/n)B$. A does, therefore so does $A + (1/n)B$, and so does $A + (2/n)B, \ldots$, and so does $A + (n/n)B = A + B$. $\blacksquare$

The above theorem fails if $a = 1$; for example, let A be an unbounded skew-adjoint operator and let $B = -A$. Then $A + B = 0|_{\mathcal{D}(A)}$ is not closed. However, the closure of $\overline{A + B}$ is m-dissipative, i.e. generates a (C_0) contraction semigroup. This is typical, according to the following result.

6.2. THEOREM. *Let A be m-dissipative and densely defined, and let B be dissipative with $\mathcal{D}(B) \supset \mathcal{D}(A)$. Assume there is a constant $b \geq 0$ such that*

$$\|Bf\| \leq \|Af\| + b\|f\|$$

for all $f \in \mathcal{D}(A)$. Finally, suppose that the adjoint B^* of B is densely defined. Then the closure $\overline{A + B}$ of $A + B$ is m-dissipative.

Proof. $A + B$ is dissipative (cf. the argument given in Theorem 6.1.). Moreover, its closure is dissipative, and Range $[I - (\overline{A + B})]$ is closed. (Compare Exercise 3.10.9.) Thus it suffices to show that $I - (A + B)$ has dense range. Choose $\phi \in \mathcal{X}^*$ such that $\langle f - (A + B)f, \phi \rangle = 0$ for all $f \in \mathcal{D}(A)$. Choose $g \in \mathcal{X}$ such that $\|g\| = \|\phi\|$, $\langle g, \phi \rangle \geq (1/2)\|\phi\|^2$. By Theorem 6.1, for $0 < t < 1$, $A + tB$ is m-dissipative and there is an f_t in $\mathcal{D}(A)$ such that

$$g = (I - A - tB)f_t.$$

By dissipativity of $A + tB$, $\|f_t\| \leq \|g\|$. Moreover, by the hypothesis of the theorem,

$$\|Bf_t\| \leq \|Af_t\| + b\|f_t\|$$
$$\leq \|(A + tB)f_t\| + \|tBf_t\| + b\|f_t\|$$
$$= \|f_t - g\| + t\|Bf_t\| + b\|f_t\|,$$

whence

$$\|(1 - t)Bf_t\| \leq \|f_t - g\| + b\|f_t\| \leq (1 + b)\|f_t\| + \|g\| \leq (2 + b)\|g\|.$$

For all ψ in the dense set $\mathcal{D}(B^*)$,

$$|\langle (1 - t)Bf_t, \psi \rangle| = (1 - t)|\langle f_t, B^*\psi \rangle| \leq (1 - t)\|g\|\|B^*\psi\| \to 0$$

as $t \to 1$. It follows that $(1 - t)Bf_t \to 0$ weakly as $t \to 1$. Consequently, since ϕ annihilates the range of $I - (A + B)$,

$$(1/2)\|\phi\|^2 \leq \langle g, \phi \rangle = \langle f_t - Af_t - tBf_t, \phi \rangle$$
$$= \langle [I - (A + B)]f_t, \phi \rangle + \langle (1 - t)Bf_t, \phi \rangle \to 0$$

as $t \to 1$, whence $\phi = 0$. ∎

6.3. COROLLARY. *The assumption in Theorem 6.3 that $\mathcal{D}(B^*)$ is dense may be dropped if $\mathcal{X}$ is reflexive.*

Proof. If B is closable and densely defined, then so is B^* as long as $\mathcal{X}$ is reflexive. ∎

6.4. THEOREM. *Let A generate a (C_0) semigroup and let $B \in \mathcal{B}(\mathcal{X})$. Then $A + B$ generates a (C_0) semigroup.*

Proof. Since $A + B = (A - \omega I) + (B + \omega I)$ we may suppose that the semigroup generated by A is uniformly bounded. Renorming $(\mathcal{X}, \|\cdot\|)$ with an equivalent norm $\||\cdot\||$ (cf. 2.13) makes A the generator of a (C_0) contraction semigroup on $(\mathcal{X}, \||\cdot\||)$. Next, $B - \||B\||I$ is dissipative and $\||(B - \||B\||I)f\|| \leq 2\||B\|| \, \||f\||$ for $f \in \mathcal{D}(A)$. Thus Theorem 6.1 applies (with $a = 0$, $b = 2\||B\||$) and tells us that $A + B - \||B\||I$ generates a (C_0) semigroup on $(\mathcal{X}, \||\cdot\||)$. Hence $A + B = (A + B - \||B\||I) + \||B\||I$ generates a (C_0) semigroup on $(\mathcal{X}, \|\cdot\|)$. ∎

6.5.   THEOREM.   *Let A generate a (C_0) semigroup T. We say that $B \in \mathscr{P}(A)$ iff*

(i) $\mathscr{D}(B) \supset \mathscr{D}(A)$ *and B is closed;*
(ii)   *for each $t > 0$ there is a $K(t) \in \mathbb{R}^+$ such that $\|BT(t)f\| \leq K(t)\|f\|$ for each $f \in \mathscr{D}(A)$;*
(iii) *$K(\cdot)$ in (ii) can be chosen so that $\int_0^1 K(t)dt < \infty$.*

If $B \in \mathscr{P}(A)$, then $A + B$ [with domain $\mathscr{D}(A)$] generates a (C_0) semigroup S given by

$$S(t) = \sum_{n=0}^{\infty} S_n(t), \quad S_0(t) = T(t),$$

$$S_{n+1}(t)f = \int_0^t T(t-s)BS_n(s)f \, ds, \, f \in \mathscr{X}.$$

The series is absolutely convergent, uniformly on compacta of $\mathbb{R}^+$.

We omit the proof (see Dunford-Schwartz [1, pp. 631–639]). Note that $\mathscr{B}(\mathscr{X}) \subset \mathscr{P}(A)$ (if $B \in \mathscr{B}(\mathscr{X})$, take $K(t) = \|B\| \cdot \|T(t)\| \leq \|B\|Me^{\omega t}$) so that Theorem 6.4 is a special case of Theorem 6.5. ∎

6.6.   THEOREM.   *Let $A \in \mathscr{GA}_b(\theta,M)$ where $\pi/2 < \theta \leq \pi$, $M \geq 1$. Let B on $\mathscr{X}$ satisfy*

(i) $\mathscr{D}(B) \supset \mathscr{D}(A)$,
(ii)   *there are constants $0 \leq a < 1/(M+1)$, $b \geq 0$ such that*

$$\|Bf\| \leq a\|Af\| + b\|f\|$$

for all $f \in \mathscr{D}(A)$. Then there are positive constants k, M' such that $A + B - kI \in \mathscr{G}A_b(\theta,M')$.

Proof.   For all $g \in \mathscr{X}$, $\lambda \in \Sigma_\theta$, $k \in \mathbb{R}^+$, we have $\lambda + k \in \Sigma_\theta$ and

$$\|B(\lambda + k - A)^{-1}g\| \leq a\|A(\lambda + k - A)^{-1}g\| + b\|(\lambda + k - A)^{-1}g\|$$

$$\leq a\|g\| + a\|(\lambda + k)(\lambda + k - A)^{-1}g\| + \frac{bM}{|\lambda + k|}\|g\|$$

$$\leq \left\{a(1 + M) + \frac{bM}{|\lambda + k|}\right\}\|g\|.$$

By choosing $k = k(\theta)$ large enough, $1/|\lambda + k|$ can be made arbitrarily small independent of $\lambda \in \Sigma_\theta$. Since $a(1 + M) < 1$, we have

$$\|B(\lambda + k - A)^{-1}g\| \leq c\|g\| \qquad (6.2)$$

for a fixed (large) k, some $c < 1$, and all $g \in \mathscr{X}$. Thus for $\lambda \in \Sigma_\theta$, $\lambda + k \in \rho(A + B)$ [i.e. $\lambda \in \rho(A + B - kI)$] and

$$[\lambda + k - (A + B)]^{-1} = (\lambda + k - A)^{-1}[I - B(\lambda + k - A]^{-1})^{-1}$$

by (6.2). Moreover, there is a positive constant C such that $|\lambda| \leq C|\lambda + k|$ for all

$\lambda \in \Sigma_\theta$. Hence for all $\lambda \in \Sigma_\theta$, $\lambda \in \rho(A + B - kI)$ and

$$\|[\lambda - (A + B - kI)]^{-1}\| = \|[\lambda + k - (A + B)]^{-1}\|$$

$$\leq \|(\lambda + k - A)^{-1}\|\|[I - B(\lambda + k - A)^{-1}]^{-1}\|$$

$$\leq \frac{M}{|\lambda + k|} \cdot \frac{1}{(1 - c)} \leq \frac{M'}{|\lambda|}$$

by (6.2), where $M' = MC/(1 - c)$. This completes the proof.   ∎

6.7. Definitions

   (i) $A \in \mathscr{GA}(\theta, M)$ if $A - kI \in \mathscr{GA}_b(\theta, M)$ for some $k \in \mathbb{R}^+$.

  (ii) $A \in \mathscr{GA}(\theta)$ if $A - kI \in \mathscr{GA}_b(\theta)$ for some $k \in \mathbb{R}^+$

 (iii) B is a *Kato perturbation* of A if $\mathscr{D}(B) \supset \mathscr{D}(A)$ and if for each $a > 0$, there is a $b = b(a) > 0$ such that

$$\|Bf\| \leq a|Af\| + b\|f\|$$

for each $f \in \mathscr{D}(A)$.

6.8. Corollary.   *If A generates (C_0) contraction semigroup and if B is a dissipative Kato perturbation of A, then $A + B$ generates a (C_0) contraction semigroup.*

6.9. Corollary.   *Let $A \in \mathscr{GA}(\theta)$ and let B be a Kato perturbation of A. Then $A + B \in \mathscr{GA}(\theta)$.*

6.10. Remark.   A finite linear combination of Kato perturbations of A is also a Kato perturbation of A.

6.11. Corollary.   *If A on a complex Hilbert space $\mathscr{H}$ is self-adjoint and if B is a symmetric Kato perturbation of A, then $A + B$ [with domain $\mathscr{D}(A)$] is self-adjoint.*

Proof.   This follows from Corollary 6.8 and Stone's theorem (Theorem 4.7).   ∎

6.12. Exercises

  1. Prove that Theorem 6.2 remains valid if the assumption that $\mathscr{D}(B^*)$ is dense is replaced by the assumption that $\mathscr{D}(A^*)$ is dense.

  2. This illustrates Theorem 6.5. Let $\mathscr{X} = BUC(\mathbb{R})$, $A = d^2/dx^2$, $\mathscr{D}(A) = \{f \in \mathscr{X} : f', f'' \in \mathscr{X}\}$. Show that A generates the (C_0) contraction semigroup T given by

$$T(t)f(x) = \frac{1}{\sqrt{4\pi t}} \int_{\mathbb{R}} e^{-y^2/4t} f(x + y)\, dy.$$

     (See Exercise 2.18.7.) Let $Bf(x) = h(x)f'(x)$, where $h \in \mathscr{X}$ and $\mathscr{D}(B) = \{f \in \mathscr{X} : f$ is C^1 in a neighborhood of each x_0 for which $h(x_0) \neq 0$, and $hf' \in \mathscr{X}\}$. Then $B \in \mathscr{P}(A)$ [and $K(t) = \|h\|_\infty/\sqrt{\pi t}$]. For hints see Dunford-Schwartz [1, pp. 639–640].

  3. Let A, B be as in Exercise 2 above. Show that $A + B$ is m-dissipative, using

Theorem 6.1. Does applying Theorem 6.1 involve less work than applying Theorem 6.6?

4. Let A be a self-adjoint operator on a complex Hilbert space $\mathscr{H}$. Let B be a symmetric operator on $\mathscr{H}$ with $\mathscr{D}(B) \supset \mathscr{D}(A)$. Suppose there is a $b > 0$ such that

$$\|Bf\| \le \|Af\| + b\|f\|$$

for all $f \in \mathscr{D}(A)$. Then the closure $\overline{A + B}$ of $A + B$ is self-adjoint.

*5. Let $\mathscr{H}$ be the complex Hilbert space $L^2(\mathbb{R}^+)$. Define $\mathscr{D}(S) = \{f \in \mathscr{H}:$ f absolutely continuous on $\mathbb{R}^+$, $f(0) = 0$, $f' \in \mathscr{H}\}$ and let $Sf = -f'$ for f $\in \mathscr{D}(S)$. S is m-dissipative, and $-S$ is dissipative but not m-dissipative. Let Rf $= f''$ on $\mathscr{D}(R) = \{f \in \mathscr{H}: f', f'' \in \mathscr{H}, f(0) = 0\}$. $\pm S$ is a Kato perturbation of R, and $\pm iR + \alpha S$ is m-dissipative for each $\alpha \in \mathbb{R}$. On the other hand, for $\beta > 0$, $(-iR - \beta S) + iR$ is not m-dissipative. Deduce that the conclusion of Theorem 6.1 can fail for any $a > 1$.

6. Let $\mathscr{H}$ be a Hilbert space. Let A be m-dissipative and B dissipative with $\mathscr{D}(B) \supset \mathscr{D}(A)$. Let there exist constants a,b with $a \le 1$ such that

$$0 \le \operatorname{Re}\langle Af, Bf \rangle + a\|Af\|^2 + b\|f\|^2$$

holds for all $f \in \mathscr{D}(A)$. Then $A + B$ is m-dissipative, and $A + B$ is m-dissipative if $a < 1$.

7. Let $\mathscr{X}$ be reflexive. Let A and B be m-dissipative. Let

$$B_n = n[(I - (1/n)B)^{-1} - I]$$

be the Yosida approximation of B. (B_n is B_λ with $\lambda = 1/n$ in the notation of 2.10.) For $f \in \mathscr{H}$ let $g_n = (I - A - B_n)^{-1}f$. If $sup_n\|B_ng_n\| < \infty$ for each $f \in \mathscr{X}$, then $A + B$ is m-dissipative.

8. Let $A + \epsilon I$ be an m-dissipative operator on a Hilbert space $\mathscr{H}$ for some $\epsilon > 0$. Let B be dissipative with $\mathscr{D}(A) \subset \mathscr{D}(B)$. Suppose there is a $c > 0$ such that for all $f \in \mathscr{D}(A)$,

$$\operatorname{Im}\langle Af, Bf \rangle \ge c\operatorname{Re}\langle Af, f \rangle. \tag{6.3}$$

Conclude that $\overline{A + B}$ is m-dissipative.

9. Prove an analogue of the above result in which the conclusion is that $\overline{A + B}$ is self-adjoint. Interpret the condition (6.3) as a condition involving commutators.

*10. Let $\{A(t): 0 \le t \le 1\}$ be a family of dissipative operators on $\mathscr{X}$ having a common domain D. Suppose that $\cap \{\mathscr{D}\{[A(t) - A(1)]^*\}: 0 \le t \le 1\}$ is dense in $\mathscr{X}^*$, and that $A(t)$ is m-dissipative for $0 \le t < 1$.
  (i) Assume
    (α) $\lim_{t \to 1}\|A(t)f - A(1)f\| = 0$ for each $f \in D$, and
    (β) there are constants C_1, C_2 such that

$$\|A(t)f - A(1)f\| \le C_1\|A(t)f\| + C_2\|f\|$$

    for all $f \in D$ and $0 \le t \le 1$.
Then $A(1)$ is m-dissipative.
  (ii) Assume that $A(1)$ is m-dissipative and (α') $A(\cdot)f \in C^1([0,1],\mathscr{X})$ for $f \in D$. Then (β) holds.
  (iii) The conclusion in (ii) is false if (α') is replaced by (α'') $A(\cdot)f \in C([0,1],\mathscr{X})$ for each $f \in D$.

7.   Approximation Theory

The approximation theorem says that a (C_0) semigroup depends continuously on its generator. We shall present several versions of this theorem.

7.1.   Definition.   Let A be a closed operator on $\mathscr{X}$. A *core* of A is a subspace $\mathscr{D} \subset \mathscr{X}$ such that A is the closure of its restriction to $\mathscr{D}$; in symbols, $\overline{A|_{\mathscr{D}}} = A$.

7.2.   Theorem.   *Let A_n $(n \in \mathbb{N}_0 = \{0,1,2,\dots\})$ generate a (C_0) semigroup T_n satisfying the "stability condition"*

$$\|T_n(t)\| \le Me^{\omega t}, n \in \mathbb{N}_0 \qquad t \in \mathbb{R}^+, \tag{7.1}$$

where M, ω are independent of n, t. Let $\mathscr{D}$ be a core for A_0. Assume that for each $f \in \mathscr{D}$,

$$f \in \liminf_{n \to \infty} \mathscr{D}(A_n)^{\dagger} \text{ and } \lim_{n \to \infty} A_n f = A_0 f.$$

Then $\lim_{n \to \infty} T_n(t)g = T_0(t)g$ for each $g \in \mathscr{X}$, uniformly for t in compact subsets of $\mathbb{R}^+$.

7.3.   Theorem (Trotter, Neveu, Kato).   *Let $A_n(n \in \mathbb{N}_0)$ generate a (C_0) semigroup T_n satisfying the stability condition (7.1) with M, ω independent of n, t.*

(i) $\lim_{n \to \infty} T_n(t)f = T_0(t)f$ *for each $t \in \mathbb{R}^+$, $f \in \mathscr{X}$ implies*
$$\lim_{n \to \infty} (\lambda - A_n)^{-1} f = (\lambda - A_0)^{-1} f$$
for each $f \in \mathscr{X}$, uniformly for λ in compact subsets of (ω, ∞).

(ii) $\lim_{n \to \infty} (\lambda - A_n)^{-1} f = (\lambda - A_0)^{-1} f$ *for each $f \in \mathscr{X}$, $\lambda > \omega$ implies $\lim_{n \to \infty} T_n(t)f = T_0(t)f$ for each $f \in \mathscr{X}$, uniformly for t in compact subsets of $\mathbb{R}^+$.*

7.4.   First we prove Theorem 7.2, assuming Theorem 7.3. We only need show that for $\lambda > \omega$,

$$\lim (\lambda - A_n)^{-1} g = (\lambda - A_0)^{-1} g \text{ for each } g \in \mathscr{X}; \tag{7.2}$$

then Theorem 7.3 (ii) implies the result. It suffices to prove (7.2) for all g of the form $g = (\lambda - A_0)f$, $f \in \mathscr{D}$, since $(\lambda - A_0)(\mathscr{D})$ is dense in $\mathscr{X}$ and $\|(\lambda - A_n)^{-1}\|$, $\|(\lambda - A_0)^{-1}\| \le M/(\lambda - \omega)$. For such g,

$$\|(\lambda - A_n)^{-1} g - (\lambda - A_0)^{-1} g\|$$
$$= \|(\lambda - A_n)^{-1}(\lambda - A_n)f + (\lambda - A_n)^{-1}(\lambda - A_0 - \lambda + A_n)f - f\|$$
$$= \|(\lambda - A_n)^{-1}(A_n - A_0)f\| \le M\|A_n f - A_0 f\|/(\lambda - \omega) \to 0$$

as $n \to \infty$, proving (7.2).   ∎

7.5.   *Proof of Theorem 7.3.*   Recall that for $\lambda > \omega$, $\lambda \in \rho(A_n)$ and

$$(\lambda - A_n)^{-1} f = \int_0^{\infty} e^{-\lambda t} T_n(t) f \, dt$$

$^{\dagger} \displaystyle\liminf_{n \to \infty} \mathscr{D}(A_n) = \bigcup_{n=1}^{\infty} \bigcup_{m=n}^{\infty} \mathscr{D}(A_m)$

for each $f \in \mathscr{X}$. Theorem 7.3(i) then follows by the dominated convergence theorem in view of the stability condition (7.1).

The idea of the proof of (ii) is to deduce the conclusion as a consequence of the semigroup generation theorem (Theorem 2.13) by viewing a convergent sequence of semigroups (or resolvents) on $\mathscr{X}$ as a single semigroup (or resolvent) on the space of convergent sequences in $\mathscr{X}$.

Let

$$X = \left\{ f = \{f_n\}_0^\infty \subset \mathscr{X} : \lim_{n \to \infty} f_n = f_0 \right\}.$$

X is a Banach space under the norm

$$|f| = \sup_n \|f_n\|.$$

Define A on X as follows. $f \in \mathscr{D}(A)$ and $Af = g$ means: $f = \{f_n\}_0^\infty$, $g = \{g_n\}_0^\infty \in X$, and for each $n \in \mathbb{N}_0$, $f_n \in \mathscr{D}(A_n)$ and $A_n f_n = g_n$. (Note that $f_n \to f_0$ and $A_n f_n = g_n \to g_0 = A_0 f_0$.) It is easy to see that $\lambda - A$ is injective for $\lambda > \omega$. For $\lambda > \omega$, Theorem 2.13 implies

$$|(\lambda - \omega)^m (\lambda - A)^{-m} f| = \sup_n \|(\lambda - \omega)^m (\lambda - A_n)^{-m} f_n\|$$

$$\leq M \sup_n \|f_n\| = M|f|.$$

Moreover, for $\lambda > \omega$ and $g \in \mathscr{X}$, let $f_n = (\lambda - A_n)^{-1} g_n$, $n \in \mathbb{N}_0$. Our hypothesis about the convergence of $(\lambda - A_n)^{-1}$ to $(\lambda - A_0)^{-1}$ together with $g_n \to g_0$ implies $f = (f_n)_0^\infty \in \mathscr{D}(A)$ and $(\lambda - A)f = g$. One can easily check that $\mathscr{D}(A)$ is dense in X (see Exercise 7.9.3). Theorem 2.13 now implies that A generates a semigroup T on X satisfying $\|T(t)\| \leq Me^{\omega t}$. For $f \in X$, $T(t)f = \{T_n(t)f_n\}_0^\infty$; whence $f_n \to f_0$ implies $T_n(t)f_n \to T_0(t)f_0$ since $T(t)f \in X$ and X is the space of convergent sequences having their limit in the zeroth coordinate. Finally, the uniformity assertion in (ii) is an easy exercise in epsilontics; we omit the argument. ∎

7.6. REMARK. *Let A_0 be any unbounded skew-adjoint operator on a complex Hilbert space $\mathscr{H}$. Then there is a sequence of skew-adjoint operators A_n ($n \in \mathbb{N} = \{1, 2, \ldots\}$) on $\mathscr{H}$ such that*

$$\lim_{n \to \infty} (\lambda - A_n)^{-1} f = (\lambda - A_0)^{-1} f$$

holds for all $f \in \mathscr{H}$ and all $0 \neq \lambda \in \mathbb{R}$, and yet

$$\mathscr{D}(A_0) \cap \bigcup_{n=1}^\infty \mathscr{D}(A_n) = \{0\}.$$

Since each A_n generates a (C_0) unitary group according to Stone's theorem (Theorem 4.7), we conclude that Theorem 7.3 is a stronger result than Theorem 7.2. In the applications, however, Theorem 7.2 is usually used.

7.7. *Proof of Remark 7.6.* Let A_0 be skew-adjoint and unbounded. According to Dixmier's generalization of a theorem of von Neumann (cf. Fillmore-Williams [1, p. 274] for a proof), there is a (C_0) unitary group $\{U(t): t \in \mathbb{R}\}$ such that

$\mathscr{D}[U(t)A_0 U(-t)] \cap \mathscr{D}(A_0) = \{0\}$ for each real $t \neq 0$. The remark follows by letting $\{t_n\}$ be a sequence in $\mathbb{R}\backslash\{0\}$ converging to zero and taking $A_n = U(t_n)A_0 U(-t_n)$. ∎

A specific example of a sequence $\{A_n\}_0^\infty$ satisfying the conditions of Remark 7.6 is discussed in Example 8.16.

7.8. Now we present a version of the approximation theorem in which $(\lambda - A_n)^{-1}f$ is assumed to be convergent to some limit which we *prove* is $(\lambda - A_0)^{-1}f$ for a suitable generator A_0. We also show that it is enough to have the convergence of $(\lambda - A_n)^{-1}$ for one value of λ. Compare this with the context of the Lumer-Phillips theorem (Theorem 3.3); for A dissipative, the fact that $\lambda - A$ is surjective for one value of $\lambda > 0$ implies that the same holds for all $\lambda > 0$.

THEOREM (Trotter-Neveu-Kato). *Let* A_n $(n \in \mathbb{N})$ *generate* (C_0) *semigroups* T_n *satisfying* $\|T_n(t)\| \leq Me^{\omega t}$ *with M and ω being independent of t and n. Let λ_0 satisfy* $\mathrm{Re}(\lambda_0) > \omega$, *and suppose*

$$\lim_{n \to \infty} (\lambda_0 - A_n)^{-1}f = Rf$$

holds for all $f \in \mathscr{X}$ and some injective operator R having dense range. Then a semigroup generator A_0 and its corresponding semigroup T_0 exist such that $R = (\lambda_0 - A_0)^{-1}$ and $\lim_{n \to \infty} T_n(t)f = T_0(t)f$ holds for all $f \in \mathscr{X}$, uniformly for t in bounded intervals in $\mathbb{R}^+$.

Proof. To begin with, we may (and do) assume $\omega = 0$ by replacing A_n by $A_n - \omega I$ for each n. For $\lambda, \mu \in \mathbb{K}$, $\mathrm{Re}(\lambda) > 0$, $\mathrm{Re}(\mu) > 0$, and $n \in \mathbb{N}$, we have $\lambda, \mu \in \rho(A_n)$ and the identity

$$(\mu - A_n)^{-1}f = (\lambda - A_n)^{-1} \sum_{m=0}^{\infty} (\lambda - \mu)^m (\lambda - A_n)^{-m} f$$

for $f \in \mathscr{X}$ [cf. equation (3.1)]. For fixed λ and ϵ with $0 < \epsilon < \mathrm{Re}(\lambda)$, this series converges uniformly in μ for $M|\mu - \lambda| \leq \mathrm{Re}(\lambda) - \epsilon$. Hence the convergence of $(\lambda - A_n)^{-1}f$ for all $f \in \mathscr{X}$ implies the convergence of $(\mu - A_n)^{-1}f$ for $M|\mu - \lambda| < \mathrm{Re}(\lambda)$. We claim that the set

$$S = \{\mu \in \mathbb{K} : \mathrm{Re}(\mu) > 0, (\mu - A_n)^{-1}f \text{ converges for all } f \in \mathscr{X}\}$$

is all of the connected set $\Sigma = \{\mu : \mathrm{Re}(\mu) > 0\}$. Indeed, S is open in Σ, since $\lambda \in S$ implies $\{\mu \in \Sigma : M|\mu - \lambda| < \mathrm{Re}(\lambda)\} \subset S$. And the complement of S is also open in Σ, since $\mu \in \Sigma\backslash S$ implies

$$\left\{\lambda \in \Sigma : |\mu - \lambda| < \frac{1}{(M+1)} \mathrm{Re}\,\mu\right\}$$

$$\subset \{\lambda \in \Sigma : M|\mu - \lambda| < \mathrm{Re}(\lambda)\} \subset \Sigma\backslash S.$$

Hence $\Sigma = S$. Define

$$R(\lambda)f = \lim_{n \to \infty} (\lambda - A_n)^{-1}f$$

for $\lambda \in \Sigma$. Clearly

$$R(\lambda) - R(\mu) = (\mu - \lambda)R(\lambda)R(\mu),$$

and $R(\mu)R(\lambda) = R(\lambda)R(\mu)$ for $\mu, \lambda \in \Sigma$. Let $\mathscr{N}$ and $\mathscr{R}$ denote, respectively, the null space and range of $R(\lambda)$. $R(\mu) = R(\lambda)[I - (\mu - \lambda)R(\lambda)]$ implies Range $[R(\mu)] \supset$ Range $[R(\lambda)]$, and so by interchanging μ and λ we conclude that $\mathscr{R}$ (and similarly $\mathscr{N}$) is independent of λ. By hypothesis on $R = R(\lambda_0)$, $\mathscr{N} = \{0\}$ and $\mathscr{R}$ is dense. Define $A_0 : \mathscr{D}(A_0) = \mathscr{R} \to \mathscr{X}$ by

$$A_0 = \lambda_0 I - R(\lambda_0)^{-1}.$$

Clearly

$$(\lambda_0 - A_0)R(\lambda_0) = R(\lambda_0)(\lambda_0 - A_0) = I$$

on $\mathscr{R}$. Moreover, for $\lambda \in \Sigma$ one has the following calculation for operators defined on $\mathscr{X}$:

$$
\begin{aligned}
(\lambda - A_0)R(\lambda) &= [(\lambda - \lambda_0) + (\lambda_0 - A_0)]R(\lambda) \\
&= [(\lambda - \lambda_0) + (\lambda_0 - A_0)]R(\lambda_0)[I - (\lambda - \lambda_0)R(\lambda)] \\
&= I + (\lambda - \lambda_0)[R(\lambda_0) - R(\lambda) - (\lambda - \lambda_0)R(\lambda)R(\lambda_0)] = I
\end{aligned}
$$

and similarly $R(\lambda)(\lambda - A_0) = I$ on $\mathscr{R}$. Consequently

$$R(\lambda) = (\lambda - A_0)^{-1}$$

for all $\lambda \in \Sigma$. The estimate

$$\|(\lambda - A_0)^{-n}f\| = \lim_{m \to \infty} \|(\lambda - A_m)^{-n}f\| \leq \frac{M\|f\|}{Re(\lambda)^n}$$

shows that A_0 generates a (C_0) semigroup. An appeal to Theorem 7.3 now completes the proof. ∎

Further results on approximation theory are given in Sections 8 and 9.

7.9. EXERCISES

1. In 7.5 we omitted the proofs of two assertions, namely the denseness of $\mathscr{D}(A)$ and the uniformity in t of the convergence. Prove these two assertions.

2. Prove the following version of the approximation theorem. (See 5.8 for the terminology and notation.) Let $0 \leq a \leq \infty$, and let $\mathscr{X}$ be complex. For $n \in \mathbb{N}$ let A_n generate T_n, an analytic contraction semigroup of type $S(a)$. Let $\mathscr{Y}$ be a closed subspace of $\mathscr{X}$, and let T_0 be an analytic contraction semigroup of type $S(a)$ on $\mathscr{Y}$. The following three statements are equivalent.

   (i) For all $g \in \mathscr{Y}$,

   $$\lim_{n \to \infty} (I - A_n)^{-1}g = (I - A_0)^{-1}g.$$

   (ii) For all $g \in \mathscr{Y}$,

   $$\lim_{n \to \infty} T_n(t)g = T_0(t)g$$

for all t in some ray in $S(a)$; i.e. for all t of the form $t = re^{i\theta}$ where $r > 0$ is arbitrary, θ is fixed, and $re^{i\theta} \in S(a)$ for all $r > 0$.

(iii)   For all $g \in \mathcal{Y}$,

$$\lim_{n \to \infty} T_n(t)g = T_0(t)g,$$

uniformly for t in bounded intervals if $a = 0$, and uniformly in bounded subsets of $S(b)$ for all $b < a$, if $a > 0$.

3. In the proof of Theorem 7.3 (see 7.5), check that $\mathcal{D}(A)$ is dense in X.

8.   Some Applications

Probably the most important applications are to initial value problems; we deal with these in Chapter II. In this section we give some applications to classical approximation theory, probability theory, and mathematical physics, including the Feynman path integral and the mean ergodic theorem.

We begin with a very special case of the approximation theorem (Theorem 7.2) which admits a particularly simple proof.

8.1.   LEMMA. *For $n \in \mathbb{N}_0$ let T_n be a (C_0) contraction semigroup with generator A_n. Suppose $A_n \in \mathscr{B}(\mathscr{X})$ for $n \geq 1$ and $A_n T_0(t) = T_0(t)A_n$ for all $n \in \mathbb{N}$ and $t \in \mathbb{R}^+$. If*

$$\lim_{n \to \infty} A_n f = A_0 f$$

for all f in a core $\mathcal{D}$ of A_0, then

$$\lim_{n \to \infty} T_n(t)f = T_0(t)f,$$

for all $f \in \mathscr{X}$ uniformly for t in bounded intervals of $\mathbb{R}^+$.

Proof.   For $f \in \mathcal{D}, n \in \mathbb{N}$,

$$T_n(t)f - T_0(t)f = -\int_0^t \frac{d}{ds}\left(T_n[t - s]T_0(s)f\right]ds$$

$$= \int_0^t T_n(t - s)(A_n - A_0)T_0(s)f\,ds$$

$$= \int_0^t T_n(t - s)T_0(s)(A_n f - A_0 f)\,ds.$$

Consequently

$$\|T_n(t)f - T_0(t)f\| \leq t\|A_n f - A_0 f\|,$$

and the result follows.   ∎

8.2.   EXAMPLE.   Let $\mathscr{X} = BUC(\mathbb{R})$. The formula

$$T_0(t)f(x) = f(x + t)$$

defines a (C_0) contraction semigroup T_0 whose generator is $A_0 = d/dx$ with domain

$$\mathscr{D}(A_0) = \{f \in \mathscr{X} : f \text{ absolutely continuous, } f' \in \mathscr{X}\}.$$

If we set

$$A_n = \frac{T_0(1/n) - I}{1/n}$$

for $n \in \mathbb{N}$, then the hypotheses of Lemma 8.1 are fulfilled because

$$\|T_n(t)\| = \left\| \exp\left\{ tn\left[T_0\left(\frac{1}{n}\right) - I \right] \right\} \right\| \le e^{-tn} \exp\left[tn \left\| T_0\left(\frac{1}{n}\right) \right\| \right] \le 1.$$

Let $f \in C[0,1]$ and regard f as (the restriction of) a member of $\mathscr{X}$. Then, uniformly for $t \in [0,1]$ and $x \in \mathbb{R}$,

$$f(x + t) = \lim_{n \to \infty} \lim_{M \to \infty} \sum_{m=0}^{M} \frac{t^m (A_n^m f)(x)}{m!}.$$

Taking $x = 0$ yields

$$f(t) = \lim_{n \to \infty} \sum_{m=0}^{M_n} \frac{t^m (A_n^m f)(0)}{m!},$$

uniformly for $t \in [0,1]$, for a suitable sequence M_n of integers. This is the classical Weierstrass approximation theorem.

8.3. EXAMPLE. In the notation of Example 8.2, let

$$\Delta(h) = \frac{T_0(h) - I}{h}.$$

$\Delta(h)$ is the difference quotient operator:

$$\Delta(h)f(x) = \frac{f(x + h) - f(x)}{h}.$$

Lemma 8.1 yields

$$f(x + t) = \lim_{h \to 0} \lim_{M \to \infty} \sum_{m=0}^{M} \frac{t^m}{m!} \Delta_h^m f(x) \tag{8.1}$$

uniformly for $x \in \mathbb{R}$ and t in compact subsets of $\mathbb{R}$. Now look at (8.1) with the limits reversed:

$$f(x + t) = \lim_{M \to \infty} \lim_{h \to 0} \sum_{m=0}^{M} \frac{t^m}{m!} \Delta_h^m f(x)$$

$$\left[= \sum_{m=0}^{\infty} \frac{t^m}{m!} f^{(m)}(x) \right].$$

This is Taylor's theorem; it's validity for all $t, x \in \mathbb{R}$ implies f is analytic on $\mathbb{R}$, and even in this case the above equality will not usually be valid for all t and x, let alone uniformly for $x \in \mathbb{R}$. Equation (8.1) is thus a remarkable result: a finite difference version of Taylor's formula which holds without any differentiability assumption on f.

8.4. For further applications we need some more machinery.

THEOREM (Chernoff's product formula).   *Let $\{V(t): t \in \mathbb{R}^+\}$ be a family of contractions on $\mathscr{X}$ with $V(0) = I$. Suppose the derivative $V'(0)f$ exists for all f in a set $\mathscr{D}$ and that the closure A of $V'(0)|_{\mathscr{D}}$ generates a (C_0) contraction semigroup T. Then for each $f \in \mathscr{X}$,*

$$\lim_{n \to \infty} V\left(\frac{t}{n}\right)^n f = T(t)f,$$

uniformly for t in compact subsets of R^+.

The proof is based on the following estimate.

8.5.   LEMMA.   *$L \in \mathscr{B}(\mathscr{X}), \|L\| \leq 1$ implies*

$$\|e^{n(L-I)}f - L^n f\| \leq \sqrt{n}\|Lf - f\|$$

for each $f \in \mathscr{X}$ and each positive integer n.

Proof. $\|e^{n(L-I)}f - L^n f\| = e^{-n}\left\| \sum_{k=0}^{\infty} \frac{n^k}{k!}(L^k - L^n)f \right\|$

$$\leq e^{-n} \sum_{k=0}^{\infty} \frac{n^k}{k!} \left\| L^k f - L^n f \right\|.$$

But

$$\|L^k f - L^n f\| \leq \|L^{|k-n|}f - f\| \quad \text{and} \quad L^m - I = \sum_{i=1}^{m}(L^i - L^{i-1}),$$

whence

$$\|L^m f - f\| \leq \sum_{i=1}^{m} \|L^i f - L^{i-1}f\| \leq m\|Lf - f\|.$$

Therefore

$$\|e^{n(L-I)}f - L^n f\| \leq e^{-n} \sum_{k=0}^{\infty} \frac{n^k}{k!}|k - n|\|Lf - f\|.$$

$$\leq e^{-n}\left\{ \sum_{k=0}^{\infty} \frac{n^k}{k!} \right\}^{1/2} \left\{ \sum_{k=0}^{\infty} \frac{n^k}{k!}(k - n)^2 \right\}^{1/2} \|Lf - f\|$$

by the Schwarz inequality. But an easy computation (cf. any elementary probability text where the mean and variance of the Poisson distribution are computed) shows that

$$\sum_{k=0}^{\infty} \frac{n^k}{k!}(k - n)^2 = ne^n$$

and the lemma follows.   ■

8.6. *Proof of Theorem* 8.4. For $t > 0$ let

$$A_n = \frac{V(t/n) - I}{t/n}$$

for $n \geq 1$ and $A_0 = A$. Then by Theorem 7.2,

$$\|e^{tA_n}f - T(t)f\| \to 0$$

as $n \to \infty$ uniformly for t in bounded intervals. Now apply Lemma 8.5 [with $L = V(t/n)$] to obtain

$$\left\|e^{tA_n}f - V\left(\frac{t}{n}\right)^n f\right\| \leq \sqrt{n}\|V(t/n)f - f\|$$

$$= \frac{t}{\sqrt{n}}\left\|\frac{V(t/n)f - f}{t/n}\right\| \to 0$$

as $n \to \infty$ for $f \in D$. The conclusion of the theorem now follows; except for the uniformity, which is left as an exercise. ∎

8.7. EXAMPLE. Let A generate a (C_0) contraction semigroup T. Letting $V(t) = (I - tA)^{-1}, D = \mathscr{D}(A)$, Chernoff's product formula yields the exponential formula

$$T(t)f = \lim_{n \to \infty} \left(I - \frac{t}{n} A\right)^{-n} f$$

for $f \in X$, uniformly for t in bounded intervals of $\mathbb{R}^+$. This representation formula for T was discussed heuristically in 2.8.

8.8. A random variable ξ on a probability space (Ω, Σ, P) is said to have the *normal distribution with parameters* $0, r$ $(r > 0)$ if its distribution function is

$$F_\xi(t) \equiv P(\xi \leq t) = F_{N(0,r)}(t)$$

$$\equiv \frac{1}{\sqrt{2\pi r}} \int_{-\infty}^t \exp(-s^2/2r) \, ds, \ t \in \mathbb{R}.$$

In this case ξ has mean zero and variance r: $E(\xi) = 0$, Var $(\xi) = r$.

THEOREM (central limit theorem). *Let $\xi_1, \xi_2, \ldots$ be a sequence of independent identically distributed random variables satisfying the normalization conditions* $E(\xi_i) = 0$, Var $(\xi_i) = 1$. *Then for all $x \in \mathbb{R}$,*

$$\lim_{n \to \infty} F_{\frac{1}{\sqrt{n}}\sum_{i=1}^n \xi_i}(x) = F_{N(0,1)}(x).$$

It is probably not an exaggeration to claim that this is the most important theorem in probability and statistics.

We are going to show how the central limit theorem follows from the Chernoff product formula.

First, let $\mathscr{F}_0$ be the set of all distribution functions. Let $\mathscr{X} = BUC(\mathbb{R})$. (We could equally well take $\mathscr{X} = C_0(\mathbb{R})$ or $C[-\infty, \infty]$.) For $F \in \mathscr{F}_0$ define $\tilde{F}: X \to X$

by the convolution formula $\tilde{F}f = f * dF$, i.e.

$$\tilde{F}f(t) = \int_{-\infty}^{\infty} f(t - s)\, dF(s)$$

for $f \in \mathscr{X}$, $t \in \mathbb{R}$. Clearly $F \in \mathscr{B}(\mathscr{X})$ and $\|\tilde{F}\| \leq 1$. The following two elementary lemmas are well-known; the proofs are left as exercises.

8.9.   LEMMA.   *Let F_n, $F \in \mathscr{F}_0$. Then F_n converges in distribution to F (i.e. $F_n(x) \to F(x)$ as $n \to \infty$ at all points x of continuity of F) if and only if $\tilde{F}_n$ converges to $\tilde{F}$ in the strong operator topology (i.e. $\|\tilde{F}_n f - \tilde{F}f\| \to 0$ as $n \to \infty$ for each $f \in \mathscr{X}$).*

8.10.   LEMMA.   *For all $F, G \in \mathscr{F}_0$, $(F * G)\tilde{} = \tilde{F}\tilde{G}$.*

8.11.   *Proof of Theorem* 8.8.   Thanks to Lemma 8.9, it is enough to prove that

$$\left\| \tilde{F}_{\frac{1}{\sqrt{n}}\sum_{i=1}^{n} \xi_i} f - \tilde{F}_{N(0,1)}f \right\| \to 0$$

as $n \to \infty$ for all f in (a dense subset of) $\mathscr{X}$. Let

$$D = \{f \in \mathscr{X} : f', f'' \in \mathscr{X}\}, ' = d/dx.$$

Let G be the common distribution function of the ξ_i. Let $G_r(x) = G(\sqrt{r}x)$ for $r > 0$, $x \in \mathbb{R}$, and let $V(t) = \tilde{G}_{1/t}$ for $t > 0$, $V(0) = I$. Finally let $A = \frac{1}{2}(d^2/dx^2)$ with domain $\mathscr{D}(A) = D$. As is well-known, A generates a (C_0) contraction semigroup T on $\mathscr{X}$ given by $T(t) = \tilde{F}_{N(0,t)}$ for $t > 0$. (In different notation, we are asserting that the unique bounded solution of $\partial u/\partial t = \frac{1}{2}\partial^2 u/\partial x^2$, $u(0,x) = f(x)$ (for $t \geq 0$, $x \in \mathbb{R}$) is $u(t,x) = 1/\sqrt{2\pi t} \int_{-\infty}^{\infty} \exp(-y^2/2t)f(x - y)\, dy$. See Exercise 2.18.7.)

   We *claim* that $\|(V(t)f - f)/t - Af\| \to 0$ as $t \to 0$ for $f \in D$. Assume this to be true. Then by Chernoff's product formula, $\|V(t/n)^n f - \tilde{F}_{N(0,t)}f\| \to 0$ as $n \to \infty$ for each $f \in X$ and $t \geq 0$. Now take $t = 1$. By Lemma 8.10,

$$V\left(\frac{1}{n}\right)^n = \tilde{G}_n^n = (G_n * \cdots * G_n)\tilde{} = \tilde{H}_n$$

where H_n is the distribution function of the sum of n independent random variables, each having G_n as its distribution function. But G_n is the distribution function of $\xi_i/\sqrt{n}$, whence

$$H_n = F \qquad\qquad ,$$

and the central limit theorem follows.

   So it remains to prove the claim. First, note that since $G_s = F_{\xi_i/\sqrt{s}}$,

$$\int_{-\infty}^{\infty} dG_s(r) = 1, \tag{8.2}$$

$$\int_{-\infty}^{\infty} r^2\, dG_s(r) = E\left(\frac{\xi_i}{\sqrt{s}}\right) = 0, \tag{8.3}$$

$$\int_{-\infty}^{\infty} r^2\, dG_s(r) = E\left[\left(\frac{\xi_i}{\sqrt{s}}\right)^2\right] = \frac{1}{s}. \tag{8.4}$$

Thus for $f \in D$,

$$
\left(\frac{V(t)f - f}{t} - Af\right)(x) = \frac{V(t)f(x) - f(x)}{t} - \frac{1}{2}f''(x)
$$

$$
= \int_{-\infty}^{\infty} \left(\frac{f(x - r) - f(x)}{t}\right) dG_{1/t}(r) - \frac{1}{2}f''(x)
$$

$$
= \int_{-\infty}^{\infty} \left(\frac{f(x - r) - f(x) + rf'(x) - \dfrac{r^2}{2}f''(x)}{t}\right) dG_{1/t}(r)
$$

$$
= \left\{\int_{|r| < \delta} + \int_{|r| \geq \delta}\right\}(\cdots)\, dG_{1/t}(r) \equiv J_1 + J_2;
$$

the integrand equals $(\cdots) = (r^2/2t)[f''(\theta) - f''(x)]$ according to Taylor's theorem, where $\theta[=\theta(x,r)]$ is between x and $x - r$. Let $\epsilon > 0$ be given. Since f'' is uniformly continuous on $\mathbb{R}$, there is a $\delta > 0$ such that $|f''(\theta) - f''(x)| < \epsilon/2$ if $\theta, x \in \mathbb{R}$ and $|\theta - x| < \delta$. Thus for this (henceforth fixed) choice of $\delta > 0$,

$$
|J_1| \leq \frac{\epsilon}{2t}\int_{|r| < \delta} r^2\, dG_{1/t}(r) \leq \frac{\epsilon}{2}
$$

by (8.4) for all $t > 0$.

Next note that $|f''(\theta) - f''(x)| \leq 2\|f''\|$. Consequently,

$$
|J_2| \leq \int_{|r| \geq \delta} \frac{2\|f''\|}{2t} r^2\, dG_{1/t}(r) = \|f''\|\int_{|r/\sqrt{t}| \geq \delta/\sqrt{t}} \left(\frac{r}{\sqrt{t}}\right)^2 dG\left(\frac{r}{\sqrt{t}}\right)
$$

$$
= \|f''\|\int_{|s| \geq \delta/\sqrt{t}} s^2\, dG(s) \to 0 \text{ as } t \downarrow 0.
$$

(This is nothing more than Chebyshev's inequality.) Thus $|J_2| < \epsilon/2$ for $0 < t < t_0(\epsilon, f)$. Combining these estimates yields

$$
\left|\left(\frac{V(t)f - f}{t} - Af\right)(x)\right| \leq |J_1| + |J_2| < \epsilon
$$

for $f \in D$, for all $x \in \mathbb{R}$, and for $0 < t < t_0(\epsilon, f)$. ∎

8.12. THEOREM (Trotter product formula). *Suppose A, B, and $\overline{A + B}$ generate (C_0) contraction semigroups T, S, U on $\mathcal{X}$. Then*

$$
\lim_{n \to \infty} \{T(t/n)S(t/n)\}^n f = U(t)f
$$

for each $f \in \mathcal{X}$, uniformly for t in compacta of $\mathbb{R}^+$.

Proof. Apply Theorem 8.4 with $\mathcal{D} = \mathcal{D}(A + B) = \mathcal{D}(A) \cap \mathcal{D}(B)$, $V(t) = T(t)S(t)$. ∎

8.13. THE FEYNMAN PATH FORMULA   First we give a little background from
quantum mechanics. Consider a spinless, nonrelativistic quantum mechanical
particle of mass m travelling in $\mathbb{R}^l$ under the influence of a potential V. (The
physically most interesting case is $l = 3$.) Here V is a real-valued (Borel) function
on $\mathbb{R}^l$. The *wave function of* the particle is the solution $u = u(t,x)$ of the
Schrödinger equation

$$\partial u/\partial t = i\left[\frac{1}{2m}\Delta u - V(x)u\right], \Delta = \sum_{j=1}^{l} \partial^2/\partial x_j^2,$$

subject to the initial condition

$$u(0,x) = f(x),\ x \in \mathbb{R}^l,$$

where $\int_{\mathbb{R}^l} |f(x)|^2\, dx = 1$. u has the following partial interpretation. $|u(t,x)|^2$ is the
position probability density, i.e. $\int_\Gamma |u(t,x)|^2\, dx$ is the probability that the vector
(in $\mathbb{R}^l$) describing the particle's position is in the Borel set Γ at time t. Similarly,
$|\hat{u}(t,\xi)|^2$ is the momentum probability density, where $\hat{u}$ is the (spatial) Fourier
transform of u:

$$\hat{u}(t,\xi) = \frac{1}{(2\pi)^{l/2}} \int_{\mathbb{R}^l} e^{-ix\cdot\xi}u(t,x)\, dx.^\dagger$$

We interpret the Schrödinger equation in the complex Hilbert space $\mathscr{X} = L^2(\mathbb{R}^l)$
by viewing $u(t) = u(t,\cdot)$ as a function from $\mathbb{R}$ to $\mathscr{X}$ satisfying

$$du/dt = (A + B)u$$

where $A = (i/2m)\,\Delta$ and B is the operator of multiplication by $-iV(x)$. A, B, and
$C = \overline{A + B}$ are all skew-adjoint operators under mild assumptions on V. (For
example, let $V = V_1 + V_2$ where $V_1 \in L^\infty(\mathbb{R}^l)$, $V_2 \in L^p(\mathbb{R}^l)$ where $p \geq 2, p > l/2$. In
this case $C = A + B$. This can be shown using Corollary 6.11.)

By Stone's Theorem (Theorem 4.7), $C = \overline{A + B}$ generates a (C_0) unitary
group, and since the Fourier transform is unitary on $\mathscr{X}$, we see that

$$\|\hat{u}(t)\|^2 = \|u(t)\|^2 = \|f\|^2 = 1,$$

explaining the significance of the assumption that $\int_{\mathbb{R}^l} |f(x)|^2\, dx = 1$ and
confirming that $|u(t,x)|^2$, $|\hat{u}(t,\xi)|^2$ are indeed probability densities. The groups
generated by A, B are given by

$$T(t)f(x) = \left(\frac{2\pi it}{m}\right)^{-l/2} \int_{\mathbb{R}^l} \exp\{im|x - y|^2/2t\}f(y)\, dy,$$

$$S(t)f(x) = \exp\{-itV(x)\}f(x).$$

The formula for S is obviously valid; the one for T follows from an elementary
analysis based on Fourier transforms, since the associated partial differential

---

† Here $u(t,\cdot) \in L^2(\mathbb{R}^l)$. If $u(t,\cdot) \notin L^1(\mathbb{R}^l)$, the integral is to be interpreted as a limit in the mean in the
usual way. For information on Fourier transforms see Section 3 of Chapter II.

equation has constatn coefficients. In view of the Trotter product formula, consider the expression

$$U_n(t) = \left[T\left(\frac{t}{n}\right) S\left(\frac{t}{n}\right) \right]^n.$$

Induction yields

$$U_n(t)f(x) = \left(\frac{2\pi it}{nm}\right)^{-l/2} \underbrace{\int_{\mathbb{R}^l} \cdots \int_{\mathbb{R}^l}}_{n} \exp\{iS(x_0,\ldots,x_n;t)\}f(x_n)\, dx_1 \cdots dx_n \quad (8.5)$$

where $x_0 = x \in \mathbb{R}^l$ and

$$S(x_0,\ldots,x_n;t) = \sum_{j=1}^{m} \left[\frac{m}{2} \frac{|x_j - x_{j-1}|^2}{(t/n)^2} - V(x_j) \right]\left(\frac{t}{n}\right).$$

Let Ω_x be the set of all continuous functions (or paths) $\omega: \mathbb{R}^+ \to \mathbb{R}^l$ such that $\omega(0) = x$. Letting $x_j = \omega(tj/n)$, we see that $S(x_0,\ldots,x_n;t)$ is a Riemann sum for the *action* integral

$$S(\omega;t) = \int_0^t \left\{ \frac{m}{2} |\dot{\omega}(s)|^2 - V[\omega(s)] \right\} ds.$$

Here the dot means differentiation with respect to time. *Formally*, letting $n \to \infty$, the right-hand side of (8.5) becomes

$$\text{constant} \int_{\Omega_x} \exp\{iS(\omega;t)\}f[\omega(t)]D\omega. \quad (8.6)$$

This is the celebrated Feynman path integral representation of the wave function $u(t,x)$. There are many mysterious aspects of the expression (8.6), however. First of all, the constant is infinite. Next, $D\omega = \prod_{0 \leq s \leq t} dx_s (= \lim \prod_{i=1}^{n} dx_i)$ doesn't make sense. Finally, to interpret (8.6) as a Wiener-type integral, we expect the a.e. differentiable paths ω to form a set of zero measure, whereas $S(\omega;t)$ fails to make sense for an ω not differentiable a.e. Conclusion: Among the ingredients of (8.6), namely, the constant, the integrand, and the measure, *none makes sense*. Nevertheless, according to the Trotter product formula, the right-hand side of (8.5) converges in the sense of the $L^2(\mathbb{R}^l)$ norm to the wave function $u(t,x) = U(t)f(x)$, where U is the group generated by C. Consequently the product formula has enabled us to give a rigorous interpretation of (8.6). Namely, the (well-defined) Riemann sums (8.5) for the (meaningless) integral (8.6) converge (not pointwise but) in $L^2(\mathbb{R}^l)$ norm to the wave function of the particle.

In the case of the heat equation with a potential, $\partial u/\partial t = \Delta u/2m - V(x)u$, Kac showed that the solution u of the initial value problem has a representation as a Wiener integral. See Exercise 8.20.24(iii) for more details.

The "mystery" of the Feynman integral formula consists of the difficulty of interpreting it rigorously. It is a tool of great power and intuitive appeal for quantum theorists; see Feynman and Hibbs [1] for the physical point of view.

8.14.   EXAMPLE.   Let $\mathcal{X} = L^p(\mathbb{R})$ $(1 \le p < \infty)$ or $C_0(\mathbb{R})$, the complex-valued continuous functions that vanish at $\pm\infty$. Let $q: \mathbb{R} \to \overline{\mathbb{R}} = [-\infty, \infty]$ be Lebesgue measurable and finite a.e. [and uniformly continuous if $\mathcal{X} = C_0(\mathbb{R})$]. Let

$$(A_q f)(x) = iq(x)f(x),$$

$\mathcal{D}(A_q) = \{f \in \mathcal{X}: qf \in \mathcal{X}\}$. A_q generates a (C_0) isometric group T (i.e. $T(t)$ is an isometry for each $t \in \mathbb{R}$) given by

$$[T(t)f](x) = e^{itq(x)}f(x).$$

Let $B = d/dx$, $\mathcal{D}(B) = \{f \in \mathcal{X}: f$ absolutely continuous on $\mathbb{R}$ and $f' \in \mathcal{X}\}$. B generates the (C_0) isometric group S of translations:

$$[S(t)f](x) = f(x + t).$$

Induction shows that

$$[T(t/n)S(t/n)]^n f(x) = \exp\left\{i \sum_{j=1}^{n} q\left(x + \frac{tj}{n}\right)\frac{t}{n}\right\}f(x + t).$$

If $q \in L^1_{\text{loc}}(\mathbb{R})^\dagger$, then for each $t \in \mathbb{R}$,

$$\lim_{n \to \infty} \sum_{j=1}^{n} q\left(x + \frac{tj}{n}\right)\frac{t}{n} = \int_0^t q(x + s)\,ds$$

in the sense of convergence in measure as functions of x on only bounded interval. This requires a computation which is left as an exercise. It follows by the dominated convergence theorem that

$$\lim_{n \to \infty} [T(t/n)S(t/n)]^n f = U(t)f,$$

for each $f \in \mathcal{X}$, where U is the (C_0) isometric group given by

$$[U(t)f](x) = \exp\left[i\int_0^t q(x + s)\,ds\right]f(x + t).$$

If $\overline{A_q + B}$ is a generator, then by Theorem 8.12, $\overline{A_q + B}$ is the generator of U. However, it is possible that $\overline{A_q + B}$ need not be a generator; in fact, it can even happen that $\mathcal{D}(A_q + B) = \{0\}$!

8.15.   EXAMPLE.   Let $\mathcal{X} = L^2(\mathbb{R})$. Let $\{r_n\}_{n=1}^{\infty}$ be an enumeration for all the rationals. Let

$$q(x) = \sum_{n=1}^{\infty} a_n|x - r_n|^\alpha.$$

Choose a_n, α such that $q \in L^1_{\text{loc}}(\mathbb{R})$, but q is not square integrable over any interval of positive length, e.g. $\alpha = -1/2$ and $a_n = 1/n!$. Then for this choice of q, $\mathcal{D}(A_q) \cap \mathcal{D}(B) = \{0\}$. Thus A_q and B are "disjoint" skew-adjoint operators.

---

† $L^p_{\text{loc}}(J)$ is the set of measurable functions on J whose restrictions belong to $L^p(K)$ for each compact subset K of J.

8.16. EXAMPLE. Let $\mathscr{X}$ be the complex space $L^2(\mathbb{R})$, let $q \in L^1_{\text{loc}}(\mathbb{R})$, and let $\tau, \sigma \in \mathbb{R}$. Define the (C_0) unitary group $U(\sigma,\tau) = \{U(\sigma,\tau;t): t \in \mathbb{R}\}$ on $\mathscr{X}$ by

$$U(\sigma,\tau;t)f(x) = f(x + \sigma t)\exp\left\{i\tau \int_0^t q(x + \sigma s)\, ds\right\}$$

for $f \in \mathscr{X}$ and $x \in \mathbb{R}$. Choosing q as in Example 8.15, one can easily show that

$$\mathscr{D}[A(1,0)] \cap \bigcup\{\mathscr{D}[A(\sigma,\tau)]: \sigma \in \mathbb{R}, \tau \in \mathbb{R}\setminus\{0\}\} = \{0\}$$

where $A(\sigma,\tau)$ denotes the generator of $U(\sigma,\tau)$. It follows that there is a sequence of generators $A_n = A(\sigma_n, \tau_n)$ with $(\sigma_0, \tau_0) = (1,0)$, $\tau_n \neq 0$ for $n \geq 1$, and (σ_n, τ_n) $\rightarrow (1,0)$, such that the hypotheses of Theorem 7.3 hold but the hypotheses of Theorem 7.2 fail to hold.

8.17. REMARK. Let A, B, C be m-dissipative on $\mathscr{X}$, and denote their respective semigroups by T, S, U. We call C the *generalized sum* (or *Lie sum*) of A and B, and write $C = A +_L B$, if

$$\lim_{n\to\infty} \left[T\!\left(\frac{t}{n}\right)S\!\left(\frac{t}{n}\right)\right]^n f = U(t)f$$

holds for all $t \in \mathbb{R}^+$ and $f \in \mathscr{X}$. If $\overline{A + B} = C$, then $A +_L B$ exists and equals C. However, $A +_L B$ can exist even when $\mathscr{D}(A) \cap \mathscr{D}(B) = \{0\}$. For a specific example take

$$A = A(1,0), \quad B = A(0,1)$$

in the notation of Example 8.16 (cf. Examples 8.14, 8.15).

8.18. Here and in 8.20 we give a version of the mean ergodic theorem, which describes the behavior of $1/\tau \int_0^\tau T(t)\, dt$ as $\tau \to \infty$, where T is a (C_0) semigroup.

THEOREM. *Let A generate a (C_0) semigroup T which is uniformly bounded. Let $\mathscr{N} = \{f \in \mathscr{D}(A): Af = 0\}$ and $\mathscr{R} = \{Af: f \in \mathscr{D}(A)\}$ denote the null space and range of A. If $f_1 \in \mathscr{N}$, $f_2 \in \mathscr{R}$, then*

$$\frac{1}{\tau}\int_0^\tau T(t)(f_1 + f_2)\, dt = f_1 + O\!\left(\frac{1}{\tau}\right)$$

as $\tau \to \infty$.

Proof. $Af_1 = 0$ implies $T(t)f_1 = f_1$ for all t. Next, since $f_2 = Af_3 \in \mathscr{R}$,

$$\frac{1}{\tau}\int_0^\tau T(t)f_2\, dt = \frac{1}{\tau}\int_0^\tau T(t)Af_3\, dt$$

$$= \frac{1}{\tau}\int_0^\tau \frac{d}{dt}[T(t)f_3]\, dt = \frac{1}{\tau}[T(t)f_3 - f_3],$$

therefore

$$\left\|\frac{1}{\tau}\int_0^\tau T(t)f_2\, dt\right\| \leq \frac{K\|f_3\|}{\tau}$$

where K is a uniform bound for $\|T(t)\| + 1$. The result follows. ∎

8.19.   LEMMA.   *Let A be a closed, densely defined operator on a reflexive space $\mathscr{X}$.*
Let

$$\mathscr{N} = \{f \in \mathscr{D}(A): Af = 0\}, \; \mathscr{R} = \{Af: f \in \mathscr{D}(A)\},$$
$$\mathscr{N}^* = \{\phi \in \mathscr{D}(A^*): A^*\phi = 0\}, \; \mathscr{R}^* = \{A^*\phi: \phi \in \mathscr{D}(A^*)\}.$$

Then

$$\mathscr{N}^\perp = \overline{\mathscr{R}^*}, \mathscr{R}^\perp = \mathscr{N}^*,$$

$$(\mathscr{N}^*)^\perp = \overline{\mathscr{R}}, (\mathscr{R}^*)^\perp = \mathscr{N}.$$

For $\mathscr{M} \subset \mathscr{X}$, by $\mathscr{M}^\perp$ we mean

$$\mathscr{M}^\perp = \{\phi \in \mathscr{X}^*: \langle f, \phi \rangle = 0 \text{ for all } f \in \mathscr{M}\}.$$

Since $\mathscr{X}$ is reflexive we identify $(\mathscr{X}^*)^*$ with $\mathscr{X}$ and conclude that the above lemma makes sense. The proof is left as a straightforward exercise in functional analysis (see Exercise 8.22.11).

8.20.   THEOREM (mean ergodic theorem).   *Let A generate a (C_0) semigroup T on a reflexive space $\mathscr{X}$, and suppose*

$$\sup\{\|T(t)\|: t \in \mathbb{R}^+\} = K < \infty.$$

Let

$$\mathscr{N} = \{f \in \mathscr{D}(A): Af = 0\}, \mathscr{R} = \{Af: f \in \mathscr{D}(A)\}.$$

Then $\mathscr{N}$ and $\overline{\mathscr{R}}$ are closed subspaces of $\mathscr{X}$, $\mathscr{N} \cap \overline{\mathscr{R}} = \{0\}$, $\mathscr{N} + \overline{\mathscr{R}} = \mathscr{X}$, and the projection P of $\mathscr{X}$ to $\mathscr{N}$ is well-defined and satisfies $\|P\| \leq K$. Moreover, each $f \in \mathscr{X}$ can be written uniquely as $f = Pf + (I - P)f \in \mathscr{N} + \overline{\mathscr{R}}$, and

$$\lim_{\tau \to \infty} \frac{1}{\tau} \int_0^\tau T(t)f \, dt = Pf$$

for each $f \in \mathscr{X}$.

Proof.   By Theorem 8.18, if $f = f_1 + f_2$ where $f_1 \in \mathscr{N}$ and $f_2 \in \overline{\mathscr{R}}$, then

$$\lim_{\tau \to \infty} \frac{1}{\tau} \int_0^\tau T(t)f \, dt = f_1.$$

It follows that $\mathscr{N} \cap \overline{\mathscr{R}} = \{0\}$, and if $P: \mathscr{N} + \overline{\mathscr{R}} \to \mathscr{X}$ is defined by $Pf = f_1$ for $f = f_1 + f_2$ as above, then

$$\|Pf\| = \lim_{\tau \to \infty} \left\| \frac{1}{\tau} \int_0^\tau T(t)f \, dt \right\| \leq K\|f\|.$$

Hence P is a bounded projection of norm $\leq K$ on $\mathscr{N} + \overline{\mathscr{R}}$. It only remains to show that $\mathscr{N} + \mathscr{R}$ is dense in $\mathscr{X}$. Suppose not. Then we can choose an $f \in \mathscr{X} \setminus \overline{\mathscr{N} + \mathscr{R}}$ and a $\phi \in \mathscr{X}^*$ such that $\phi \in (\mathscr{N} + \mathscr{R})^\perp$ but $\langle f, \phi \rangle = 1$. If $\mathscr{R}^*, \mathscr{N}^*$ are as in Lemma 8.19, then $\phi \in \mathscr{N}^\perp = \overline{\mathscr{R}^*}$ and $\phi \in \mathscr{R}^\perp = \mathscr{N}^*$, whence $\phi \in \mathscr{N}^* \cap \overline{\mathscr{R}^*}$. But the above argument which gave $\mathscr{N} \cap \overline{\mathscr{R}} = \{0\}$ also gives $\mathscr{N}^* \cap \overline{\mathscr{R}^*} = \{0\}$, and so $\phi = 0$, contradicting $\langle f, \phi \rangle = 1$.   ∎

8.21.   REMARK.   Let $M = \{M(t) : t \in \mathbb{R}\}$ be a one-parameter group of measure preserving transformations on a finite measure space (Ω, Σ, μ); measure preserving means $\mu(M(t)\Gamma) = \mu(\Gamma)$ for each $t \in \mathbb{R}$ and $\Gamma \in \Sigma$. Define T on $\mathscr{X} = L^2(\Omega, \Sigma, \mu)$ by

$$T(t)f(x) = f[M(t)x]$$

for $x \in \Omega$, $f \in \mathscr{X}$, and $t \in \mathbb{R}$. Then T is a (C_0) unitary group if M satisfies a suitable continuity condition. M is called *metrically transition* (or *ergodic*) if, whenever $\Gamma \in \Sigma$ and $\mu\{[M(t)\Gamma \backslash \Gamma] \cup [\Gamma \backslash M(t)\Gamma]\} = 0$ for all $t \in \mathbb{R}$, then either $\mu(\Gamma) = 0$ or $\mu(\Omega \backslash \Gamma) = 0$; in other words, modulo sets of measure zero, $M(t)\Gamma = \Gamma$ for all t iff $\Gamma = \emptyset$ or Ω. In this case the null space of the generator A of T consists of the constant functions, and the associated projection P is given by

$$Pf(x) = \frac{1}{\mu(\Omega)} \int_\Omega f(y)\mu(dy).$$

The conclusion of the mean ergodic theorem is that the *time average*

$$\frac{1}{\tau} \int_0^\tau f[M(t)x]\, dt$$

converges to the *space average*

$$\frac{1}{\mu(\Omega)} \int_\Omega f(y)\mu(dy)$$

in the sense of the L^2 norm as $\tau \to \infty$.

8.22.   We shall give a refinement of the mean ergodic theorem. But first we prepare a lemma. The existence of $1/\tau \int_0^\tau u(t)\, dt$ can be interpreted to mean that $u(t)$ converges (as $t \to \infty$) in a certain generalized sense. We can rewrite this as: $\lim_{\tau \to \infty} \int_0^\infty p(\tau, t)u(t)\, dt$ exists where $p(\tau, t) = \tau^{-1}$ or 0, according as $t \leq \tau$ or $t > \tau$. Here $p \geq 0$, $\int_0^\infty p(\tau, t)\, dt = 1$, and the support of $p(\tau, \cdot)$ drifts toward infinity as $\tau \to \infty$. The same is true for $q(\tau, t) = \tau^{-1}e^{-t/\tau}$, and q turns out to be equivalent to p in the following sense.

LEMMA.   *Let $u : \mathbb{R}^+ \to \mathscr{X}$ be continuous and bounded. Then*

$$\lim_{\tau \to \infty} \frac{1}{\tau} \int_0^\tau u(t)\, dt = \lim_{\lambda \to 0^+} \lambda \int_0^\infty e^{-\lambda t}u(t)\, dt$$

in the sense that one limit exists iff the other does, and then they are equal.

The verification is omitted (cf. Exercise 8.24.16).

8.23.   THEOREM.   *Let A generate a uniformly bounded semigroup T on $\mathscr{X}$ and let $\mathscr{N}, \mathscr{R}$, be as in Theorem 8.20. Then the following four conditions are equivalent.*

(i) $\mathscr{X} = \mathscr{N} + \bar{\mathscr{R}}$.
(ii) $\lim_{\tau \to \infty} 1/\tau \int_0^\tau T(t)f\, dt$ exists for all $f \in \mathscr{X}$.
(iii) $\lim_{\lambda \to 0^+} \lambda(\lambda I - A)^{-1}f$ exists for all $f \in \mathscr{X}$.
(iv) $\lim_{\lambda \to 0^+} (\lambda I - A)^{-1}f$ exists for all $f \in \mathscr{R}$.

Proof. (i) implies (ii): By the proof of Theorem 8.20, $1/\tau \int_0^\tau T(t)f \, dt$ converges as $\tau \to \infty$ for all $f \in \mathcal{N} + \bar{\mathcal{R}} = \mathcal{X}$.

(ii) implies (iii): This holds by Lemma 8.22 since $\lambda(\lambda I - A)^{-1}f = \lambda \int_0^\infty e^{-\lambda t} T(t)f \, dt$.

(iii) implies (iv): This follows from

$$(\lambda I - A)^{-1} Af = \lambda(\lambda I - A)^{-1}f - f \tag{8.7}$$

for $f \in \mathcal{D}(A)$.

(iv) implies (ii): By (8.7), $\lambda(\lambda I - A)^{-1}f$ converges for all $f \in \mathcal{D}(A)$. This implies (ii) by Lemma 8.22.

(ii) implies (i). Define P on $\mathcal{X}$ to $\mathcal{X}$ by

$$Pf = \lim_{\tau \to \infty} \frac{1}{\tau} \int_0^\tau T(t)f \, dt, \, f \in \mathcal{X}.$$

One routinely checks that $\mathcal{X} = \mathcal{R}(P) + \mathcal{R}(I - P)$, $PT(t)f = T(t)Pf = Pf$ for all $f \in \mathcal{X}$ and $t \in \mathbb{R}^+$, $P^2 = P$, $\mathcal{R}(P) = \mathcal{N}$, $\mathcal{R}(I - P) = \mathcal{N}(P)$, and $\bar{\mathcal{R}} \subset \mathcal{N}(P)$. The only nontrivial thing to check is that $\mathcal{N}(P) \subset \bar{\mathcal{R}}$. To that end let $f \subset \mathcal{N}(P)$. Then (8.7) and (iii) imply

$$\lim_{\lambda \to 0^+} A(\lambda I - A)^{-1}f = Pf - f = -f,$$

whence $f \in \bar{\mathcal{R}}$.   ∎

8.24. EXERCISES

1. For $n \in \mathbb{N}_0$ let T_n be a (C_0) contraction semigroup with generator A_n. Suppose $A_n \in \mathcal{B}(\mathcal{X})$ for $n \in \mathbb{N}$ and $\lim_{n \to \infty} A_n f = A_0 f$ for all f in a core $\mathcal{D}$ of A_0 such that $\{f \in \mathcal{D}: A_0 f \in \mathcal{D}\}$ is dense in $\mathcal{X}$. Then $\lim_{n \to \infty} T_n(t)f = T_0(t)f$ for each $f \in \mathcal{X}$, uniformly for t in compact subsets of $\mathbb{R}^+$. Fill in the details of the following proof of this result, which is a special case of Theorem 7.2. *Case 1*: A_n and $T_0(t)$ commute. (This is Lemma 8.1.) *Case 2*: Sup$\{\|A_n\|: n \in \mathbb{N}_0\} < \infty$. Use the proof of Lemma 8.1 together with the bounded convergence theorem of Lebesgue. For the general case, write $T_n(t)f - T_0(t)f = \sum_{i=1}^{3} J_i$ where $J_1 = T_n(t)f - \exp[tA_n(h)]f$, $J_2 = \exp[tA_n(h)f - \exp[tA_0(h)]f$, $J_3 = \exp[tA_0(h)]f - T_0(t)f$, where $A_n(h) = A_n(I - hA_n)^{-1}$. Control J_1 and J_3 by case 1 and J_2 by case 2.

2. Show that the (weaker) result of Exercise 1 above is an adequate substitute for Theorem 7.2 in the proof of the Chernoff product formula (Theorem 8.4).

3. Use the method of 8.8 to prove the Weak Law of Large Numbers: Let $\xi_1, \xi_2, \ldots$ be a sequence of independent identically distributed random variable having mean μ. Then for each $\epsilon > 0$.

$$\lim_{n \to \infty} P\left\{ \left| \frac{1}{n} \sum_{i=1}^{n} \xi_i - \mu \right| \geq \epsilon \right\} = 0.$$

Hints. Let G be the distribution function of $\xi_i - \mu$, and let $G_r(x) = G(rx)$, $V(t) = \tilde{G}_{1/t}$. Apply the Chernoff product formula with $A = 0$, $T(t) = I$. Finally note that $1/n \sum_{i=1}^{n} \xi_i$ converges to μ in distribution iff it does so in probability.

4. Verify the details of Example 8.14.

5. Verify the details of Example 8.15.

6. Verify the details of Example 8.16.

7. Verify the last sentence of Remark 8.17.
8. Let A, B be bounded skew-adjoint operators on a Hilbert space $\mathscr{H}$, and let T, S denote the associated unitary group. Let U denote the unitary group generated by $C = [A,B] = AB - BA$. Show that for all $t \in \mathbb{R}^+$,

$$\lim_{n \to \infty} \left[T\left(\frac{t}{n}\right) S\left(\frac{t}{n}\right) T\left(-\frac{t}{n}\right) S\left(-\frac{t}{n}\right) \right]^{n^2} = U(t^2)$$

9. Formulate and prove a version of Exercise 8 when A and B unbounded.
10. Use the result of Exercise 9 to formulate a definition of the *generalized* (or *Lie*) *commutator* of two skew-adjoint operators. Construct skew-adjoint operators A and B such that their generalized commutator exists but $\mathscr{D}(AB) \cap \mathscr{D}(BA) = \{0\}$.
11. Prove Lemma 8.19.
12. Let A generate a (C_0) semigroup T. Let $\lambda \in \mathbb{K}$ and $f \in \mathscr{X}$. Show that $Af = \lambda f$ iff $T(t)f = e^{\lambda t}f$ for all $t \in \mathbb{R}^+$.
13. Let $\Omega = \{z \in \mathbb{C}: |z| = 1\}$ be the unit circle. Let $\theta \in \mathbb{R}$ and let $M(t)z = e^{2\pi i t \theta}z$ for $t \in \mathbb{R}$ and $z \in \Omega$. This defines a group M of measure preserving transformations on Ω. For which values of θ is M metrically transitive?
14. Show that in the commutative case of the Chernoff product formula (i.e. when $V(t) V(s) = V(s) V(t)$ for all $t,s \in \mathbb{R}^+$), the simple result Lemma 8.1 can be substituted for the complicated result Theorem 7.2 in the proof. Conclude that the approximation theorem enters into the proof of the central limit theorem only through Lemma 8.1.
15. Establish the uniformity assertion of Theorem 8.4.
*16. Prove Lemma 8.22.
17. Let T be an analytic semigroup of type (α, M) on a reflexive space $\mathscr{X}$. Let $\mathscr{N}$, $\mathscr{R}$ be as in Lemma 8.19. Then $\mathscr{N}$ and $\bar{\mathscr{R}}$ are closed subspaces, $\mathscr{N} \cap \mathscr{R} = \{0\}$, $\mathscr{N} + \bar{\mathscr{R}} = \mathscr{X}$, and the projection P of $\mathscr{X}$ onto $\mathscr{N}$ along $\bar{\mathscr{R}}$ is well-defined. Moreover,

$$\lim_{t \to \infty} T(t)f = Pf$$

for each $f \in \mathscr{X}$.

9. Further Developments and Applications

9.1. In this section we treat fractional powers (e.g. square roots) of semigroup generators, spectral mapping theorems, applications to classical inequalities, the connections between semigroups and Markov processes, potential theory, semigroups on lattices, approximation theory motivated by finite difference equation approximations to the differential equation $du/dt = Au$, and analytic vectors.

9.2. FRACTIONAL POWERS OF GENERATORS
Let $\mathscr{X}$ be a complex Banach space. Let T be a uniformly bounded (C_0) semigroup on $\mathscr{X}$ with generator A. For $t > 0$, $\lambda > 0$, and $0 < \alpha < 1$ introduce the function

$$g_{t,\alpha}(\lambda) = \frac{1}{2\pi i} \int_{\sigma - i\infty}^{\sigma + i\infty} \exp(z\lambda - tz^\alpha) \, dz;$$

here the branch of z^α is chosen so that $\operatorname{Re}(z^\alpha) > 0$ for $\operatorname{Re}(z) > 0$, and the integral is independent of $\sigma > 0$ (by Cauchy's theorem). The formula

$$T_\alpha(t)f = \int_0^\infty g_{t,\alpha}(\lambda)T(\lambda)f\,d\lambda$$

defines a uniformly bounded (C_0) semigroup T_α which has an analytic extension into a sector of the complex plane. If A_α denotes the generator of T_α, then the following formulas are valid:

$$(\mu - A_\alpha)^{-1}f = \frac{\sin\alpha\pi}{\pi}\int_0^\infty \frac{\lambda^\alpha(\lambda - A)^{-1}f}{\mu^2 - 2\mu\lambda^\alpha\cos\alpha\pi + \lambda^{2\alpha}}\,d\lambda$$

for $f \in \mathscr{X}$, and for $f \in \mathscr{D}(A)$,

$$A_\alpha f = \frac{\sin\alpha\pi}{\pi}\int_0^\infty \lambda^{\alpha-1}(\lambda - A)^{-1}\,Af\,d\lambda$$

$$= \frac{1}{\Gamma(-\alpha)}\int_0^\infty t^{-\alpha-1}\bigl(f - T(t)f\bigr)\,dt.$$

Other properties are valid such as

$$A_\alpha A_\beta f = -A_{\alpha+\beta}f \tag{9.1}$$

for $f \in \mathscr{D}(A^2)$ and $0 < \alpha,\beta$ with $\alpha + \beta < 1$,

$$\lim_{\alpha \to 1-} \|A_\alpha f - Af\| = 0$$

for $f \in \mathscr{D}(A)$, etc.

We mention two ways of viewing fractional powers. First, the spectral theorem (see Section 7 of Chapter II) enables us to define general functions of self-adjoint or skew-adjoint operators on complex Hilbert spaces. Semigroup theory tells us how to take exponential functions of certain other operators, but it is desirable to enlarge the class of functions of semigroup generators which make sense. Secondly, the operators A_α provide us with a large class of examples of analytic semigroups.

To see why the curious minus sign appears in (9.1), consider the calculus formula

$$a^\alpha = \frac{\sin\pi\alpha}{\pi}\int_0^\infty \lambda^{\alpha-1}(\lambda + a)^{-1}a\,d\lambda$$

for $a > 0$ and $0 < \alpha < 1$. This suggests that we can take roots of "positive" operators which are themselves "positive." It is natural to interpret dissipative to stand for negative and accretive to stand for positive. Consequently, A_α is $-(-A)^\alpha$.

For a concrete example, let $\mathscr{X} = BUC(\mathbb{R})$ or $L^p(\mathbb{R})$, $1 \le p < \infty$, and

$$T(t)f(x) = \frac{1}{\sqrt{4\pi t}}\int_{-\infty}^\infty \exp[-(x - y)^2/4t]f(y)\,dy;$$

T is a (C_0) semigroup with generator $A = d^2/dx^2$ (see Exercise 2.18.7). The semigroup associated with the square root of A is

$$T_{1/2}(t)f(x) = \frac{t}{\pi}\int_{-\infty}^{\infty}\frac{f(y)}{t^2+(x-y)^2}\,dy.$$

Its generator is given by the singular integral operator $A_{1/2}$ defined by

$$A_{1/2}f(x) = \lim_{h\to 0}\frac{1}{\pi}\int_{-\infty}^{\infty}\frac{f(x-y)-f(x)}{y^2+h^2}\,dy.$$

Note that $A_{1/2}$ has nothing to do with d/dx. It is easiest to see why in an L^2 setting (see Chapter II, Section 4). In the Fourier transform representation, A becomes multiplication by $-\xi^2$, $A_{1/2} = -(-A)^{1/2}$ becomes multiplication by $-|\xi|$, and d/dx becomes multiplication by $-i\xi$. Since $-|\xi|$ is not a polynomial, $A_{1/2}$ is not a differential operator; it is a *pseudo-differential* operator.

Let $\qquad\qquad v(t) = T_{1/2}(t)f$ for $f \in \mathscr{D}(A)$.

Then $\qquad\qquad v''(t) = (A_{1/2})^2 v(t) = -Av(t) \qquad v(0) = f,$

whence $v(t) = T(t)f$. For $A = d^2/dx^2$, v thus satisfies the *Laplace equation*

$$v_{tt} + v_{xx} = 0 \qquad (x\in\mathbb{R}, t\geq 0)$$

with boundary value $v(0,x) = f(x)$ and v is given by the *Poisson integral formula*

$$v(t,x) = \frac{t}{\pi}\int_{-\infty}^{\infty}\frac{f(y)}{t^2+(x-y)^2}\,dy$$

as noted above. Thus fractional power considerations enable us to solve certain two-point boundary value problems for abstract "elliptic" problems (such as $u_{tt} = -Au$ with A dissipative). The boundary conditions here are that $v = f$ at $t = 0$ and v is bounded at $t = \infty$.

9.3. SPECTRAL MAPPING THEOREMS. Let A generate a (C_0) semigroup T. Since we think of $T(t)$ as "e^{tA}", can we conclude that the spectrum of $T(t)$ satisfies $\sigma[T(t)] = e^{t\sigma(A)}$?

9.4. THEOREM. *Let A generate a (C_0) semigroup T on $\mathscr{X}$. Then*

$$\sigma[T(t)] \supset e^{t\sigma(A)} = \{e^{t\lambda}:\lambda\in\sigma(A)\}$$

for each $t \in \mathbb{R}^+$.

Proof. It is enough to show that

$$\rho(A) \supset \{\lambda: e^{t\lambda}\in\rho[T(t)]\}.$$

For $f \in \mathscr{D}(A)$ and $\lambda \in \mathbb{K}$,

$$[e^{t\lambda} - T(t)]f = -\int_0^t \frac{d}{ds}[e^{(t-s)\lambda}T(s)f]\,ds$$

$$= \int_0^t e^{(t-s)\lambda}T(s)(\alpha - A)f\,ds.$$

The formula

$$Sf = \int_0^t e^{(t-s)\lambda} T(s)f \, ds$$

defines a bounded operator S from $\mathscr{X}$ to $\mathscr{D}(A)$, equipped with its graph norm. Consequently $[e^{t\lambda} - T(t)]f = S(\lambda - A)f$ for all $f \in \mathscr{D}(A)$, and $[e^{t\lambda} - T(t)]f = (\lambda - A)Sf$ for all $f \in \mathscr{X}$. Therefore $e^{t\lambda} \in \rho[T(t)]$ implies $\lambda \in \rho(A)$ and $(\lambda - A)^{-1} = S[e^{t\lambda} - T(t)]^{-1}$.   ■

9.5. Let $\sigma_p(B)$, the *point spectrum* of an operator B, be the set of eigenvalues of B.

THEOREM.   *Let A generate a (C_0) semigroup T on a complex space $\mathscr{X}$. Then*

$$\exp[t\sigma_p(A)] \subset \sigma_p[T(t)] \subset \exp[t\sigma_p(A)] \cup \{0\}$$

for $t \in \mathbb{R}^+$.

Proof.   The first containment is easy; see Exercise 8.24.12. For the second containment, let $t > 0$ and $e^{\lambda t} \in \sigma_p[T(t)]$. Define the (C_0) semigroup S by $S(s) = e^{-\lambda s}T(s)$. The generator B of S is $A - \lambda I$ and $S(t)f = f$. Since $S(\cdot)f \in C(\mathbb{R}^+, \mathscr{X})$ is nonzero and periodic of period t, it has a nonzero Fourier coefficient, i.e. for some integer n,

$$0 \neq g = t^{-1} \int_0^t e^{-as} S(s)f \, ds$$

where $a = 2\pi int^{-1}$. Writing $k(s) = e^{-as} S(s)f$ we obtain, for $h > 0$,

$$h^{-1}[S(h) - I]g = (th)^{-1}e^{ah} \int_h^{t+h} k(s) \, ds - (th)^{-1} \int_0^t k(s) \, ds$$

$$= h^{-1}(e^{ah} - 1)t^{-1} \int_h^t k(s) \, ds + t^{-1}e^{ah}h^{-1} \int_t^{t+h} k(s) \, ds$$

$$-t^{-1}h^{-1} \int_0^h k(s) \, ds \to ag + t^{-1}e^{-at}S(t)f - tf = ag$$

as $h \to 0^+$. Thus $g \in \mathscr{D}(B) = \mathscr{D}(A)$ and

$$0 = (aI - B)g = [(a + \lambda)I - A]g,$$

whence $a + \lambda = 2\pi int^{-1} + \lambda \in \sigma_p(A)$, i.e. $e^{\lambda t} \in \exp[t\sigma_p(A)]$.   ■

9.6.   EXAMPLE.   Let $\mathscr{X}$ be the complex space $\{f \in C[0,1]: f(0) = 0\}$. Let $Af = f'$ on $\mathscr{D}(A) = \{f \in \mathscr{X}: f' \in \mathscr{X}\}$. The resolvent of A is given by

$$(\lambda - A)^{-1}f(x) = \int_0^x e^{\lambda(x-s)} f(s) \, ds$$

for all $x \in [0,1]$ and $f \in \mathscr{X}$; and A generates a (C_0) contraction semigroup T given by

$$T(t)f(x) = f(x - t) \text{ or } 0,$$

according as $t \le x$ or $x \le t$ with $x,t \in [0,1]$. Note that $\sigma(A) = \emptyset$, $\rho(A) = \mathbb{C}$, and $T(t) = 0$ for $t \ge 1$.

9.7. REMARK. Let A generate a (C_0) semigroup T on a complex space $\mathscr{X}$. Then $T(t) = 0$ for all $t \ge t_0$, where $t_0 > 0$, iff $\sigma(A) = \emptyset$ and there is constant M such that

$$\|(\alpha + i\beta - A)^{-1}\| \le M \max(1, e^{-\alpha t_0})$$

for all $\alpha + i\beta \in \mathbb{C}$.

9.8. Next we prove two theorems that generalize and simplify some classical inequalities of Hardy, Landau, and Littlewood. The classical inequalities are discussed in Example 9.10.
THEOREM. *Let A generate a (C_0) contraction semigroup T. Then for all $f \in \mathscr{D}(A^2)$,*

$$\|Af\|^2 \le 4\|A^2 f\| \|f\|.$$

Proof. It is easy to establish the Taylor formula

$$T(t)f = f + tAf + \int_0^t (t - s)T(s)A^2 f \, ds.$$

It follows that

$$\|Af\| \le t^{-1}(\|T(t)f\| + \|f\|) + \int_0^t (t - s)\|T(s)A^2 f\| \, ds$$

$$\le \frac{2}{t}\|f\| + \frac{t}{2}\|A^2 f\|. \tag{9.2}$$

If $A^2 f = 0$, letting $t \to \infty$ in (9.2) shows $Af = 0$. If $A^2 f \ne 0$, we minimize the right-hand side of (9.2) over t. The minimum occurs at $t = 2\|f\|^{1/2}\|A^2 f\|^{-1/2}$; inserting this into (9.2) yields $\|Af\| \le 2\|A^2 f\|^{1/2}\|f\|^{1/2}$. ∎

9.9. THEOREM. *Let A generate a (C_0) contraction semigroup on a Hilbert space $\mathscr{H}$. Then for all $f \in \mathscr{D}(A^2)$,*

$$\|Af\|^2 \le 2\|A^2 f\| \|f\|;$$

equality holds iff

$$A^2 f + \lambda Af + \lambda^2 f = 0 \text{ and } \mathrm{Re}\langle A^2 f, f \rangle = 0$$

for some $\lambda \in \mathbb{R}^+$.

Proof.   Let $C_\lambda = (\lambda + A)(\lambda - A)^{-1}$ for $\lambda > 0$ be the Cayley transform of A. Then $\|C_\lambda\| \leq 1$ (cf. Remarks 3.8). Using the identity

$$-2\lambda A f = (A^2 f + \lambda^2 f) + C_\lambda (A^2 f - \lambda^2 f)$$

and the parallelogram law we obtain

$$\|Af\|^2 \leq \frac{1}{4\lambda^2} (\|A^2 f + \lambda^2 f\| + \|C_\lambda\| \|A^2 f - \lambda^2 f\|)^2$$

$$\leq \frac{1}{2\lambda^2} (\|A^2 f + \lambda^2 f\|^2 + \|A^2 f - \lambda^2 f\|^2)$$

$$= \frac{1}{\lambda^2} \|A^2 f\|^2 + \lambda^2 \|f\|^2.$$

$A^2 f = 0$ implies $Af = 0$ (let $\lambda \to 0$). If $A^2 f \neq 0$, choosing $\lambda^2 = \|A^2 f\| \|f\|^{-1}$ (i.e. minimizing) yields $\|Af\|^2 \leq 2\|A^2 f\| \|f\|$. The issue of when equality holds is left as an exercise.   ∎

9.10.   *Example.*   Take $A = d/dx$ on $\mathscr{X} = L^p[0,\infty[$, $1 \leq p < \infty$ or $BUC(\mathbb{R}^+)$. The two theorems above imply

$$\|f'\|_p^2 \leq C_p \|f''\|_p \|f\|_p$$

for all $f \in \mathscr{X}$ such that $f'' \in \mathscr{X}$, where $1 \leq p \leq \infty$, and $C_p = 4$ for $p \neq 2$, $C_2 = 2$. The best values of C_p are not known for $1 < p < \infty, p \neq 2$. The best value of C_∞ is 4, while the best value of C_1 is 5/2. It is known that the best value of C_p is a continuous function of p.

9.11.   Feller-Markov Processes and Semigroups
Classical probability theory was concerned with sequences of independent random variables. Applications made it clear that it was necessary to consider random variables indexed by a real parameter (time) and to relax the notion of independence. Markov dependence is a notion that has two particularly significant features: (1) it leads to important applications, and (2) it leads to a mathematically rich theory involving semigroups of operators in a crucial way. We shall indicate how these features evolved in the following, rather informal, introduction to Markov processes.

A *Markov process* is a family $\xi = \{\xi_t : t \in \mathbb{R}^+\}$ of random variables (i.e. measurable functions) from a probability space (Ω, Σ, p) to a measurable space $(S, \mathscr{B})$ such that the Markov property is satisfied. Intuitively, the Markov property says that, given the present, the future is independent of the past. Alternatively, one could say that, intuitively, a Markov process is a model for a system evolving randomly in time and having no memory of the past. This property is usually expressed in terms of conditional expectations, which we avoid going into in detail; indeed, the modern definition of a Markov process is quite complicated (see, for example, Dynkin [2] or Blumenthal–Getoor [1]). We shall continue informally. Consider the probability p that ξ_t is in Γ, given that $\xi_s = x$; here $\Gamma \in \mathscr{B}$

and $0 \leq s < t < \infty$. We assume that the process ξ is temporally homogeneous, which means that p depends on t and s only through $t - s$. Denote p by $P(t - s, x, \Gamma)$; as a function of Γ this is a probability measure. When does ξ_{t+s} belong to $\Gamma \in \mathscr{B}$? $\xi_{t+s}(\omega)$ is in Γ iff $\xi_t(\omega)$ is somewhere in S and the function $\xi_{t+r}(\omega)$ goes from $\xi_t(\omega)$ when $r = 0$ to a point in Γ when $r = s$. Using the Markov property, we get the Chapman-Kolmogorov equation

$$P(t + s, x, \Gamma) = \int_S P(t, x, dy) P(s, y, \Gamma); \qquad (9.3)$$

here y is thought of as $\xi_t(\omega)$, and we have multiplied the probabilities of independent events to get the probability of their intersection. Intuitively the right-hand side of (9.3) is the sum over all $y \in S$ of the probability that $\xi_u(\omega)$ goes from x, which it is at time $u = 0$, to y at time $u = t$, and then to a value in Γ t units of time later. Now let $\mathscr{X}$ be the Banach space (under supremum norm) of all bounded measurable functions on S. Define $T = \{T(t): t \in \mathbb{R}^+\}$ by

$$T(t)f(x) = \int_S P(t, x, dy) f(y) \qquad (9.4)$$

for $f \in \mathscr{X}$, $t \in \mathbb{R}^+$, and $x \in S$. Because of (9.3), T satisfies $T(t + s) = T(t)T(s)$ for all $t, s \in \mathbb{R}^+$. Clearly also $T(0) = I$ and $\| T(t) \| \leq 1$ for each $t \geq 0$. In general, however, $T(\cdot)f$ need not be continuous for $f \in \mathscr{X}$.

Now assume that S is a compact metric space and $\mathscr{B}$ is its collection of Borel sets. A typical case is that S is a closed subset of either $[-\infty, \infty]$ or the one-point compactification of $\mathbb{R}^n$. We assume that T is a (C_0) semigroup on the space $C(S)$ of continuous real functions on S. Then we shall call T a *Feller-Markov semigroup* and ξ a *Feller-Markov process*. In general, a *Feller-Markov semigroup* is a (C_0) contraction semigroup on $C(S)$, where S is a compact metric space, such that

$$T(t)1 = 1,$$

$$f \geq 0 \text{ implies } T(t)f \geq 0,$$

for each $t \in \mathbb{R}^+$. Here 1 denotes the constant function with value 1. Clearly T defined by (9.4) satisfies these two properties.

We next give an alternative expression for T. ξ determines a family of measures $\{P_\mu\}$ on the sample path space $\tilde{\Omega}$, which consists of *all* functions from $\mathbb{R}^+$ to S. These measures are indexed by μ, which runs over all probability measures on $(S, \mathscr{B})$. If δ_x denotes the point mass measure at $x \in S$, we shall denote P_{δ_x} by P_x. This gives the probability of an event for a path which starts from x. The corresponding expectation (or integral) operator is E_x defined by

$$E_x(\eta) = \int_\Omega \eta(\omega) P_x(d\omega)$$

defined for real functions on Ω. Thus $E_x(\eta)$ is the (conditional) expected value of η along a random path starting at x. Then (9.4) is equivalent to

$$T(x)f(x) = E_x[f(\xi_t)],$$

which is the conditional expectation of $f(\xi_t)$, given that $\xi_0 = x$. Because of this formula, T is sometimes called the *expectation semigroup* of the Markov process ξ.

The above discussion shows how a (stationary) Markov process leads to a semigroup. We now show the converse construction. Let S be a compact metric space and let T be a Feller-Markov semigroup on $C(S)$. Let $\mathscr{B}$ denote the Borel sets of S. By the Riesz Representation theorem (see Dunford-Schwartz [1, p. 265]) for each $t \in \mathbb{R}^+$ and $x \in S$ there is a unique regular probability measure $P(t,x,\cdot)$ on $\mathscr{B}$ such that

$$T(t)f(x) = \int_S P(t,x,dy)f(y)$$

for each $f \in C(S)$. Let $S_\alpha = S$ for all $\alpha \in \mathbb{R}^+$ and let Ω be the product space $\Omega = \Pi_{\alpha \in \mathbb{R}^+} S_\alpha$. Ω is nothing more than what was called $\tilde{\Omega}$ above. Ω is a compact Hausdorff space. Let $\mathscr{A}$ be the algebra of all continuous functions on Ω of the form

$$\phi(\omega) = F[\omega(t_1),\dots,\omega(t_n)] \qquad \omega \in \Omega, \tag{9.5}$$

where $n \in \mathbb{N}$, $0 \le t_1 < t_2 < \dots < t_n < \infty$, and $F \in C(S^n)$. The Stone-Weierstrass theorem (see Dunford-Schwartz [1, p. 272]) implies that $\mathscr{A}$ is dense in $C(\Omega)$. Thus there is a bijective correspondence between the set of bounded, positivity preserving linear functionals on $\mathscr{A}$ and the set of regular measures on the Borel sets of Ω. An example of such a functional is

$$L_\mu(\phi) = \int_{S^{n+1}} F(x_1,\dots,x_n)\mu(dx_0)P(t_1,x_0,dx_1)\dots P(t_n - t_{n-1},x_{n-1},dx_n)$$

for $\phi \in \mathscr{A}$ as in (9.5) and μ a probability measure on S. Let P_μ denote the corresponding probability measure on Ω. If δ_x denotes the unit mass measure at $x \in S$, then one can check that

$$T(t)f(x) = E_x[f(\xi_t)] \qquad (x \in S, t \in \mathbb{R}^+)$$

holds for all $f \in C(S)$, where $\xi_t : \Omega \to S$ is defined by $\xi_t(\omega) = \omega(t)$, and E_x is the expectation corresponding to the probability measure P_{δ_x}. We have recovered the process ξ from T. Standard arguments enable us to extend the domain of $T(t)$ to the bounded measurable functions on S. Then we can recover $P(t,x,\Gamma)$ via

$$P(t,x,\Gamma) = T(t)X_\Gamma(x),$$

where X_Γ is the characteristic (or indicator) function of Γ.

Once we have the measures $\{P_\mu\}$, sample function properties, limit theorems, and other things of interest to probabilists can be studied. The main point of our discussion is that the starting point of such a study can be taken to be the theory of Feller-Markov semigroups.

9.12. EXAMPLES. We give three examples of Feller-Markov semigroups.

1. Let $S = \{1,\dots,n\}$. Identify $\mathscr{X} = C(S)$ with $\mathbb{R}^n$ and $\mathscr{B}(\mathscr{X})$ with all $n \times n$ real matrices. A matrix $A = (a_{ij})$ generates a Feller-Markov semigroup iff (i) $a_{ii} \le 0$

for $1 \le i \le n$, (ii) $a_{ij} \ge 0$ for $1 \le i, j \le n$ with $i \ne j$, and (iii) $\sum_{j=1}^{n} a_{ij} = 0$. The corresponding process is called a continuous time finite state stationary Markov chain.

A specific example is given by

$$A = \begin{pmatrix} -\alpha & \alpha \\ \alpha & -\alpha \end{pmatrix}, \quad T(t) = \frac{1}{2} \begin{pmatrix} 1 + e^{-2\alpha t} & 1 - e^{-2\alpha t} \\ 1 - e^{-2\alpha t} & 1 + e^{-2\alpha t} \end{pmatrix}$$

for any $\alpha > 0$.

2. (n dimensional Brownian motion). Let $\dot{\mathbb{R}}^n$ denote the one-point compactification of $\mathbb{R}^n$. Let $\mathscr{X} = C(\dot{\mathbb{R}}^n)$ and for $f \in \mathscr{X}$, $x \in \mathbb{R}^n$, $t > 0$ let

$$T(t)f(x) = (4\pi t)^{-n/2} \int_{\mathbb{R}^n} \exp(-|x - y|^2/4t)f(y)\, dy,$$

$$T(t)f(\infty) = f(\infty).$$

(Here $|\cdot|$ is the Euclidean norm in $\mathbb{R}^n$.) This defines a (C_0) semigroup T on $\mathscr{X}$ whose generator is given by $A = \Delta = \sum_{i=1}^{n} \partial^2/\partial x_i^2$ on $\mathscr{D}(A) = \{f \in \mathscr{X}$: the distributional Laplacian Δf exists and is in $\mathscr{X}\}$. This is the higher dimensional version of Exercise 2.18.7.

3. (n-dimensional Cauchy process). Let $\mathscr{X} = C(\dot{\mathbb{R}}^n)$ as above. For $f \in \mathscr{X}$, $x \in \mathbb{R}^n$, $t > 0$ let

$$T(t)f(x) = d_n \int_{\mathbb{R}^n} t(t^2 + |x - y|^2)^{-(n+1)/2} f(y)\, dy,$$

$$T(t)f(\infty) = f(\infty)$$

where $d_n = \Gamma\left(\frac{n+1}{2}\right)\pi^{-(n+1)/2}$. This defines a Feller-Markov semigroup T whose generator A is given on functions f of the form $g + c$ (where g is a $C^\infty(\mathbb{R}^n)$ function with compact support and $c \in \mathbb{R}$) by

$$Af(x) = \lim_{t \to 0} d_n \int_{\mathbb{R}^n} [f(y) - f(x)](t^2 + |x - y|^2)^{-(n+1)/2}\, dy.$$

This is the n-dimensional generalization of Exercise 2.18.8. Moreover, if A,T denote the generator and semigroup of this example, and if B,S denote the generator and semigroup of Example 2 above, then $T = S_{1/2}$ and $A = B_{1/2}$ in the notation of Section 9.2. In the language of stochastic processes, the Cauchy process (i.e. T) is said to be *subordinate* to the Brownian motion process (S). Subordinate just means that the generator of the subordinate process is a fractional power of the generator of the other process.

9.13. SEMIGROUPS ON LATTICES.   A key property of Feller-Markov semigroups is that they are positivity preserving. We state here a generation theorem for general positivity preserving semigroups on a Banach lattice. Let $\mathscr{X}$ be a Banach lattice, i.e. $\mathscr{X}$ is a Banach space and a vector lattice for which the order relation satisfies

$$|f| \leq |g| \text{ implies } \|f\| \leq \|g\|.$$

When $\mathscr{X}$ is a real space, $|f| = f^+ + f^-$ and $f^+ = f \vee 0$, $f^- = -f \vee 0$. When $\mathscr{X}$ is complex we assume that $\mathscr{X}$ comes equipped with a complex conjugation $f \to \bar{f}$ so that $f = \operatorname{Re} f + i \operatorname{Im} f$ with $\operatorname{Re} f = 2^{-1}(f - \bar{f})$, $\operatorname{Im} f = (2i)^{-1}(f - \bar{f})$, and $|f| = [(\operatorname{Re} f)^2 + (\operatorname{Im} f)^2]^{1/2}$. A is *dispersive* if

$$\operatorname{Re}\langle Af, \phi \rangle \leq 0 \text{ for } f \in \mathscr{D}(A), \phi \in \mathscr{J}(f^+).$$

A is *m-dispersive* if, in addition, Range $(I - \alpha A) = \mathscr{X}$ for some $\alpha > 0$.

THEOREM. *A is the generator of a (C_0) positivity preserving contraction semigroup on a Banach lattice $\mathscr{X}$ iff A is densely defined and m-dispersive.*

Positivity preserving means $f \geq 0$ implies $T(t)f \geq 0$ for each $f \in \mathbb{R}^+$. As a corollary, *m-dispersive* implies *m-dissipative*.

9.14. THEOREM. *Let T be a uniformly bounded (C_0) semigroup on $\mathscr{X}$ with generator A. Then A has a densely defined inverse V (i.e. A is injective with dense range) iff*

$$\lim_{\lambda \to 0^+} \int_0^\infty e^{-\lambda t} T(t)f \, dt = \lim_{\lambda \to 0^+} (\lambda - A)^{-1}f = Vf \qquad (9.6)$$

holds for all $f \in \mathscr{R}(A)$.

The proof is left as an exercise.

Let A, T be as in the above theorem. When (9.6) holds, we define the *potential operator associated with A* to be $V = -A^{-1}$.

When T is the Brownian motion semigroup (i.e. A is the Laplacian on $\mathbb{R}^n$, see Example 9.12.2), then the associated potential operator V is the integral operator whose kernel is the classical Newtonian potential.

9.15. Let Ω be a locally compact separable metric space. Let $\mathscr{X} = C_0(\Omega)$ be the real continuous functions on Ω which vanish at ∞. Recall that $C_c(\Omega)$ consists of the functions in $\mathscr{X}$ having compact support. Let T be a positivity preserving (C_0) contraction semigroup on $\mathscr{X}$. T can be viewed as a restriction of a Feller-Markov semigroup (see Exercise 9.21.5). We can check that, under some additional (weak) assumptions on T, the conditions of Theorem 9.14 hold and that the potential operator V satisfies the following three properties.

(i) $C_c(\Omega) \cap \mathscr{D}(V)$ is dense in $\mathscr{X}$.
(ii) V is *codissipative*, i.e.

$$\|f - \lambda Vf\| \geq \|\lambda Vf\|$$

holds for all $f \in \mathscr{D}(V)$ and all $\lambda > 0$.

(iii) V satisfies the *complete principle of the maximum*: for $f, g \in C_c(\Omega) \cap \mathscr{D}(V)$ with $f \geq 0$, $g \geq 0$, we have $Vf \geq 0$, $Vg \geq 0$, and the condition

$$c + Vf(x) \geq Vg(x)$$

for all $x \in \Omega$ such that $g(x) > 0$ (and some $c \in \mathbb{R}$) implies

$$c + Vf(x) = Vg(x)$$

for all $x \in \Omega$.

Each of these last two properties characterizes V, as the following theorem shows.

HUNT'S THEOREM.   *Let* $\mathscr{X} = C_0(\Omega)$, *where* Ω *is a locally compact separable metric space. Let* $\mathscr{D}(V)$ *be dense in* $C_c(\Omega)$ *and let* $V: \mathscr{D}(V) \to \mathscr{X}$ *be a linear operator having dense range. If either*
   (i)   *V is codissipative, or*
   (ii) *V satisfies the complete principle of the maximum,*

then there is a unique Feller-Markov semigroup T on $C(\dot{\Omega})$, $\dot{\Omega}$ being the one-point compactification of Ω, such that for all $f \in \mathscr{D}(V)$,

$$Vf = \lim_{\lambda \to 0^+} \int_0^\infty e^{-\lambda t} T(t) f \, dt.$$

In the context of Feller-Markov semigroups, the generator, the semigroup, the potential operator, and the Markov process all determine one another. We can mimic classical potential theory and associate a potential theory with a complete principle of the maximum, superharmonic and harmonic functions, balayages, réduites, etc. with each Feller-Markov semigroup (or process). This theory is very rich and has close connections with many branches of mathematics.

We refer the interested reader to the books of Meyer [1], Cornea-Licea [1], and M. Rao [1].

9.16.   The approximation theorems given in Section 7 all assume that the approximating semigroups are defined on the same space as the limit semigroups. This need not be the case.

THEOREM.   *For each $n \in \mathbb{N}$ let A_n generate a (C_0) contraction semigroup T_n on a Banach space $\mathscr{X}_n$. Let $\mathscr{X}_0$ be a Banach space and suppose for each $n \in \mathbb{N}$ there is a $P_n \in \mathscr{B}(\mathscr{X}_0, \mathscr{X}_n)$ such that $\sup_n \|P_n\| < \infty$ and $\lim_{n \to \infty} \|P_n f\| = \|f\|$ for each $f \in \mathscr{X}_0$. Write $f \in \mathscr{D}(A_0)$ and $A_0 f = g$ to mean there is a sequence $\{f_n\}_1^\infty$ with $f_n \in \mathscr{D}(A_n)$ such that $\lim_{n \to \infty} \|f_n - P_n f\| = 0$ and $\lim_{n \to \infty} \|A_n f_n - P_n g\| = 0$. Then there is a (C_0) contraction semigroup T_0 on $\mathscr{X}_0$ such that*

$$\lim_{n \to \infty} \| T_n(t) P_n f - P_n T_0(t) f \| = 0$$

for all $f \in \mathscr{X}_0$ and all $t \in \mathbb{R}^+$ iff $\mathscr{D}(A_0)$ is dense in $\mathscr{X}_0$ and the range of $\lambda I - A_0$ is dense in $\mathscr{X}_0$ for some $\lambda > 0$. In this case the generator of T_0 is the closure of A_0 and

$$\limsup_{n \to \infty \; 0 \le s \le t} \| T_n(s) P_n f - P_n T(s) f \| = 0$$

for every $f \in \mathscr{X}_0$ and each $t > 0$.

9.17. DEFINITIONS. Let $\mathscr{X}_n (n \in \mathbb{N}_0)$ be a Banach space. Suppose that for each $n \in \mathbb{N}$ there is a $P_n \in \mathscr{B}(\mathscr{X}_0, \mathscr{X}_n)$ such that $\sup_n \|P_n\| < \infty$, $\|P_n f\| \to \|f\|$ for each $f \in \mathscr{X}_0$, and each $g \in \mathscr{X}_n$ is of the form $g = P_n f$ for some $f \in \mathscr{X}_0$, where $\|f\| \le C\|g\|$, C being some absolute constant. We shall say that $\mathscr{X}_n$ *approximates* $\mathscr{X}_0$ when these conditions hold.

Let A_0 generate a (C_0) semigroup T_0 on $\mathscr{X}_0$. Now let $\{T_n\}_{n=1}^\infty$ be a sequence of discrete semigroups, $T_n = \{T_n^k : k \in \mathbb{N}_0\}$, in $\mathscr{B}(\mathscr{X}_n)$. We associate with T_n a "time unit" $\tau_n > 0$. (The role of τ_n will become clear momentarily, in Section 9.19.) We say that $\{T_n\}$ *approximates* T_0 if, for all $f \in \mathscr{X}$ and all $t \in \mathbb{R}^+$.

$$\lim_{n \to \infty} \|T_n^{k_n} P_n f - P_n T_0(t) f\| = 0$$

whenever $\{k_n\} \subset \mathbb{N}$ is such that $\lim_{n \to \infty} k_n \tau_n = t$.

9.18. THEOREM. *Let $\mathscr{X}_n$ approximate $\mathscr{X}_0$ as above. For $n \in \mathbb{N}$ let T_n be a discrete semigroup with time unit $\tau_n > 0$. Let $A_n = \tau_n^{-1}(T_n - I)$, so that $T_n^k = (I + \tau_n A_n)^k$ for $k, n \in \mathbb{N}$. Suppose that the stability condition*

$$\|T_n^k\| \le M e^{\omega t}$$

holds, where $t = k\tau_n$ and M, ω are independent of n and k. Then $\{T_n\}$ approximates T_0 if, for each f in some core of A_0,

$$\lim_{n \to \infty} \|A_n P_n f - P_n A_0 f\| = 0.$$

9.19. THE LAX EQUIVALENCE THEOREM. Consider the differential equation

$$du/dt = A_0 u$$

subject to the initial condition $u(0) = f$. If A_0 generates a (C_0) semigroup T_0 on $\mathscr{X}_0$ and if $f \in \mathscr{D}(A_0)$, then the solution is $u(t) = T_0(t)f$. Suppose we want to approximate this problem by a difference equation for $u^n = u(n\Delta t)$ for some $\Delta t > 0$. If the difference scheme is $u^{n+1} = C(\Delta t)u^n$, then the solution is $u^n = C(\Delta t)^n u^0$. (Here u^0 should approximate f in some sense.) The operators $C(\Delta t)$ are said to satisfy the *consistency condition* if for all g in a core of A_0,

$$\left\| \left[\frac{C(\Delta t) - I}{\Delta t} \right] T_0(t)g - A_0 T_0(t)g \right\| \to 0$$

as $\Delta t \to 0$, uniformly for t in bounded subsets of $\mathbb{R}^+$. Here we have assumed that $C(\Delta t)$ acts on the same space $\mathscr{X}_0$ as T_0 does. If not, consider a sequence $\Delta_n t \to 0^+$ and let $C(\Delta_n t)$ act on $\mathscr{X}_n$, where $\mathscr{X}_n$ approximates $\mathscr{X}_0$ in the sense of Definitions 9.17. The consistency condition can be defined in this context as well.

We shall write $\Delta_n t$ as τ_n and $C(\Delta_n t)$ as T_n. The difference schemes $\{T_n\}$ are said to be *convergent* provided that

$$\lim_{n \to \infty} \|T_n^{k_n} g - T(t)g\| = 0$$

where $g \in \mathscr{X}$ and $\lim_{n \to \infty} \tau_n k_n = t$. Again, the P_n's must be brought in to modify this definition when $\mathscr{X}_n$ differs from $\mathscr{X}_0$. Finally, the difference schemes are said to be *stable* if

$$\sup\{\|T_n^{k_n}\|: 0 < \Delta_n t \le \tau_1, k_n\Delta_n t \le \tau_2, n \in \mathbb{N}\} < \infty$$

for each $\tau_1, \tau_2 > 0$. This notion of stability is a property of the sequence of difference schemes, and makes no reference to the underlying differential equation.

Now let A_0 generate a (C_0) semigroup T_0, and let $\{T_n[=C(\Delta_n t)]\}$ be a sequence of difference schemes that satisfies the consistency condition. Then, the *Lax equivalence theorem* states that the difference schemes are convergent iff they are stable.

It should be noted that an unstable sequence of difference schemes may yield convergence for certain special choices of initial data.

9.20. ANALYTIC VECTORS. Let A be a linear operator in $\mathscr{X}$. Vectors in $\bigcap_{n=1}^{\infty} \mathscr{D}(A^n)$ will be called C^∞ *vectors* for A. (When $A = d/dx$, the C^∞ vectors for A are simply C^∞ functions.) An *analytic vector* for A is a C^∞ vector f such that the power series

$$\sum_{n=0}^{\infty} \frac{t^n}{n!} \|A^n f\|$$

has a positive radius of convergence.

THEOREM. *Let A be a dissipative operator on a Hilbert space $\mathscr{H}$ and suppose that A has a dense set of analytic vectors. Then $\bar{A}$ is m-dissipative.*

Proof. Let B be a maximal dissipative extension of A. B generates a (C_0) contraction semigroup T on $\mathscr{H}$ (see Remark 3.8). Let $g \in \mathscr{H}$ be orthogonal to the range of $I - A$. It is enough to show $g = 0$. Let $\mathscr{D}$ be the set of all analytic vectors of A. $f \in \mathscr{D}$ implies $\langle f - Af, g \rangle = 0$ or $\langle Af, g \rangle = \langle f, g \rangle$. Since $f \in \mathscr{D}$ implies $A^n f \in \mathscr{D}$, we get $\langle A^{n-1}f, g \rangle = \langle A^n f, g \rangle$, and so $\langle A^n f, g \rangle = \langle f, g \rangle$ for all $n \in \mathbb{N}$. Let $F(t) = \langle T(t)f, g \rangle$. Since f is an analytic vector, F is an analytic function on $\mathbb{R}^+$. Since $F^{(n)}(0) = \langle A^n f g \rangle = \langle f, g \rangle = F(0)$ for all n, it follows that $F(t) = F(0)e^t$. But $\|T(t)\| \le 1$ implies that $|F(t)|$ is bounded by $\|f\|\|g\|$. Consequently, $F(0) = 0$. Since $F(0) = \langle f, g \rangle$, g is orthogonal to each analytic vector f of A. The density of $\mathscr{D}$ implies $g = 0$. ∎

9.21. EXERCISES

1. In the notation at the end of Section 9.2, verify that $A_{1/2}$ is the generator of $T_{1/2}$.
2. Let T be a (C_0) semigroup with generator A. Show that $T(t)$ is compact for each $t > 0$ iff $(\lambda - A)^{-1}$ is compact for one (hence all) λ in $\rho(A)$ and $\lim_{t \to s} \|T(t) - T(s)\| = 0$ for all $s > 0$.
3. Let A generate a (C_0) semigroup on a complex Banach space, and suppose the conditions of Exercise 2 are satisfied. Which subsets of $\mathbb{C}$ can $\sigma(A)$ be? What about $\sigma_p(A)$?
*4. Prove Remark 9.7.
5. Let S_0 be a locally compact metric space and let $C_0(S_0) = \{f \in C(S_0): f$ vanishes at $\infty\}$ be the closure of the continuous functions having compact

support in S_0. Let T_0 be a (C_0) contraction semigroup on $C_0(S_0)$ satisfying: $f \geq 0$ implies $T_0(t)f \geq 0$. Adjoin a "dead point" Δ to S_0 and let $S = S_0 \cup \{\Delta\}$. (Take Δ to be the point at infinity if S_0 is not compact; otherwise take Δ to be an isolated point.) Define T on $C(S)$ by

$$T(t)f(x) = \begin{cases} T_0(t)[f - f(\Delta)](x) + f(\Delta) \text{ if } x \in S_0 \\ f(\Delta) \qquad\qquad\qquad\qquad \text{ if } x = \Delta. \end{cases}$$

Show that T is a Feller-Markov semigroup on $C(S)$. {Thus sub-Markov semigroups on locally compact spaces can be extended to be Feller-Markov semigroups. Intuitively, when the process reaches Δ it stays there; Δ can therefore be thought of as a cemetery.}

6. Let $\mathscr{B}$ denote the Borel sets of $\mathbb{R}$. Let $P \colon \mathbb{R}^+ \times \mathbb{R} \times \mathscr{B} \to [0,1]$ satisfy
    (i) $P(t,x,\Gamma)$ is a Borel function in x for fixed t,Γ;
    (ii) $P(t,x,\Gamma)$ is a probability measure in Γ for fixed t,x;
    (iii) $P(t + s,x,\Gamma) = \int_{-\infty}^{\infty} P(t,x,dy)P(s,y,\Gamma)$ for all t,s,x,Γ;
    (iv) $P(0,x,\Gamma) = 1$ or 0 according as $x \in \Gamma$ or $x \notin \Gamma$;
    (v) $P(t,x,\Gamma)$ is absolutely continuous in Γ for fixed t,x; define $p = \mathbb{R}^+ \times \mathbb{R} \times \mathbb{R} \to [0,\infty)$ by

$$P(t,x,\Gamma) = \int_{\Gamma} p(t,x,y)\, dy.$$

Assume further that
    (vi) $\partial p/\partial x$ and $\partial^2 p/\partial y^2$ exist and are continuous in $\mathbb{R}^+ \times \mathbb{R} \times \mathbb{R}$;
    (vii) $\lim_{t \to 0} (1/t) \int_{|y-x| \geq \delta} p(t,x,y)\, dy = 0$ for each $\delta > 0$, uniformly $x \in \mathbb{R}$;
    (viii) $\lim_{t \to \infty} (1/t) \int_{|y-x| < \delta} (y - x)p(t,x,y)\, dy = m(x)$ and

$$\lim_{t \to 0} \frac{1}{t} \int_{|y-x| < \delta} (y - x)^2 p(t,x,y)\, dy = \sigma^2(x)$$

exist uniformly in $x \in \mathbb{R}$, for each $\delta > 0$, and are continuous.

Then $u(t,x) = p(t,x,y)$ is a solution of the partial differential equation

$$\frac{\partial u}{\partial t} = \frac{\sigma^2(x)}{2} \frac{\partial^2 u}{\partial x^2} + m(x) \frac{\partial u}{\partial x}.$$

(In fact, $p(t,x,y)$ is the fundamental solution of this equation.)

7. In the above exercise, let $\mathscr{X} = C(\dot{\mathbb{R}})^\dagger$, where $\dot{\mathbb{R}}$ is the one-point compactification of $\mathbb{R}$, and define

$$T(t)f(x) = \int_{-\infty}^{\infty} f(y)p(t,x,y)\, dy$$

for $f \in \mathscr{X}$, $t \in \mathbb{R}^+$, and $x \in \mathbb{R}$. Show that T is a (C_0) Feller-Markov semigroup whose generator is an extension of

$$\frac{\sigma^2(x)}{2} \frac{d^2}{dx^2} + m(x) \frac{d}{dx}$$

defined on the C^2 functions having compact support in $\mathbb{R}$.

$\dagger$ $f \in \mathscr{X}$ iff $f \in C[-\infty,\infty]$ and $f(-\infty) = f(+\infty)$.

8. Verify the details of Example 9.12.1
*9. Prove Theorem 9.13.
10. Let A be m-dissipative, let $0 < \alpha < 1$, and let $A_\alpha = -(-A)^\alpha$ be the operator constructed in 9.2. If B is a closed operator and $\mathscr{D}(B) \supset \mathscr{D}(A)$, then B is a Kato perturbation of A. Prove this, and deduce some perturbation theorems as consequences.
11. If T is an analytic contraction semigroup of type $S(a)$ (where $a > 0$) on a uniformly convex space $\mathscr{X}$, then
$$\limsup_{t \to 0} \|T(t) - I\| < 2.$$
(This provides a partial converse to Exercise 5.10.5.)
12. Let T be a (C_0) semigroup on $\mathscr{X}$ with generator A. Show that
$$\{f \in \mathscr{X} : \text{weak } \lim_{t \to 0} t^{-1}[T(t)f - f] \text{ exists}\} = \mathscr{D}(A).$$
13. Let A be a densely defined dissipative operator on $\mathscr{X}$. If $\mathscr{D}(A^*)$ is dense in $\mathscr{X}^*$, the $\bar{A}$ is m-dissipative.
14. Let A generate a (C_0) group which has an analytic extension into a sector of the complex plane. Then A is bounded.
15. Let T be a (C_0) semigroup such that $T(t) - I$ is compact for some $t > 0$. Then T can be embedded in a (C_0) group.
16. Let $\mathscr{X}$ be the continuous functions on the unit circle, or, equivalently, the continuous 2π-periodic function on $\mathbb{R}$. Show that $A = d^3/dx^3$ cannot generate a Feller-Markov semigroup on $\mathscr{X}$. Hint: If it does and if $f \in \mathscr{D}(A)$ attains its minimum at x_0, then $Au(x_0) \geq 0$. Consider $f(x) = 2\sin^2 x - \sin^3 x$.
17. Prove the following variant of Theorem 9.20. If A is a symmetric operator on a Hilbert space, and if A has a dense set of analytic vectors, then $\bar{A}$ is self-adjoint.
18. A *quasi-analytic* vector for a linear operator A on $\mathscr{X}$ is an C^∞ vector f such that the least nonincreasing majorant of the series $\Sigma \|A^n f\|^{-1/n}$ diverges, i.e. $\|A^n f\|^{1/n} = O(m_n)$ implies $\sum_{n=1}^\infty (\inf\{m_k : k \geq n\})^{-1} = \infty$. Show that an analytic vector is quasianalytic. Prove the versions of Theorem 9.20 and Exercise 9.21.18 obtained by replacing the assumption that the set of analytic vectors is dense with the assumption that the set of quasianalytic vectors has dense span.
19. Show that the converse of Theorem 9.20 is false.
20. Show that A generates a (C_0) group of isometries iff both A and $-A$ generate (C_0) contraction semigroups on $\mathscr{X}$. For such an A and for $f \in \mathscr{D}(A^2)$, show that
$$\|Af\|^2 \leq 2\|A^2 f\|\|f\|.$$
Show that the constant 2 can be replaced by 1 when $\mathscr{X}$ is a Hilbert space.
*21. Let T be a Feller-Markov semigroup on $\mathscr{X} = C([r_1, r_2])$ where $-\infty \leq r_1 < r_2 \leq \infty$. Let $P(t,x,\Gamma) = T(t)\lambda_\Gamma(x)$ be the associated transition function (cf. 9.11). T is said to satisfy the *Lindeberg condition* if, for all $x \in]r_1, r_2[$ and all $\epsilon > 0$,
$$\lim_{t \to 0^+} \left\|\frac{1}{t} T(t)\chi_\Gamma\right\|_\infty = 0$$
where $\Gamma = \{y \in [r_1, r_2] : |y - x| \geq \epsilon\}$. The Lindeberg condition implies the following three conditions.

(i) $A1 = 0$, where A is the generator of T and 1 is the constant function with value 1.

(ii) A is *local* in the sense that if $f \in \mathscr{D}(A)$ vanishes in an open interval J, then
      $Af|_J = 0$.

(iii) If $f \in \mathscr{D}(A)$ has a relative maximum at x_0, then $Af(x_0) \le 0$.

Now let A_0 be densely defined on $\mathscr{X} = C([r_1,r_2])$ and satisfy (i), (ii), (iii) together with the nondegeneracy conditions:

(iv) There is an $f_0 \in \mathscr{D}(A_0)$ such that $A_0 f_0(x) > 0$ for all x in $]r_1,r_2[$.

(v)   There is a solution v of $A_0 v = 0$ which is nonconstant on each open interval of $]r_1,r_2[$.

Show that A_0 is a generalized second-order differential operator in the following sense: For $f \in \mathscr{D}(A)$, $x \in]r_1,r_2[$,

$$Af(x) = D_m^+ D_s^+ f(x). \tag{9.7}$$

Here, for q a strictly increasing function on $]r_1,r_2[$,

$$D_q^+ g(x) = \lim_{y \to x^+} \frac{g(y) - g(x)\mathord{\restriction}}{q(y) - q(x)\mathord{\restriction}}$$

or $D_q^+ g(x) = \dfrac{g(x+0) - g(x-0)}{q(x+0) - q(x-0)}$ according as q is continuous or not at x. The functions m and s that appear in (9.7) come about in the following way: s is any strictly increasing solution of $A_0 s = 0$.

$$m(x) = \int_a^x \frac{1}{(A_0 f_0)(y)}\, d[D_s^+ f_0(y)],$$

where $a \in]r_1,r_2[$ is fixed.

(Hints: First show that if $u \in \mathscr{D}(A_0)$ and $Au = 0$, then $u(x_1) = u(x_2)$ for $r_1 < x_1 < x_2 < r_2$ implies $u \equiv$ constant on $[x_1,x_2]$. Use (iii) and (iv) for this. Get s from this assertion and (v). Reparametrize by $y = s(x)$, so that s becomes the identity function. Show that if $A_0 h(x) > 0$ for all x in $]x_1,x_2[\subset]r_1,r_2[$, then h is strictly convex on $[x_1,x_2[$. Conclude that $f \in \mathscr{D}(A)$ satisfies $f = (f + \delta f_0) - (\delta f_0)$ is the difference of two convex functions, and so locally f is absolutely continuous and f' is of bounded variation. Argue that $A_0(f - t_0 f_0)(x) > 0$ for $x \in]x_1,x_2[$ if

$$t_0 < \inf\left\{\frac{A_0 f(x)}{A_0 f_0(x)} : x_1 < x < x_2\right\} = \alpha;$$

conclude that

$$\alpha[D_s^+ f_0(x_2) - D_s^+ f_0(x_1)]$$

$$\le D_s^+ f(x_2) - D_s^+ f(x_1)$$

$$\le \sup\left\{\frac{A_0 f(x)}{A_0 f_0(x)} : x_1 < x < x_2\right\}[D_s^+ f_0(x) - D_s^+ f_1(x_1)].$$

This implies

$$D_s^+ f(x_2) - D_s^+ f(x_1) = \int_{x_1}^{x_2} \frac{A_0 f(x)}{A_0 f_0(x)}\, d[D_s^+ f_0(x)],$$

and the desired conclusion follows.

*22. Let A_0, r_1, r_2, m, and s be as in the preceding problem. Formula (9.7) tells how A_0 acts in the interior $]r_1,r_2[$ of $[r_1,r_2]$. Let

$$\sigma_1 = \iint_{r_1<y<x<t_1} dm(x)\,ds(y), \quad \mu_1 = \iint_{r_1<y<x<t_1} ds(\lambda)\,dm(y),$$

where t_1 is any number in $]r_1,r_2[$. r_1 is called a *regular*, *exit*, *entrance*, or *natural boundary point* according as $\{\sigma_1,\mu_1 < \infty\}$, $\{\sigma_1 < \infty = \mu_1\}$, $\{\mu_1 < \infty = \sigma_1\}$, or $\{\sigma_1 = \mu_1 = \infty\}$. Similarly for r_2. Establish the following.

(i) Let $A_0 = d^2/dx^2$, $r_1 = -\infty$, $r_2 = +\infty$. Take $s(x) = m(x) = x$. Both r_1 and r_2 are natural boundary points.

(ii) Let $A_0 = d^2/dx^2$, $r_1 = -\infty$, $r_2 = 0$. Take $s(x) = m(x) = x$. r_1 is a natural boundary point whereas r_2 is a regular boundary point.

(iii) Let $A_0 = x^2 d^2/dx^2 - d/dx$, $r_1 = 0$, $r_2 = \infty$. Take $s(x) = \int_1^x e^{-1/y}\,dy$, $m(x) = \int_1^x y^{-2} e^{1/y}\,dy$. Then r_1 is an exit boundary point and r_2 is a natural boundary point.

(iv) Let $A_0 = x^2 d^2/dx^2 + d/dx$, $r_1 = 0$, $r_2 = 2$. Then r_1 is an entrance boundary point while r_2 is a regular boundary point.

{Here is a partial interpretation of these notions. Let a particle diffuse according to the Markov process described by A_0. Suppose the particle starts in $]r_1,r_2[$. Let p be the probability that the particle reaches the boundary point r_i $(i = 1,2)$ in a finite time. Then $p > 0$ if r_i is either a regular or exit boundary point while $p = 0$ if r_i is an entrance or natural boundary point.}

*23. Let $\{\mu_t : t \geq 0\}$ be a family of probability measures on the Borel sets of $\mathbb{R}$. Suppose that the formula

$$T(t)f(x) = (f * d\mu_t)(x) = \int_{-\infty}^{\infty} f(x - y)\,\mu_t(dy) \tag{9.8}$$

defines a (C_0) contraction semigroup T on $BUC(\mathbb{R})$. Define the Fourier transform of μ_t to be

$$\phi_t(\xi) = \int_{-\infty}^{\infty} e^{is\xi}\,\mu_t(ds). \tag{9.9}$$

Show that a constant $\gamma \in \mathbb{R}$ and a nonnegative measure v exist on the Borel sets of $\mathbb{R}$ such that

$$\phi_t(\xi) = \exp\left\{ t\left[i\gamma + \int_{-\infty}^{\infty} \left(e^{i\xi s} - 1 - \frac{i\xi}{1+s^2} \right)\left(\frac{1+s^2}{s^2} \right) v(ds) \right] \right\} \tag{9.10}$$

for $t > 0$ and $\xi \in \mathbb{R}$. Moreover, γ and v are uniquely determined. Conversely, if ϕ_t is given by (9.10) and μ_t by (9.9), then T given by (9.8) is a (C_0) contraction semigroup on $BUC(\mathbb{R})$.

*24. Let $\{\xi_t : t \in \mathbb{R}^+\}$ be a one-dimensional Brownian motion process. Then the formula

$$T(t)f(x) = E_x[f(\xi_t)] \qquad (x \in \mathbb{R}, t \in \mathbb{R}^+)$$

defines a (C_0) contraction semigroup on $BUC(\mathbb{R})$, which is called the *Brownian motion semigroup*.

(i)   Let $\mathscr{X} = BUC(\mathbb{R}^+)$ and define

$$S(t)f(x) = E_x[f(|\xi_t|)] \qquad (x \in \mathbb{R}^+, t \in \mathbb{R}^+).$$

Then S is a (C_0) semigroup on $\mathscr{X}$, the *reflected Brownian notion semigroup*.

(ii) Let $\mathscr{Y} = \{f \in BUC(\mathbb{R}): f \text{ is odd}, f(0) = 0\}$. The formula

$$U(t)f(x) = E_x[f(\xi_t)]$$

defines a (C_0) semigroup on $\mathscr{Y}$, the *Brownian motion semigroup with absorbing barrier at* 0.

(iii) Let $\mathscr{X} = C(\dot{\mathbb{R}}^n)$ (see Example 9.12.2 for the notation). Let $V: \mathbb{R}^n \to \mathbb{R}$ be bounded and measurable. The (Feynman-Kac) formula

$$W(t)f(x) = E_x\left\{f(\xi_t)\exp\left[-\int_0^t V(\xi_s)\, ds\right]\right\} \qquad (t \in \mathbb{R}^+, x \in \mathbb{R}^+)$$

defines a (C_0) semigroup W on $\mathscr{X}$ with generator A given (formally) by $Af = 1/2\,\Delta f - Vf$.

25. Let A generate a uniformly bounded (C_0) semigroup T. Then the semigroup $T_{1/2}$ generated by $A_{1/2} = -(-A)^{1/2}$ is given by

$$T_{1/2}(t)f = \frac{t}{\sqrt{4\pi}}\int_0^\infty s^{-3/2}e^{-t^2/4s}T(s)f\, ds.$$

*26. Let A generate a (C_0) semigroup T on $\mathscr{X}$. Let D be a dense subspace of $\mathscr{D}(A)$ such that $T(t)(D) \subset D$ for each $t > 0$. Then $A^* = (A|_D)^*$ and D is a core for A.

27. Give a variant of Example 9.6 involving $f(x + t)$ instead of $f(x - t)$ and the boundary condition $f(1) = 0$ rather than $f(0) = 0$.

*28. Establish the calculus formula

$$e^{-xy} = 2^{-1}\pi^{1/2}y\int_0^\infty \exp(-y^2/4\alpha)\exp(-\alpha x^2)\alpha^{-3/2}\, d\alpha$$

for $x > 0$, $y > 0$ and discuss its connection with self-adjoint and m-dissipative operators. (What is the connection between e^{tA} and $\exp\{t[-(-A)^{1/2}]\}$?) Note that the above calculus formula corresponds to the one for $T_{1/2}(t)f$ given at the beginning of 9.2. Compare this with Exercise 25 above.

*29. Prove Theorem 9.15, using the sequence space approach of 7.5.

*30. Let A generate a (C_0) semigroup T on $\mathscr{X}$. Let $\mathscr{D}$ be a dense linear set in $\mathscr{D}(A)$ such that $T(t)(D) \subset D$ for almost every $t > 0$. Then D is a core for A.

*31. There exists a (C_0) group T on $\mathscr{H}$ with generator A such that $\sigma(A) \subset \mathbb{R}$, $(\lambda I - A)^{-1}$ is compact for each $\lambda \in \mathbb{C}\backslash\mathbb{R}$, and $\|T(t)\| = e^{|t|} \in \sigma[T(t)]$ for each real t. Compare this with Theorem 9.4.

10.   Historical Notes and Remarks

Some historical remarks are given in the Preface. For more comments on the early history of the subject, see Hille and Phillips [1] and Dunford and Schwartz [1], who, incidentally, refer to Stone's 1930 paper [1] as the first one on "modern" semigroup theory.

Many books deal with the topic of semigroup theory. These include Bellini-Morante [1], Butzer and Berens [1], Davies [8], Dunford and Schwartz [1],

Dynkin [2], Fattorini [14], A. Friedman [1], Hille and Phillips [1], T. Kato [12], S. Kreĭn [1,2], Ladas and Lakshmikantham [1], Pazy [5], Reed and Simon [1], Showalter [11], Tanabe [5], J. Walker [1], Yosida [10], and others. Some of these authors also give historical notes.

The theory of (C_0) semigroups has many extensions, none of which are discussed in this volume. The extension from a Banach space context to that of a locally convex space was first given by L. Schwartz [1]. The first extension to a distribution context was proposed by J.-L. Lions [2]. Investigators have also looked at semigroups of unbounded operators. Much of their work is included in the References. Prominent contributors include Bababola, Barbu, Chazarain, Ciorănescu, DaPrato, Dembart, Fujiwara, Guillement, Hille, Hughes, Komatsu, T. Kōmura, S. Kreĭn, Lai, Miyadera, Mosco, Nussbaum, Ôharu, Ōuchi, Okazawa, Singbal-Vedek, Sunouchi, Ushijima, M. Watanabe, Yosida, and others.

Other generalizations replace the functional equation $T(t + s) = T(t)T(s)$ by a different one. Interested readers can browse through the References and find titles dealing with these and other generalizations.

The references on semigroups of nonlinear operators are deferred to the next volume.

The following discussion presents a section-by-section review of additional references on the topics of interest.

Section 1. Most textbooks on functional analysis deal with closed unbounded operators. Kato's treatment [12] is especially nice.

Section 2. Proposition 2.5 is due (independently) to Nathan [1], Nagumo [1], and Yosida [1]. Our simple proof of Proposition 2.5 comes from Bardos [personal communication] and Pazy [4]. Theorem 2.6 is due (independently) to Hille [1] and Yosida [2]. The operator A_λ introduced in 2.10 is widely known as the *Yosida approximation* of A. Theorem 2.13 is due, independently, to Feller [1], Miyadera [1], and Phillips [3]. The change of norm tricks used in the proof are those of Feller [1], although these ideas originated much earlier, cf. Eilenberg [1]. The counterexample mentioned in Exercise 2.18.3 is that of C. Fefferman [personal communication]. The convolution semigroups of 2.18.4, 7, 8, can be given a unified theory. See Feller [10] and Davies [8, pp. 55–59].

Exercise 2.18.11 is nontrivial to prove. (For a complete proof see Hille and Phillips [1].) Here is a simple proof in the case when $\mathscr{X}$ is reflexive. A uniform boundedness principle argument (see Lemma 2.12) gives the estimate $\|T(t)\| \le Me^{\omega t}$ where we assume without loss of generality that $\omega > 0$. Thus for $f \in \mathscr{X}, \phi \in \mathscr{X}^*, t > 0$,

$$\left| \int_0^t \langle T(s)f, \phi \rangle \, ds \right| \le M(e^{\omega t} - 1) \|f\| \|\phi\|.$$

Fix f and t and vary ϕ. By reflexivity there is an $f_t \in \mathscr{X}^{**} = \mathscr{X}$ such that

$$\langle f_t, \phi \rangle = t^{-1} \int_0^t \langle T(s)f, \phi \rangle \, ds$$

for all $\phi \in \mathcal{X}^*$. Thus for $h > 0$,

$$|\langle T(h)f_t - f_t, \phi \rangle| = |\langle f_t, T(h)^*\phi \rangle - \langle f_t, \phi \rangle|$$

$$= \left| t^{-1} \int_0^t \langle T(s)f, T(h)^*\phi \rangle \, ds - t^{-1} \int_0^t \langle T(s), \phi \rangle \, ds \right|$$

$$= \left| t^{-1} \left\{ \int_h^{t+h} - \int_0^t \right\} \langle T(\tau)f, \phi \rangle \, d\tau \right|$$

$$= \left| t^{-1} \left\{ \int_t^{t+h} - \int_0^h \right\} \langle T(\tau)f, \phi \rangle \, d\tau \right|$$

$$\leq t^{-1} \|t\| \|\phi\| M\omega^{-1}[(e^{\omega(t+h)} - e^{\omega t}) + (e^{\omega h} - 1)].$$

Taking the supremum over $\|\phi\| \leq 1$ yields

$$\|T(h)f_t - f_t\| \to 0$$

as $h \to 0$. The set $F = \{g \in \mathcal{X} : \lim_{h \to \infty} T(h)g = g\}$ is a closed subspace of $\mathcal{X}$ containing each f_t; but since weak $\lim_{t \to 0} f_t = f$ and $f \in \mathcal{X}$ is arbitrary, it follows that $F = \mathcal{X}$. ∎

In the nonreflexive case the Kreĭn-Šmulian theorem can be used. We omit the details.

The example in 2.19(1) is Trotter's [2]; the one in 2.19(2) is that of Packel [1]. Chernoff [8] modified Packel's example to produce a (C_0) semigroup T on a complex Hilbert space $\mathcal{H}$ such that for each $\alpha > 0$ and every inner product norm $\|\|\cdot\|\|$ on $\mathcal{H}$ equivalent to the given one.

$$\lim_{t \to 0_+} \sup \|\|e^{-\alpha t} T(t)\|\| > 1.$$

This shows that, in a certain sense, the theory of (C_0) semigroups on Hilbert space does not reduce to the contraction case. The result of 2.19(3) is due to Sz.-Nagy [1]. For Banach limits see Dunford and Schwartz [1, p. 73].

Section 3.   Theorem 3.3 is due to Lumer and Phillips [1]. Earlier the Hilbert space case was obtained by Phillips [8]. As far as we know, Proposition 3.9 has not appeared before in the literature. The inequalities in Exercise 3.10.3 are due to Clarkson [1]. Semi-inner products (Exercise 3.10.2) are due to Lumer [1]. The proof outlined in Exercise 3.10.2 was suggested by Davies [personal communication].

Section 4.   For more on adjoint semigroups see Hille and Phillips [1]. The classical results for adjoint semigroups are due to Phillips. Theorem 4.7 is Stone's [3]. Our proof is patterned after the one given by Lax and Phillips [2].

Section 5.   Theorem 5.2 and 5.3 are due to Hille [1]. (See also Hille [2].) Cannon [1] devised the pretty formulation of Theorem 5.9. Exercise 5.10.3 is due to Gel'fand [1], and 5.10.5 is due to Neuberger [3]. For an extension see Kato [17].

Section 6.   Theorem 6.1 is due to Trotter [1], Nelson [5] (in special cases), and Gustafson [1]. The idea behind it goes back to work of F. Rellich on self-adjoint operators in the thirties. Theorem 6.2 is due to Chernoff [2]; see also Wüst [1]. Theorems 6.4 and 6.5, which predate Theorems 6.1 and 6.2, are those of Phillips [4]. Theorem 6.6 is Hille's; cf. Hille and Phillips [1]. The term *Kato perturbation* is not yet standard but it deserves to be. Exercises 6.12.5, 10 seem to be new.

Section 7. Theorems 7.3, 7.8 are due, independently, to Neveu [1] (in a special case) and Trotter [1]. Trotter's proof contained a gap which was filled by Kato [5], whence the name Trotter-Neveu-Kato theorem. The sequence space idea of the proof goes back to Kisyński [3]. Many interesting variants of Theorem 7.3, 7.8 (motivated by probabilistic applications) appear in the work of Kurtz. Example 7.6 is that of Goldstein [11] and 7.9.2 is Cannon's [1].

Section 8. Lemma 8.1 can be found in Goldstein [14]. Examples 8.2 and 8.3 are due to Hille [1]. Theorem 8.4 and Lemma 8.5 are Chernoff's [1]. The proof of Theorem 8.8 is that of Goldstein [14]. The Trotter product formula (Theorem 8.12) is due to Trotter [2], although the idea behind it was discovered by S. Lie in the nineteenth century. Ideas close to those in Theorems 8.4 and 8.12 appeared in the work of Lax in the fifties; see Lax and Richtmeyer [1]. For the Feynman path integral and its physical background see Feynman [1] and Simon [15]. The beautiful approach of 8.14 is that of Nelson [5]. The definition of *generalized sum* given in 8.17 is due to Chernoff in his 1968 thesis; see [6] which contains a wealth of results. The proof of the ergodic theorem (Theorem 8.20) given is due to Goldstein, Radin, and Showalter [1]. For more history on ergodic theory see Dunford and Schwartz [1, pp. 783–784]. For more on ergodic theory and semigroups, see the works of Akcoglu, Davies, Ellis and Pinsky, Kakutani, Kubokawa, C. Lin, M. Lin, McGrath, R. Sato, Shaw, Yosida, and others listed in the References. Exercise 8.22.3 is due to Goldstein [21] and 8.22.8 is due to Goldstein [7] and Nelson [8].

Section 9. Fractional powers were introduced by Balakrishnan [2]. They have been systematically studied by many authors, including Kato [9,10] and Komatsu [2–6]. For spectral mapping theorems (such as Theorems 9.4, 9.5) see Hille and Phillips [1]. Theorem 9.8 is Kallman and Rota's [1] and 9.9 is Kato's [18]. Some of the generalizations include those of Certain and Kurtz [1], Gindler and Goldstein [1,2], and Holbrook [1]. For more information on Example 9.10 see the articles by Kwong and Zettl. For more on Feller-Markov processes and semigroups see the books of Blumenthal and Getoor [1], Dynkin [2], Fukushima [6], Silverstein [2,3], or others. Theorem 9.13 is due to Phillips [9]. At the moment, many investigators are actively working on positivity-preserving semigroups (or positive semigroups, for short). This list includes, among others, Arendt, Batty, Bratelli, Davies, Derndinger, D. Evans, Greiner, Kato, Kishimoto, Nagel, Robinson, K. Sato, Schaefer, and B. Simon. The article of Arendt, Chernoff, and Kato [1] is quite pretty. Theorem 9.14 is Yosida's [15]. For Hunt's theorem, see Hunt [2] and the books cited at the end of 9.15. The approximation theorems with varying spaces (Theorem 9.16 and its variants) go back to Trotter [1]. The Lax equivalence theorem can be found in Lax and Richtmeyer [1]. Modern versions are due to the Aachen group; cf. for example, Butzer, Dickmeis,

Hahn, and Nessel [1]. See also the nonlinear analogue in our forthcoming volume. Theorem 9.20 is Nelson's [4]; see also Chernoff [6]. Exercise 9.21.11 is due to Pazy [5], Exercise 9.21.15 is Cuthbert's [2], and 9.21.20 is Ditzian's [3]. The theory of Exercise 9.21.21–22 was devised by Feller; cf. Yosida [10] for further discussion. Formula (9.10) in Exercise 9.21.23 is the celebrated Levy-Khintchine representation formula. See Feller [10] and Davies [8].

Chapter II

Linear Cauchy Problems

1. Homogeneous and Inhomogeneous Equations

Let A be a densely defined operator on $\mathscr{X}$. Consider the *abstract Cauchy problem* (or *initial value problem*)

$$du(t)/dt = Au(t) \ (t \in \mathbb{R}^+), \ u(0) = f \in \mathscr{D}(A). \tag{1.1}$$

1.1. DEFINITION. The problem (1.1) is called *well-posed* if $\rho(A) \neq \phi$ and for each $f \in \mathscr{D}(A)$ there is a unique solution $u: \mathbb{R}^+ \to \mathscr{D}(A)$ of (1.1) in $C^1(\mathbb{R}^+, \mathscr{X})$.

"Well-posedness" involves more than existence and uniqueness, namely, continuous dependence on the initial data. Existence and uniqueness alone are not sufficient to guarantee that (1.1) is governed by a (C_0) semigroup. The simplest additional condition that ensures this is the technical condition $\rho(A) \neq \phi$. The following result is fundamental.

1.2. THEOREM. *The problem* (1.1) *is well-posed iff* A *generates a* (C_0) *semigroup* T *on* $\mathscr{X}$. *In this case the solution of* (1.1) *is given by* $u(t) = T(t)f, t \in \mathbb{R}^+$.

Proof. Assume A generates a (C_0) semigroup T. Then for $f \in \mathscr{D}(A)$, $u(\cdot) = T(\cdot)f \in C^1(\mathbb{R}^+, \mathscr{X})$ is $\mathscr{D}(A)$-valued and (1.1) holds. It remains to check the uniqueness of the solution. Let v be any solution of (1.1). For $0 \leq s \leq t < \infty$,

$$\frac{d}{ds}[T(t-s)v(s)] = T(t-s)Av(s) - T(t-s)Av(s) = 0,$$

whence $T(t-s)v(s)$ is independent of s; setting $s = 0$, $s = t$ yields $v(t) = T(t)v(0) = T(t)f$.

We sketch the sufficiency. Choose $\lambda \in \rho(A)$. We may assume $\lambda = 0$. [Otherwise replace A by $B = A - \lambda I$. Then u solves (1.1) iff $v(t) = e^{-\lambda t}u(t)$ solves $dv/dt = Bv$, $v(0) = f$; and $0 \in \rho(B)$.] Define Banach spaces $\mathscr{Y}, \mathscr{Z}$ as follows:

$$\mathscr{Y} = C^1([0,1], \mathscr{X}), \ Z = [\mathscr{D}(A), \text{graph norm}]$$

with norms

$$\|u\|_{\mathscr{Y}} = \sup\{\|u(t)\| + \|u'(t)\| : 0 \leq t \leq 1\}, \|f\|_{\mathscr{Z}} = \|Af\|.$$

Define a linear operator $L: \mathscr{Z} \to \mathscr{Y}$ by $Lf = u$ where u is the unique solution of (1.1). Clearly $\mathscr{D}(L) = \mathscr{Z}$. We claim that L is closed. To see why, suppose $\|f_n\|_{\mathscr{Z}} \to 0$ and $\|u_n - u\|_{\mathscr{Y}} \to 0$. Then for $0 \leq t \leq 1$,

$$u_n(t) \to u(t), \ Au_n(t) = u'_n(t) \to u'(t)$$

in $\mathscr{X}$, so by the closedness of A, $u(t) \in \mathscr{D}(A)$ and $u'(t) = Au(t)$. Since $u(0) = \lim_{n \to \infty} u_n(0) = 0$ also, we must have $u \equiv 0$. The closed graph theorem implies that L is bounded, i.e.

$$\sup\{\|u(t)\| + \|u'(t)\| : 0 \le t \le 1\} \le \|L\|\,\|Af\|$$

holds for the solution of (1.1). The family $T = \{T(t) : t \ge 0\}$ is clearly a semigroup, strongly continuous on $\mathscr{D}(A)$. That it is a (C_0) semigroup follows easily from the estimate $\|T(t)\| \le \|L\|$ for $0 \le t \le 1$. ∎

1.3. Consider the (abstract) Cauchy problem

$$du(t)/dt = Au(t) + f(t) \qquad (t \in \mathbb{R}^+), \tag{1.2}$$

$$u(0) = u_0. \tag{1.3}$$

THEOREM. *Suppose A generates a (C_0) semigroup on $\mathscr{X}$ and $u_0 \in \mathscr{D}(A)$. Assume either*

(i) *$f \in C(\mathbb{R}^+, \mathscr{X})$ takes values in $\mathscr{D}(A)$ and $Af \in C(\mathbb{R}^+, \mathscr{X})$, or*
(ii) *$f \in C^1(\mathbb{R}^+, \mathscr{X})$.*

Then (1.2), (1.3) has a unique solution $u \in C^1(\mathbb{R}^+, \mathscr{X})$ with values in $\mathscr{D}(A)$.

Proof. First we establish the uniqueness. If u, v are two solutions of (1.2), (1.3), then $w = u - v$ satisfies

$$dw(t)/dt = Aw(t) \qquad (t \in \mathbb{R}^+),\ w(0) = 0.$$

Thus $w \equiv 0$ by the uniqueness proof of Theorem 1.2.

For the existence, we seek a solution of the form

$$u(t) = T(t)u_0 + \int_0^t T(t-s)f(s)\,ds. \tag{1.4}$$

To "derive" this formula, note that

$$\frac{d}{ds}[T(t-s)u(s)] = T(t-s)f(s)$$

if (1.2) holds; integrate from $s = 0$ to $s = t$ and use (1.3) to obtain (1.4). Even more simply, pretend that A is a real number, solve the trivial problem (1.2), (1.3) by freshman calculus, and write $T(t)$ in place of e^{tA} in the formula for the solution (1.4). Equation (1.4) is called the *variation of parameters formula*. Let

$$v(t) = \int_0^t T(t-s)f(s)\,ds.$$

Clearly $u \in C^1(\mathbb{R}^+, \mathscr{X})$ iff $v \in C^1(\mathbb{R}^+, \mathscr{X})$, in which case

$$u'(t) = AT(t)u_0 + v'(t);$$

in particular, since (1.3) holds, (1.2) holds iff

$$v'(t) = Av(t) + f(t). \tag{1.5}$$

Case (I): Assume (i) holds. Then $v(t) \in \mathscr{D}(A)$ by Lemma I.1.5, and

$$Av(t) = \int_0^t T(t-s)Af(s)\, ds.$$

Also, $v \in C^1(\mathbb{R}^+, \mathscr{X})$ and

$$v'(t) = T(0)f(t) + \int_0^t \frac{\partial}{\partial t} T(t-s)f(s)\, ds$$

$$= f(t) + \int_0^t T(t-s)Af(s)\, ds$$

$$= f(t) + Av(t),$$

proving (1.5).

Case (II): Assume (ii) holds. Recall that for all $g \in \mathscr{X}$,

$$\int_a^b T(t)g\, dt \in \mathscr{D}(A) \text{ and } A \int_a^b T(t)g\, dt = T(b)g - T(a)g \qquad (1.6)$$

(cf. equations (2.1), (2.2) of Chapter I).

$$v(t) = \int_0^t T(t-s)\left[f(0) + \int_0^s f'(r)\, dr \right] ds$$

$$= \int_0^t T(t-s)f(0)\, ds + \int_0^t \int_r^t T(t-s)f'(r)\, ds\, dr$$

$$= \int_0^t T(\tau)f(0)\, dt + \int_0^t \int_0^{t-r} T(\tau)f'(r)\, d\tau\, dr \in \mathscr{D}(A) \qquad (1.7)$$

by (1.6) and Lemma I. 1.5; also $v \in C^1(\mathbb{R}^+, \mathscr{X})$ and

$$v'(t) = \frac{d}{dt} \int_0^t T(\tau)f(t-\tau)\, d\tau$$

$$= T(t)f(0) + \int_0^t T(\tau)f'(t-\tau)\, d\tau$$

$$= T(t)f(0) + \int_0^t T(t-s)f'(s)\, ds; \qquad (1.8)$$

$$Av(t) = T(t)f(0) - f(0) + \int_0^t [-f'(r) + T(t-r)f'(r)]\, dr$$

by (1.6), (1.7)

$$= T(t)f(0) - f(0) - f(t) + f(0) + \int_0^t T(t-r)f'(r)\, dr. \qquad (1.9)$$

Combining (1.8) with (1.9) yields (1.5). ∎

1.4. THEOREM. *Let $A \in \mathcal{GA}(\theta)$ (see Section I.6.8). Let f be Hölder continuous: for each $\tau \in \mathbb{R}^+$ there are numbers $K = K(\tau) > 0$, $k = k(\tau) \in (0,1]$ such that*

$$\|f(t) - f(s)\| \leq K|t - s|^k \quad \text{for} \quad 0 \leq s, t \leq \tau. \tag{1.10}$$

Then the problem (1.2), (1.3) has a unique solution $u \in C(\mathbb{R}^+, \mathcal{X}) \cap C^1[(0,\infty), \mathcal{X}]$ for each $u_0 \in \mathcal{X}$; moreover, $u \in C^1(\mathbb{R}^+, \mathcal{X})$ if $u_0 \in \mathcal{D}(A)$.

Proof. If T denotes the analytic semigroup generated by A, then $T(t)(\mathcal{X}) \subset \mathcal{D}(A)$ for $t > 0$ by Theorem I.5.3. Because of this and the proof of Theorem 1.3, it suffices to show that $v: \mathbb{R}^+ \to \mathcal{D}(A)$ and $Av \in C(\mathbb{R}^+, \mathcal{X})$ where

$$v(t) = \int_0^t T(t - s)f(s)\, ds, \quad t \in \mathbb{R}^+.$$

Write $v = v_1 + v_2$ where

$$v_1(t) = \int_0^t T(t - s)f(t)\, ds = \int_0^t T(r)f(t)\, dr,$$

$$v_2(t) = \int_0^t T(t - s)[f(s) - f(t)]\, ds.$$

By (2.2) of Chapter I, $v_1(t) \in \mathcal{D}(A)$, and Av_1 is continuous on $\mathbb{R}^+$ (since $Av_1(t) = [T(t) - I]f(t)$). To reach the same conclusions for v_2 we first approximate v_2 by w_ϵ where

$$w_\epsilon(t) = \begin{cases} \displaystyle\int_0^{t-\epsilon} T(t - s)[f(s) - f(t)]\, ds & \text{for } t \geq \epsilon \\ 0 & \text{for } 0 \leq t \leq \epsilon \end{cases}$$

where $\epsilon > 0$. Clearly $\lim_{\epsilon \to 0} w_\epsilon(t) = v_2(t)$, uniformly for t on bounded subsets of $\mathbb{R}^+$.

Since

$$AT(t - s)[f(s) - f(t)] = [AT(\epsilon)]\{T(t - s - \epsilon)[f(s) - f(t)]\}$$

we have $w_\epsilon(t) \in \mathcal{D}(A)$ and Lemma I.1.5 implies

$$Aw_\epsilon(t) = \int_0^{t-\epsilon} AT(t - s)[f(s) - f(t)]\, ds.$$

From the estimates (1.10) and $\|AT(t - s)\| \leq C/(t - s)$ for $0 \leq s < t \leq \tau$ and some constant C depending on $\tau > 0$ [cf. Definition 1.5.2(iv)], it follows that $\lim_{\epsilon \to 0} Aw_\epsilon(t)$ exists and equals the convergent integral $\int_0^t AT(t - s)[f(s) - f(t)]\, ds$. Since A is closed it follows that $v_2(t) \in \mathcal{D}(A)$ and

$$Av_2(t) = \int_0^t AT(t - s)(f(s) - f(t))\, ds.$$

It remains only to show that $Av_2 \in C(\mathbb{R}^+, \mathcal{X})$. It is continuous at $t = 0$. For $t > 0$

write

$$Av_2(t) = \left\{ \int_0^\delta + \int_\delta^t \right\} AT(t - s)[f(s) - f(t)]\, ds = J_1(t) + J_2(t)$$

where $0 < \delta < t$. $J_2 \in C(\mathbb{R}^+, \mathcal{X})$ for each fixed $\delta > 0$, and $\|J_1(t)\| = O(\delta)$, uniformly for t in bounded subsets of $\mathbb{R}^+$. The continuity of Av_2 follows. Alternatively, Av_2 is the uniform limit as $\epsilon \to 0$ of the continuous functions Aw_ϵ on $[\delta, 1/\delta]$ for each $\delta > 0$. ∎

1.5. EXERCISES

1. Fill in the details of the sufficiency proof of Theorem 1.2.
*2. Suppose that A is a densely defined operator such that for all $\lambda >$ some λ_0, $\lambda \in \rho(A)$ and

$$\limsup_{\lambda \to \infty} \frac{1}{\lambda} \log \|(\lambda - A)^{-1}\| = 0.$$

Show that the initial value problem $du/dt = Au$, $u(0) = f$ has at most one solution.
*3. Let A generate a (C_0) semigroup T such that $T(t)(\mathcal{X}) \subset \mathcal{D}(A)$ for each $t > 0$. Let $u_0 \in \mathcal{X}$ and let $f : [0,R] \to \mathcal{X}$ be of bounded variation. Define u by

$$u(t) = T(t)u_0 + \int_0^t T(t - s)f(s)\, ds, \qquad 0 \le t \le R.$$

Show that u satisfies the following four conditions.

(i) $u(t) \in \mathcal{D}(A)$ for $0 < t \le R$ and $Au \in C(]0,R], \mathcal{H})$.
(ii) $d^+u(t)/dt = Au(t) + f(t^+)$ for $t \in]0,R[$, and d^+u/dt is continuous from the right on $]0,R[$.
(iii) $d^-u(t)/dt = Au(t) + f(t^-)$ for $t \in]0,R]$ and d^-u/dt is continuous from the left on $]0,R]$.
(iv) $du(t)/dt = Au(t) + f(t)$ holds and du/dt is continuous except at countably many points of $[0,R]$.

*4. Show that Theorem 1.2 remains valid if, in the definition of well-posedness for Cauchy problem (1.1) (Definition 1.1), the condition that $\rho(A) \ne \emptyset$ is replaced by the assumption that there is a real number ω such that for the unique solution u of (1.1),

$$\|u(t)\| \le Me^{\omega t}\|u(0)\|$$

for $t \in \mathbb{R}^+$, and A is a closed operator.

2. Nonlinear Equations

Consider the (nonlinear) Cauchy problem

$$du(t)/dt = Au(t) + f[t, u(t)] \qquad (t \in \mathbb{R}^+), \tag{2.1}$$

$$u(0) = u_0, \tag{2.2}$$

where A generates a (C_0) semigroup T on $\mathcal{X}$.

We shall solve this by successive approximations. Specifically, the solution of

$$du_n(t)/dt = Au_n(t) + f[t,u_{n-1}(t)], \qquad u(0) = u_0$$

is given formally by

$$u_n(t) = T(t)u_0 + \int_0^t T(t - s)f[s,u_{n-1}(s)]\, ds$$

according to results of the previous section. We want to let $n \to \infty$ and hope that $u_n(t)$ converges to the solution $u(t)$ of (2.1), (2.2) for $0 \le t < c$ (possibly $c = \infty$). We may have to take $c < \infty$ as the following example shows.

2.1. EXAMPLE.   Take $\mathscr{X} = \mathbb{R}$, $A = 0$, $f(t,x) = x^2$, $u_0 = 1/c$ where $c > 0$. Then $u(t) = 1/(c - t)$ is the (unique) solution of

$$u'(t) = u(t)^2, u(0) = 1/c \qquad (0 \le t < c).$$

This problem does not have a solution on $\mathbb{R}^+$; moreover, by taking c small, the subinterval of $\mathbb{R}^+$ on which the solution exists can be made arbitrarily small.

2.2 THEOREM.   (Picard-Banach fixed point theorem).   *Let $\mathscr{M}$ be a complete metric space with metric d. Let $\alpha < 1$ and let $S: \mathscr{M} \to \mathscr{M}$ satisfy*

$$d(S^n x, S^n y) \le \alpha d(x,y)$$

for some positive integer n and all $x,y \in \mathscr{M}$. Then S has a unique fixed point in $\mathscr{M}$, i.e. there is a unique $x_0 \in \mathscr{M}$ such that $Sx_0 = x_0$.

Proof.   This is a familar result if $n = 1$. For all $z \in \mathscr{M}$, $\{S^m z\}_{m=1}^\infty$ is a Cauchy sequence tending to a fixed point of S. For the uniqueness, if x_0, x_1 are fixed points of S, then $d(x_0,x_1) = d(Sx_0,Sx_1) \le \alpha d(x_0,x_1)$, whence $x_0 = x_1$.

In the general case ($n \ge 1$), S^n has a *unique* fixed point x_0. Then $S^n(Sx_0) = S(S^n x_0) = Sx_0$, whence $Sx_0 = x_0$. On the other hand, any fixed point of S is also a fixed point of S^n; thus x_0 is the unique fixed point of S.   ∎

2.3. DEFINITION.   Note that any solution u of (2.1), (2.2) satisfies the *integral equation*

$$u(t) = T(t)u_0 + \int_0^t T(t - s)f[s,u(s)]\, ds, \qquad (2.3)$$

but not conversely since a solution of (2.3) is not necessarily differentiable. We shall refer to a continuous solution of (2.3) as a *mild solution* of (2.1), (2.2); a mild solution is thus a kind of generalized solution.

2.4. THEOREM (Local Existence theorem).   *Let $\Omega \subset \mathscr{X}$ be open and $u_0 \in \Omega$. Let $f: \mathbb{R}^+ \times \Omega \to \mathscr{X}$ be jointly continuous and satisfy the Lipschitz condition: for each $\tau \in \mathbb{R}^+$ there is a $K = K(\tau)$ such that*

$$\|f(t,x) - f(t,y)\| \le K\|x - y\|$$

whenever $0 \le t \le \tau$, $x,y \in \Omega$. Then for $\tau > 0$ sufficiently small, there is a unique mild solution of (2.1), (2.2) defined $[0,\tau)$.

Proof. Let $\tau > 0$. Let $\mathscr{Y} = C([0,\tau],\mathscr{X})$. Let E be a closed neighborhood of u_0 in Ω. Define S by

$$(Su)(t) = T(t)u_0 + \int_0^t T(t-s)f[s,u(s)]\,ds$$

for $0 \le t \le \tau$ and $u \in \mathscr{M} = \{v \in \mathscr{Y}: v(0) = u_0, v([0,\tau]) \subset E\}$. Note that $\mathscr{M}$ is a complete metric space and $Su \in \mathscr{Y}$. Let M,ω be such that $\|T(t)\| \le Me^{\omega t}$. Then

$$\|Su - Sv\|_{\mathscr{Y}} = \sup_{0 \le t \le \tau} \|Su(t) - Sv(t)\|$$

$$= \sup_{0 \le t \le \tau} \left\|\int_0^t T(t-s)[f[s,u(s)] - f[s,v(s)]]\,ds\right\|$$

$$\le Me^{\omega\tau} \int_0^\tau \|f[s,u(s)] - f[s,v(s)]\|\,ds$$

$$\le Me^{\omega\tau}K(\tau) \int_0^\tau \|u(s) - v(s)\|\,ds$$

$$\le Me^{\omega\tau}K(\tau)\cdot\tau\|u - v\|_{\mathscr{Y}}.$$

And $Me^{\omega\tau}K(\tau)\tau \to 0$ as $\tau \to 0^+$ (since $K(\tau)$ can be assumed to be bounded by $K(1)$ for $\tau < 1$). Thus once we establish that $S(\mathscr{M}) \subset \mathscr{M}$, it follows that for τ sufficiently small, Theorem 2.2 implies the desired conclusion since u is a mild solution of (2.1), (2.2) iff u is a fixed point of S.

Finally, to show that S maps into $\mathscr{M}$, assume without loss of generality that E is bounded.

$$\|Su - u_0\|_{\mathscr{Y}} \le \sup_{0 \le t \le \tau} \|T(t)u_0 - u_0\|$$

$$+ \sup_{0 \le t \le \tau} \left\|\int_0^t T(t-s)f[s,u(s)]\,ds\right\| = J_1(\tau) + J_2(\tau).$$

$J_1(\tau) \to 0$ as $\tau \to 0$, while

$$J_2(\tau) \le \tau Me^{\omega\tau} \sup_{0 \le t \le \tau} \|f[t,u(t)]\|$$

$$\le \tau Me^{\omega\tau}\left\{\sup_{0 \le t \le \tau} \|f(t,u_0)\| + K(\tau) \sup_{0 \le t \le \tau} \|u(t) - u_0\|\right\}$$

$$\to 0 \text{ as } \tau \to 0^+.$$

Hence $S(\mathscr{M}) \subset \mathscr{M}$ provided τ is sufficiently small. ∎

2.5. THEOREM (Global Existence theorem). *Let $u_0 \in \mathscr{X}$. Let $f: \mathbb{R}^+ \times \mathscr{X} \to \mathscr{X}$ satisfy: for each $\tau > 0$ there is a constant $K = K(\tau)$ such that*

$$\|f(t,x) - f(t,y)\| \le K(\tau)\|x - y\|$$

whenever $x, y \in \mathscr{X}$, $0 \le t \le \tau$. Then (2.1), (2.2) has a unique mild solution on $\mathbb{R}^+$.

Proof. Let S, M, ω be as in the proof of Theorem 2.4. We *claim*:

$$\|S^n u(t) - S^n v(t)\| \leq [MK(t)e^{\omega t}t]^n \sup_{0 \leq s \leq t} \|u(s) - v(s)\|/n! \qquad (2.4)$$

for all $t > 0$, $u, v \in C([0,t], \mathscr{X})$. We may (and do) assume that $t \to K(t)$ is monotone nondecreasing. For $n = 1$, (2.4) holds by the simple calculation given in the proof of Theorem 2.4. Assume that (2.4) is true for $n = m$. Then

$$\|S^{m+1} u(t) - S^{m+1} v(t)\|$$

$$= \left\| \int_0^t T(t-s)\{f[s, S^m u(s)] - f[s, S^m v(s)]\} \, ds \right\|$$

$$\leq Me^{\omega \tau} \int_0^t K(s)[Me^{\omega s}K(s)s]^m \sup_{0 \leq r \leq s} \|u(r) - v(r)\| \, ds/m!$$

$$\leq [Me^{\omega \tau}K(t)]^{m+1} \sup_{0 \leq r \leq t} \|u(r) - v(r)\| \int_0^t s^m \, ds/m!,$$

and (2.4) follows with $n = m + 1$. By induction, then, (2.4) holds for all positive integers n. Now let $\tau > 0$ be fixed but otherwise arbitrary. Choose n so large that

$$\alpha = [Me^{\omega \tau}K(\tau)\tau]^n/n! < 1.$$

Then by (2.4),

$$\|S^n u - S^n v\|_{\mathscr{Y}} \leq \alpha \|u - v\|_{\mathscr{Y}}$$

for all $u, v \in \mathscr{Y} = C([0,\tau], \mathscr{X})$. Hence S has a unique fixed point in $\mathscr{Y}$, and so the problem (2.1), (2.2) has a unique continuous mild solution on $[0,\tau]$. The result follows since $\tau > 0$ is arbitrary. ∎

2.6. THEOREM. *Suppose $f: \mathbb{R}^+ \times \mathscr{X} \to \mathscr{X}$ is continuous and satisfies this condition: for each $c > 0$ there is a constant $K(c)$ such that*

$$\|f(t,x) - f(t,y)\| \leq K(c)\|x - y\|$$

whenever $0 \leq t \leq c$, $\|x\| \leq c$, $\|y\| \leq c$. Let $u \in C([0,\tau], \mathscr{X})$ be a mild solution of (2.1), (2.2) (for some $u_0 \in \mathscr{X}$) on $[0,\tau]$, where $\tau < \infty$. Then:

(a) *(Uniqueness) For each $\tau > 0$ there is at most one mild solution on $[0,\tau]$.*

(b) *(Extension or blow-up) Either (i) there is a mild solution on $\mathbb{R}^+$ or (ii) there is a $\tau = \tau_{\max} > 0$ such that there is a mild solution u on $[0,\tau)$ satisfying $\lim_{t \to \tau^-} \|u(t)\| = \infty$.*

Thus if a solution does not exist for all positive time, then it "blows up," and Example 2.1 therefore exhibits the "worst" behavior that a mild solution can have.

Proof. By Theorem 2.4 we know that for each $u_0 \in \mathscr{X}$ there is a unique mild solution on $[0,\epsilon)$ for some $\epsilon = \epsilon(u_0) > 0$. This proves (a).

Let $\tau = \sup\{t: \text{a mild solution exists on } [0,t)\}$. If $\tau = \infty$ we are finished. Assume $\tau < \infty$ and let u be the mild solution on $[0,\tau)$. Then u is not a solution on

$[0,\tau]$, for if it were, then by (a) we could solve $v' = Av + f(t,v)$, $v(0) = u(\tau)$ for $[0,\epsilon_0)$, and letting $u(t + \tau) = v(t)$ we would have a mild solution u of (2.1), (2.2) on $[0,\tau + \epsilon_0)$, contradicting our choice of τ.

Suppose $\lim_{t \to \tau^-} \|u(t)\| = \infty$ is false. Then there is a sequence $\{t_n\}_{n=1}^{\infty}$ converging to τ from below such that $C = \sup_n \|u(t_n)\| + 1 < \infty$. There is a K such that $\|f(t,x) - f(t,y)\| \leq K\|x - y\|$ if $\|x\|, \|y\| \leq D = Me^{\omega}C + C$ and $t \leq \tau + 1$. Here M, ω are as before. Also,

$$\|f(t,x)\| \leq \|f(t,0)\| + \|f(t,0) - f(t,x)\|$$

$$\leq \sup_{0 \leq t \leq \tau + 1} \|f(t,0)\| + KD \equiv L < \infty$$

if $\|x\| \leq D$. Hence, as in the proof of Theorem 2.4,

$$S_n v(t) = T(t)u(t_n) + \int_{t_n}^{t_n + t} T(t_n + t - s)f[s,v(s)]\, ds$$

defines a mapping from $C([t_n, t_n + \delta], \{x \in \mathscr{X}: \|x\| \leq D\}) = \mathscr{M}_n$ to itself if $\delta > 0$ is sufficiently small, namely if $0 < \delta < \delta_0 (M,\omega,L,C,u_0,\tau)$, which does not depend on n. (Compare the proof of Theorem 2.4.) By Theorem 2.2, S_n has a unique fixed point $v_n \in \mathscr{M}_n$.

$$v(t) = \begin{cases} u(t) & \text{for } 0 \leq t \leq t_n \\ v_n(t) & \text{for } t_n \leq t \leq t_n + \delta \end{cases}$$

is a mild solution on $[0, t_n + \delta]$; and $t_n + \delta > \tau$ for n sufficiently large, and $u(t) = v(t)$ for $0 \leq t < \tau$ by uniqueness. The theorem follows. ∎

2.7. EXERCISES

1. Find an example of a (C_0) contraction semigroup T on $\mathscr{X}$ and an $f \in C(\mathbb{R}^+, \mathscr{X})$ such that

$$v(t) = \int_0^t T(t - s)f(s)\, ds$$

is nowhere differentiable. Conclude that a mild solution of (2.1), (2.2) need not be a solution in the usual sense.

2. Can the Lipschitz condition of Theorem 2.4 (or 2.5) be weakened to a Hölder condition if A is assumed to generate an analytic semigroup?

3. Let A generate a (C_0) semigroup T on $\mathscr{X}$ satisfying $\|T(t)\| \leq Me^{\omega t}$ for $t \in \mathbb{R}^+$. Let $g: \mathbb{R}^+ \times \mathscr{X} \to \mathscr{X}$ be continuous and satisfy

$$\|g(t,x) - g(t,y)\| \leq [g_1(t) + g_2(t)]\|x - y\|$$

for all $t \in \mathbb{R}^+$, $x, y \in \mathscr{X}$, where $g_1 \in L^1(\mathbb{R}^+) \cap L^{\infty}(\mathbb{R}^+)$, $g_2 \in L^{\infty}(\mathbb{R}^+)$. Then the Cauchy problem

$$u'(t) = Au(t) + g[t,u(t)] \qquad (t \in \mathbb{R}^+), \; u(0) = f \qquad (2.5)$$

has a unique mild solution for each $f \in \mathscr{X}$. Let $\alpha = \omega + M\|g_2\|_{\infty}$. Let u_i be the mild solution of (2.5) with initial data f_i, $i = 1,2$. If $\alpha < 0$, then $\lim_{t \to \infty} \|u_1(t) - u_2(t)\| = 0$, exponentially fast. If $\alpha = 0$, then given $\epsilon > 0$ there is a $\delta > 0$ such that $\|f_1 - f_2\| < \delta$ implies $\|u_1(t) - u_2(t)\| < \epsilon$ for all $t \in \mathbb{R}^+$.

4. Explicitly construct the $\delta_0(M,\omega,L,C,u_0,\tau) > 0$ of the proof of Theorem 2.6.

3. Fourier Transforms, Partial Differential Operators, and Unitary Equivalence

We assume that the reader is at least vaguely familiar with the material of this section.

3.1. DEFINITION. The *Schwartz space* $\mathscr{S}(\mathbb{R}^n)$ (of "rapidly decreasing smooth functions") consists of all $f \in C^\infty(\mathbb{R}^n)$ satisfying

$$\lim_{|x| \to \infty} P(x) \frac{\partial^{\alpha_1 + \cdots + \alpha_n} f}{\partial x_1^{\alpha_1} \cdots \partial x_n^{\alpha_n}}(x) = 0$$

for each polynomial P and each partial derivative as indicated above ($\alpha_1, \ldots \alpha_n \in \{0, 1, 2, \ldots\}$).

3.2. REMARKS. $C_c^\infty(\mathbb{R}^n) \subset \mathscr{S}(\mathbb{R}^n)$. Here $f \in C_c^\infty(\mathbb{R}^n)$ iff $f \in C^\infty(\mathbb{R}^n)$ and f has compact support. If $a: \mathbb{R} \to \mathbb{R}$ is defined by $a(x) = e^{1/x}$ or 0 according as $x < 0$ or $x \geq 0$, then $a \in C^\infty(\mathbb{R})$; it follows that $f_a \in C_c^\infty(\mathbb{R}^n)$ where $f_a(x) = a(|x|^2 - 1)$. (Here $|x|^2 = x_1^2 + \cdots + x_n^2$.) $C_c^\infty(\mathbb{R}^n)$ is dense in $L^p(\mathbb{R}^n)$, $1 \leq p < \infty$, and in $C_0(\mathbb{R}^n)$, the continuous functions on $\mathbb{R}^n$ that vanish at infinity; hence $\mathscr{S}(\mathbb{R}^n)$ is also dense in these spaces. If $\alpha > 0$, $f(x) = \exp(-\alpha|x|^2)$ defines a function in $\mathscr{S}(\mathbb{R}^n)$.

3.3 DEFINITION. The *Fourier transform* $\hat{f}$ of $f \in \mathscr{S}(\mathbb{R}^n)$ is defined as

$$\hat{f}(\xi) = (2\pi)^{-n/2} \int_{\mathbb{R}^n} e^{ix \cdot \xi} f(x) \, dx \qquad \xi \in \mathbb{R}^n.$$

Here $x \cdot \xi = \sum_{j=1}^n x_j \xi_j$. The map $\mathscr{F} : f \to \hat{f}$ is called the *Fourier tranformation* (or the *Fourier tranform*).

3.4. REMARKS. The following properties of the Fourier transform are elementary, and can be proven by integrating by parts, differentiating under the integral sign, using Fubini's theorem, etc.

(1) $\mathscr{F} : \mathscr{S}(\mathbb{R}^n) \to C^\infty(\mathbb{R}^n)$ is linear.
(2) Define $g = f(\cdot + y)$ by $g(x) = f(x + y)$; then $[f(\cdot + y)]\hat{\ }(\xi) = e^{-iy \cdot \xi} \hat{f}(\xi)$ for all $\xi, y \in \mathbb{R}^n, f \in \mathscr{S}(\mathbb{R}^n)$.
(3) Define $M_j : \mathscr{S}(\mathbb{R}^n) \to \mathscr{S}(\mathbb{R}^n)$ by $(M_j f)(x) = x_j f(x)$; then $(M_j f)\hat{\ }(\xi) = -i(\partial \hat{f}/\partial \xi_j)(\xi)$ for $\xi \in \mathbb{R}^n, f \in \mathscr{S}(\mathbb{R}^n)$.

3.5. NOTATION. α is a *multi-index* if $\alpha = (\alpha_1, \ldots, \alpha_n)$, where $\alpha_1, \ldots, \alpha_n \in \{0, 1, 2, \ldots\}$. Define $|\alpha| = \sum_{j=1}^n \alpha_j$, $x^\alpha = x_1^{\alpha_1} \ldots x_n^{\alpha_n}$, and $D^\alpha = D_1^{\alpha_1} \ldots D_n^{\alpha_n}$, where $D_j = i\partial/\partial x_j$. If P is the polynomial defined by $P(\xi) = \sum_{|\alpha| \leq m} a_\alpha \xi^\alpha$, then define $P(D) = \sum_{|\alpha| \leq m} a_\alpha D^\alpha$.

3.6. Here are two more elementary properties of $\mathscr{F}$.

(4) For all $f \in \mathscr{S}(\mathbb{R}^n)$, all multi-indices α, and all $\xi \in \mathbb{R}^n$, $[P(D)f]\hat{\ }(\xi) = P(\xi)\hat{f}(\xi)$.

(5) For all $f \in \mathscr{S}(\mathbb{R}^n)$, all multi-indices α, and all $\xi \in \mathbb{R}^n$, $(M_P f)\hat{}(\xi) = (\check{P}(D)\hat{f})(\xi)$, where $(M_P f)(x) = P(x)f(x)$ and $\check{P}(\xi) = P(-\xi)$.

3.7. DEFINITION.   For $f, g \in \mathscr{S}(\mathbb{R}^n)$, the *convolution* $f * g$ is defined by

$$(f * g)(x) = \int_{\mathbb{R}^n} f(x - y)g(y) \, dy \qquad x \in \mathbb{R}^n.$$

$f * g \in \mathscr{S}(\mathbb{R}^n)$ and for all $\xi \in \mathbb{R}^n$,

$$(f * g)\hat{}(\xi) = (2\pi)^{n/2} \hat{f}(\xi)\hat{g}(\xi).$$

3.8. LEMMA.   *If* $f(x) = \exp(-|x|^2/2)$, *then* $\hat{f} = f$.

Sketch of proof.   Suppose $n = 1$. $\hat{f}$ satisfies $\hat{f}'(\xi) = -\xi\hat{f}(\xi)$, $\hat{f}(0) = (2\pi)^{-1/2} \int_{\mathbb{R}} \exp(-x^2/2) \, dx = 1$. Hence $\hat{f}(\xi) = \exp(-\xi^2/2)$. For general n,

$$\hat{f}(\xi) = \prod_{j=1}^{n} \left(\frac{1}{\sqrt{2\pi}} \int_{\mathbb{R}} \exp(ix_j\xi_j) \exp(-x_j^2/2) \, dx_j \right) = \exp(-|\xi|^2/2). \qquad \blacksquare$$

3.9.   The next three properties of $\mathscr{F}$ can be easily verified.

(6) If $f(x) = (2m)^{-n/2} \exp(-|x|^2/4m)$, then $\hat{f}(\xi) = \exp(-m|\xi|^2)$.
(7) $\int_{\mathbb{R}} \hat{f}(x)g(x) \, dx = \int_{\mathbb{R}^n} f(x)\hat{g}(x) \, dx$ for all $f, g \in \mathscr{S}(\mathbb{R}^n)$.
(8) For each $f \in \mathscr{S}(\mathbb{R}^n)$ and each $y \in \mathbb{R}^n$,

$$f(y) = \lim_{m \to \infty} (m/\pi)^{n/2} \int_{\mathbb{R}^n} \exp(-m|x - y|^2)f(x) \, dx.$$

3.10.   THEOREM (Inversion formula).   *For each* $f \in \mathscr{S}(\mathbb{R}^n)$ *and each* $\xi \in \mathbb{R}^n$,

$$f(\xi) = (2\pi)^{-n/2} \int_{\mathbb{R}^n} e^{-ix \cdot \xi} \hat{f}(x) \, dx.$$

Thus if $Rf(x) = f(-x)$ *(so that* R *reverses the sign), then* $\mathscr{F}R\mathscr{F}^{-1} = $ *identity.*

Proof. $\displaystyle f(y) = \lim_{m \to \infty} \left(\frac{m}{\pi} \right)^{n/2} \int_{\mathbb{R}_n} \exp(-m|x|^2)f(x + y) \, dx$ by (8)

$\displaystyle \qquad = \lim_{m \to \infty} \left(\frac{m}{\pi} \right)^{n/2} \int_{\mathbb{R}^n} [(2m)^{-n/2} \exp(-|\cdot|^2/4m)]\hat{}(x)f(x + y) \, dx$ by (6)

$\displaystyle \qquad = \lim_{m \to \infty} (2\pi)^{-n/2} \int_{\mathbb{R}_n} \exp(-|x|^2/4m)[f(\cdot + y)]\hat{}(x) \, dx$ by (7)

$\displaystyle \qquad = \lim_{m \to \infty} (2\pi)^{-n/2} \int_{\mathbb{R}^n} \exp(-|x|^2/4m)e^{-iy \cdot x}\hat{f}(x) \, dx$ by (2)

$\displaystyle \qquad = (2\pi)^{-n/2} \int_{\mathbb{R}^n} e^{-iy \cdot x}\hat{f}(x) \, dx. \qquad \blacksquare$

3.11   THEOREM.   *For all* $f \in \mathscr{S}(\mathbb{R}^n)$, $\|f\|_2 = \|\hat{f}\|_2$.

Sketch of proof. Let $\tilde{f}(x) = \overline{f(-x)}$, $g = -\tilde{f}*f$. Then $\hat{g} = (2\pi)^{n/2}|\hat{f}|^2$. The inversion formula implies

$$(2\pi)^{-n/2}\int_{\mathbb{R}^n} e^{-ix\cdot\xi}\hat{g}(\xi)\,d\xi = g(x) = \int_{\mathbb{R}^n} \overline{f(y-x)}f(y)\,dy.$$

Setting $x = 0$ yields $\int_{\mathbb{R}^n}|\hat{f}(\xi)|^2\,d\xi = \int_{\mathbb{R}^n}|f(y)|^2\,dy.$ ∎

3.12. REMARK. $\mathscr{F}$ extends by continuity to a unitary map from the complex Hilbert space $L^2(\mathbb{R}^n)$ to itself. (We denote the extension by $\mathscr{F}$; this should cause no confusion.)

3.13. REMARK. (This section will not be used in the sequel.) One can topologize $\mathscr{S}(\mathbb{R}^n)$ so that it becomes a Fréchet space (i.e. a topological vector space whose topology is given by a complete metric ρ_0 which satisfies $\rho_0(f,g) = \rho_0(f - g,0)$ for all f,g in the space). In fact, define the norms

$$\|f\|_{\alpha,\beta} = \sup_{x\in\mathbb{R}^n}|x^\alpha D^\beta f(x)|.$$

Let $\{\|\cdot\|_{(m)} : m = 1, 2,\ldots\}$ be $\{\|\cdot\|_{\alpha,\beta} : \alpha,\beta \text{ arbitrary multi-indices}\}$ written as a sequence. The metric ρ_0 for $\mathscr{S}(\mathbb{R}^n)$ is

$$\rho_0(f,g) = \sum_{m=1}^{\infty} 2^{-m}\|f - g\|_{(m)}/(1 + \|f - g\|_{(m)}).$$

It is easy to check that $\mathscr{F}:\mathscr{S}(\mathbb{R}^n) \to \mathscr{S}(\mathbb{R}^n)$ is a linear homeomorphism.

3.14. DEFINITIONS. If P is a polynomial (with real coefficients), then $P(D)$ is a *linear partial differential operator with constant coefficients.* The *order* of $P(D)$ is the degree of the polynomial P. A *homogeneous polynomial* is a polynomial of the form $P(\xi) = \sum_{|\alpha|=m} a_\alpha \xi^\alpha$, so that all monomials which make up P are of the same degree. A homogeneous polynomial of degree $2m$ is called *weakly elliptic* [or *strictly elliptic*] iff $P(\xi) \le 0$ [or $P(\xi) < 0$] for all $\xi \in \mathbb{R}^n\backslash\{0\}$. (Note that a nonzero homogeneous polynomial which is weakly elliptic is necessarily of even order.) If P is a strictly elliptic homogeneous polynomial of degree $2m$, then P is negative on the unit sphere $\{\xi \in \mathbb{R}^n : |\xi| = 1\}$, which is compact. It follows that there is a number $c_0 > 0$ such that $P(\xi) \le -c_0|\xi|^{2m}$ for all $\xi \in \mathbb{R}^n$.

3.15. DEFINITION. Let $\mathscr{H}_1$, $\mathscr{H}_2$ be Hilbert spaces and let S_j be an operator on $\mathscr{H}_j$, $j = 1, 2$. S_1 is *unitarily equivalent* to S_2 iff there is a unitary $U \in \mathscr{B}(\mathscr{H}_1,\mathscr{H}_2)$ such that $S_1 = U^{-1}S_2 U$.

3.16. LEMMA

   (i)   *Unitary equivalence is an equivalence relation.*
   (ii) *If S_1 on $\mathscr{H}_1$ and S_2 on $\mathscr{H}_2$ are unitarily equivalent, then $\rho(S_1) = \rho(S_2)$, and $(\lambda - S_1)^{-1}$ and $(\lambda - S_2)^{-1}$ are unitarily equivalent for each $\lambda \in \rho(S_1)$.*
   (iii) *Two bounded unitarily equivalent operators have the same norm.*

Proof. (i) is trivial. For (ii), $\lambda - S_1$ and $\lambda - S_2$ are unitarily equivalent. Hence one of them is bijective with a continuous inverse iff the other is. Moreover, if $S_1 = U^1 S_2 U$, then $(\lambda - S_1)^{-1} = U^{-1}(\lambda - S_2)^{-1}U$ for each $\lambda \in \rho(S_1)$. For (iii), let

$S_j \in \mathcal{B}(\mathcal{H}_j), j = 1, 2$, and let $U \in \mathcal{B}(\mathcal{H}_1, \mathcal{H}_2)$ be unitary and satisfy $S_1 = U^{-1} S_2 U$. Then $\|S_1\| \le \|U^{-1}\| \|S_2\| \|U\| = \|S_2\|$; and similarly $S_2 = U S_1 U^{-1}$ implies $\|S_2\| \le \|S_1\|$; hence $\|S_2\| = \|S_1\|$. ∎

3.17. LEMMA.   *Let $T_j = \{T_j(t) : t \in \mathbb{R}^+\}$ be a (C_0) semigroup on $\mathcal{H}_j$ with generator A_j, $j = 1, 2$, and let $U \in \mathcal{B}(\mathcal{H}_1, \mathcal{H}_2)$ be unitary. Then $T_1(t) = U^{-1} T_2(t) U$ for all $t \in \mathbb{R}^+$ iff $A_1 = U^{-1} A_2 U$.*

In other words, two (C_0) semigroups are unitarily equivalent iff their generators are. The trivial proof is left as an exercise. The same result holds with "semigroups" replaced by "groups."

3.18. EXERCISES

1. Give detailed proofs of the asserted properties of the Fourier transform.
2. Let $\mathcal{X} = L^p(\Omega, \Sigma, \mu)$ where (Ω, Σ, μ) is a measure space and $1 \le p \le \infty$. Let $m : \Omega \to \mathbb{K}$ be Σ-measurable and suppose $\operatorname{Re} m(x) \le 0$ for each $x \in \Omega$. Define the multiplication operator M_m by $(M_m f)(x) = m(x) f(x)$ for $f \in \mathcal{D}(M_m) = \{g \in \mathcal{X} : mg \in \mathcal{X}\}$. Show that M_m is m-dissipative. Moreover, when m is real-valued and $p = 2$, M_m is self-adjoint. When $\Omega = \mathbb{R}$, $p = 2$, and m is a real polynomial, what is $\mathcal{F}^{-1} M_m \mathcal{F}$? When $p = 1$ is M_m densely defined? What about when $p = \infty$?

4.   Parabolic Equations

4.1. THEOREM.   *Let P be a weakly elliptic homogeneous polynomial of degree $2m$ $(m = 0, 1, 2, \dots)$ with real coefficients. Let $\mathcal{X} = L^2(\mathbb{R}^n)$. Then "$P(D)$" $\in \mathcal{GA}_b(\pi/2)$, so that "$P(D)$" generates an analytic semigroup of type $(\pi/2)$.*

Proof.   First we must specify exactly what "$P(D)$" means; it will be a certain extension of the differential operator $P(D)$ acting on $\mathcal{S}(\mathbb{R}^n)$.

Let $\mathcal{X}$ be the complex space $L^2(\mathbb{R}^n)$; denote the norm in $\mathcal{X}$ by $\|\cdot\|$. Define the multiplication operator M_p on $\mathcal{X}$ by $(M_p f)(x) = P(x) f(x)$ for $x \in \mathbb{R}^n$ and

$$f \in \mathcal{D}(M_p) = \left\{ g \in \mathcal{X} : \int_{\mathbb{R}^n} |P(x) g(x)|^2 \, dx < \infty \right\}.$$

Define "$P(D)$" $= \mathcal{F}^{-1} M_p \mathcal{F}$. We have

$$\mathcal{D}(\text{"}P(D)\text{"}) = \left\{ f \in \mathcal{X} : \int_{\mathbb{R}^n} |P(\xi) \hat{f}(\xi)|^2 \, d\xi < \infty \right\}.$$

Note that $\mathcal{S}(\mathbb{R}^n) \subset \mathcal{D}(\text{"}P(D)\text{"})$ since $\mathcal{S}(\mathbb{R}^n) \subset \mathcal{D}(M_p)$ and since $\mathcal{F}$ is a bijection on $\mathcal{S}(\mathbb{R}^n)$ to itself. From now on we shall not bother to write quotation marks around $P(D)$, so that $P(D) = \mathcal{F}^{-1} M_p \mathcal{F}$.

$$(\lambda - M_p) f = g \quad \text{iff} \quad f = M_{1/(\lambda - p)} g;$$

hence if $\lambda \in \mathbb{C} \setminus \{x \in \mathbb{R} : x \le 0\}$, $\lambda \in \rho(M_p)$ and so $\lambda \in \rho[P(D)]$ by Lemma 3.16 since M_p and $P(D)$ are unitarily equivalent. Let $\operatorname{Re} \lambda > 0$. Then

$$|\lambda - P(\xi)| \ge |\lambda|,$$

whence (using Lemma 3.16 again),

$$\|[\lambda - P(D)]^{-1}\| = \|(\lambda - M_p)^{-1}\| \leq 1/|\lambda|. \tag{4.1}$$

Now let $\lambda \in \Sigma_{\pi-\epsilon} \cap \{z: \text{Re } z \leq 0\}$ where $\epsilon > 0$. Then

$$\left|\frac{\lambda}{\text{Im } \lambda}\right| = \frac{1}{|\sin(\arg \lambda)|} \leq \frac{1}{\sin \epsilon};$$

hence

$$\|[\lambda - P(D)]^{-1}\| = \|(\lambda - M_p)^{-1}\| \leq \frac{1}{|\text{Im } \lambda|} \leq \frac{(\sin \epsilon)^{-1}}{|\lambda|}.$$

It follows that

$$\|[\lambda - P(D)]^{-1}\| \leq \frac{(\sin \epsilon)^{-1}}{|\lambda|}$$

for all $\lambda \in \Sigma_{\pi-\epsilon}$. It follows that $P(D) \in \mathcal{GA}_b(\pi/2)$ since $\epsilon > 0$ is arbitrary. The theorem now follows from Theorem I.5.4. ∎

4.2. REMARKS

(i)   One can easily show that

$$P(D) = \overline{P(D)|_{\mathscr{S}(\mathbb{R}_n)}}.$$

(ii) Equation (4.1) shows that $P(D)$ generates an analytic contraction semigroup of type $\mathscr{S}(\infty)$ (see Section 5 of Chapter I).

(iii) A vast generalization of Theorem 4.1 will be given in Section 6 (Theorem 6.12).

4.3. EXAMPLE. Let $P(\xi) = -|\xi|^2 = -\sum_{j=1}^n \xi_j^2$; P satisfies the hypotheses of Theorem 4.1. Let $T = \{T(t): t \in \Sigma_{\pi/2} \cup \{0\}\}$ denote the semigroup generated by $P(D)$, the "Laplacian." We have $T(t) = \mathscr{F}^{-1}S(t)\mathscr{F}$ where S is the semigroup generated by M_p. Since $S(t)f(x) = \exp(-t|x|^2)f(x)$, we obtain, using Definition 3.7, Theorem 3.10, and property (6) of the preceding section,

$$T(t)f(x) = (4\pi t)^{-1/2} \int_{\mathbb{R}^n} \exp(-|x-y|^2/4t)f(y)\, dy \tag{4.2}$$

for all t,x, and $f \in \mathscr{S}(\mathbb{R}^n)$, and hence for all $f \in L^2(\mathbb{R}^n)$. Alternatively, if one defines T by (4.2), then one readily checks that T is an analytic semigroup of type $(\pi/2)$, and the calculation

$$\|t^{-1}[T(t)f - f] - g]\|^2 = \|\mathscr{F}\{t^{-1}[T(t)f - f] - g\}\|^2$$

$$= \int_{\mathbb{R}^n} |t^{-1}(\exp(-t|\xi|^2) - 1)\hat{f}(\xi) - \hat{g}(\xi)|^2\, d\xi$$

shows that the generator of T is $\Delta = \mathscr{F}^{-1}\mathscr{M}_{-|\xi|^2}\mathscr{F}$,

$$\mathscr{D}(\Delta) = \left\{f \in L^2(\mathbb{R}^n): \int_{\mathbb{R}^n} \left||\xi|^2\hat{f}(\xi)\right|^2\, d\xi < \infty\right\}.$$

4.4. If $P(D)$ is an elliptic homogeneous partial differential operator of order $2m$, then the Cauchy problem

$$u' = P(D)u \qquad u(0) = u_0$$

is an example of a *parabolic partial differential equation*. (The use of the term "parabolic" will be discussed in Section 11.) One can rewrite this as

$$\frac{\partial u}{\partial t} = P(D)u \qquad u(0,x) = u_0(x),$$

where $P(D)$ acts on the "spatial variables" x.

The next result uses a perturbation theorem to enable us to solve Cauchy problems for parabolic equations having variable coefficients.

An important feature of parabolic problems is that they are governed by analytic semigroups. This fact enables one to prove "regularity" theorems, i.e. theorems which state that $u(t)$ is actually a smooth function of x, not just a function in $L^2(\mathbb{R}^n)$ (cf. Section 5).

4.5. If $Q(x,\xi) = \Sigma b_\alpha(x)\xi^\alpha$ is a polynomial in $\xi \in \mathbb{R}^n$ whose coefficients are functions of $x \in \mathbb{R}^n$, then the *partial differential operator with variable coefficients* $Q(\cdot,D)$ is defined by

$$[Q(\cdot,D)f](x) = \Sigma b_\alpha(x)D^\alpha f(x);$$

in other words, $Q(\cdot,D) = \Sigma M_{b_\alpha}D^\alpha$.

THEOREM. *Let P be a strictly elliptic real homogeneous polynomial of order $2m$ ($m = 1,2,\dots$), and let $Q(x,\xi) = \sum_{|\alpha| \le 2m-1} b_\alpha(x)\xi^\alpha$ where $b_\alpha \in L^\infty(\mathbb{R}^n)$, $|\alpha| \le 2m - 1$ (the b_α can be complex-valued). Then $P(D) + Q(\cdot,D) \in \mathcal{GA}(\pi/2)$, so that $P(D) + Q(\cdot,D)$ generates a semigroup analytic in the right half plane.*

Proof. By Theorem 4.1, Remark I.6.10, and Corollary I.6.11, it suffices to prove that $M_{b_\alpha}D^\alpha$ is a Kato perturbation of $P(D)$ whenever $b_\alpha \in L^\infty(\mathbb{R}^n)$ and $|\alpha| \le 2m - 1$. Let $\epsilon > 0$. Then there is a $d_\epsilon > 0$ such that

$$|\xi^\alpha|^2 \le d_\epsilon + \epsilon|\xi|^{4m} \text{ for all } \xi \in \mathbb{R}^n$$

(here $|\alpha| \le 2m - 1$). Thus

$$\|D^\alpha f\|^2 = \int_{\mathbb{R}^n} |\xi^\alpha|^2 |\hat{f}(\xi)|^2 \, d\xi$$

$$\le d_\epsilon\|\hat{f}\|^2 + \epsilon \int_{\mathbb{R}^n} |\xi|^{4m}|\hat{f}(\xi)|^2 \, d\xi$$

$$\le d_\epsilon\|f\|^2 + \frac{\epsilon}{c_0}\|P(D)f\|^2$$

where $c_0 > 0$ is such that $P(\xi) \le -c_0|\xi|^{2m}$, $\xi \in \mathbb{R}^n$; such a c_0 exists since P is strictly elliptic. Next,

$$\|M_{b_\alpha}(D^\alpha f)\| \le \|b_\alpha\|_\infty \|D^\alpha f\|$$

$$\le \epsilon\|b_\alpha\|_\infty c_0^{-1}\|P(D)f\| + \|b_\alpha\|_\infty d_\epsilon\|f\|, \, f \in \mathscr{D}[P(D)].$$

98   Linear Cauchy Problems

Since $\epsilon > 0$ is arbitrary, it follows that $M_{b_\alpha} D^\alpha$ is a Kato perturbation of $P(D)$.   ∎

4.6. EXERCISES

1. State and prove a version of Theorem 4.1 in which P is allowed to have complex coefficients.
*2. Example 4.3 gives a rigorous $L^2(\mathbb{R}^n)$ treatment of the semigroup associated with n dimensional Brownian motion (cf. Example I.9.12.2). Give the analogous arguments in the case of an n dimensional Cauchy process (cf. Example I.9.12.3) in the $L^2(\mathbb{R}^n)$ setting.

5. Regularity for Parabolic Problems

5.1. THEOREM.   *Let P, Q be as in Theorem 4.5, and let $u_0 \in \mathscr{X} = L^2(\mathbb{R}^n)$. Then the Cauchy problem*

$$u_t = P(D)u + Q(\cdot,D)u \qquad (t \in \mathbb{R}^+), \tag{5.1}$$

$$u(0) = u_0 \in \mathscr{X}$$

has a unique solution

$$u \in C(\mathbb{R}^+,\mathscr{X}) \cap C^1\{(0,\infty),\mathscr{D}[P(D)]\};$$

moreover,

$$u \in C^1\{(0,\infty),\mathscr{D}[P(D)^k]\}$$

if $b_\alpha \in C^{(k-1)2m}(\mathbb{R}^n)$ and $D^\beta b_\alpha \in L^\infty(\mathbb{R}^n)$ for $|\alpha| \leq 2m-1, |\beta| \leq (k-1)2m$. Finally, $u \in C^1(\mathbb{R}^+,\mathscr{X})$ if $u_0 \in \mathscr{D}[P(D)]$.

Proof.   The first part of the theorem follows immediately from Theorem 4.5 and the results of Section I.5. The second part follows from Exercise I.5.10.2 as soon as we show

$$\mathscr{D}[P(D)^k] = \mathscr{D}\{[P(D) + Q(\cdot,D)]^k\};$$

this is a straightforward consequence of the smoothness condition on the coefficients b_α.   ∎

5.2. THEOREM.   *Let the hypotheses of Theorem 5.1 hold. If the coefficients b_α satisfy $b_\alpha \in C^\infty(\mathbb{R}^n)$ and $D^\beta b_\alpha \in L^\infty(\mathbb{R}^n)$ for all α, β, then for $t > 0$, the solution $u(t)$ (at time t) belongs to $\bigcap_{\nu=0}^\infty H^\nu(\mathbb{R}^n)$ where*

$$H^\nu(\mathbb{R}^n) = \left\{ f \in L^2(\mathbb{R}^n): \int_{\mathbb{R}^n} (1 + |\xi|^2)^\nu |\hat{f}(\xi)|^2 \, d\xi < \infty \right\}.$$

(Here we take $\nu = 0,1,2,\ldots$ although we could equally well take $\nu \in \mathbb{R}$.) $H^\nu(\mathbb{R}^n)$ is a Hilbert space with inner product

$$\langle f,g \rangle_\nu = \int_{\mathbb{R}^n} (1 + |\xi|^2)^\nu \hat{f}(\xi)\overline{\hat{g}(\xi)} \, d\xi$$

and norm $|f|_v = \langle f,f \rangle_v^{1/2}$. Clearly $H^v(\mathbb{R}^n) \subset H^\mu(\mathbb{R}^n)$ if $v \geq \mu$ and $\mathscr{S}(\mathbb{R}^n)$ $[\subset H^v(\mathbb{R}^n)]$ is dense in $H^v(\mathbb{R}^n)$. The spaces $H^v(\mathbb{R}^n)$ are called *Sobolev spaces*.

Proof of Theorem 5.2. Given n, l there exist positive constants c, C (depending on n, l) such that

$$c(1 + |\xi|^2)^l \leq \sum_{|\alpha| \leq l} |\xi^\alpha|^2 \leq C(1 + |\xi|^2)^l \tag{5.2}$$

for all $\xi \in \mathbb{R}^n$. It follows that

$$\mathscr{D}[P(D)^k] = \mathscr{D}(\mathscr{F}^{-1} M_{pk} \mathscr{F})$$

$$= \{f \in L^2(\mathbb{R}^n) : \int_{\mathbb{R}_n} |P^k(\xi)\hat{f}(\xi)|^2 \, d\xi < \infty\}$$

$$= H^{2mk}(\mathbb{R}^n)$$

by (5.2) and by the strict ellipticity of P. ∎

5.3. THEOREM. $\mathscr{S}(\mathbb{R}^n) \subset \bigcap_{v=0}^\infty H^v(\mathbb{R}^n) \subset C_0^\infty(\mathbb{R}^n)$.

Here $C_0^\infty(\mathbb{R}^n)$ is the set of all $f \in C^\infty(\mathbb{R}^n)$ such that $D^\alpha f \in C_0(\mathbb{R}^n)$ (i.e. vanishes at ∞) for all multi-indices α.

Before proving Theorem 5.3 we note that Theorem 5.1, 5.2, and 5.3 together yield the following regularity theorem.

5.4. THEOREM. *Let P, Q be as in Theorem 4.5. Suppose $b_\alpha \in C^\infty(\mathbb{R}^n)$ and $D^\beta b_\alpha \in L^\infty(\mathbb{R}^n)$ for all α, β. Then any solution u of (5.1) satisfies $u(t) \in C_0^\infty(\mathbb{R}^n)$ for each $t > 0$.*

5.5. Proof of Theorem 5.3. The first containment is clear. For the second we prepare three lemmas.

5.6. LEMMA (Sobolev's inequality). *If $N > n/2$, then there is a constant $A = A(n,N)$ such that*

$$\|f\|_\infty \leq A \sum_{|\alpha| \leq N} \|D^\alpha f\|$$

for all $f \in \mathscr{S}(\mathbb{R}^n)$.

Proof. Let $B = \int_{\mathbb{R}^n}(1 + |\xi|^2)^{-N} \, d\xi$. $B < \infty$ since $2N > n$. For $f \in \mathscr{S}(\mathbb{R}^n)$, by Theorem 3.10 we have

$$f(x) = (2\pi)^{-n/2} \int_{\mathbb{R}^n} e^{-ix \cdot \xi} \hat{f}(\xi) \, d\xi;$$

thus

$$\|f\|_\infty^2 \leq (2\pi)^{-n}\left\{\int_{\mathbb{R}^n} |\hat{f}(\xi)| \, d\xi\right\}^2$$

$$\leq (2\pi)^{-n} \int_{\mathbb{R}^n} |\hat{f}(\xi)|^2(1 + |\xi|^2)^N \, d\xi \cdot \int_{\mathbb{R}^n} (1 + |\xi|^2)^{-N} \, d\xi$$

by the Schwarz inequality

$$\leq B(2\pi)^{-n}c^{-1} \sum_{|\alpha| \leq N} \int_{\mathbb{R}^n} |\xi^\alpha \hat{f}(\xi)|^2 \, d\xi \text{ by (5.2)}$$

$$= B(2\pi)^{-n}c^{-1} \sum_{|\alpha| \leq N} \|D^\alpha f\|^2.$$

The result follows.   ■

5.7. LEMMA (Sobolev's inequality). *If $N > n/2$, $k \geq 0$, there is a constant $B = B(n,N,k)$ such that*

$$\sum_{|\alpha| \leq k} \|D^\alpha \hat{f}\|_\infty \leq B \sum_{|\alpha| \leq k+N} \|D^\alpha f\|$$

for all $f \in \mathscr{S}(\mathbb{R}^n)$.

Proof. This is an easy consequence of the preceding lemma.   ■

5.8. LEMMA. $H^{N+k}(\mathbb{R}^n) \subset C_0^k(\mathbb{R}^n)$ *if* $N > n/2$, $k \geq 0$.

Proof. This means that if $f \in H^{N+k}(\mathbb{R}^n)$, then (after correction on a null set) $f \in C^k(\mathbb{R}^n)$ and $D^\alpha f \in C_0(\mathbb{R}^n)$ for $|\alpha| \leq k$.

Let $f \in H^{k+N}(\mathbb{R}^n)$. There is a sequence $\{f_\mu\}_{\mu=1}^\infty$ in $\mathscr{S}(\mathbb{R}^n)$ such that

$$\lim_{\mu \to \infty} \int_{\mathbb{R}^n} |\xi^\alpha [\hat{f}_\mu(\xi) - \hat{f}(\xi)]|^2 \, d\xi = 0,$$

$|\alpha| \leq N + k$. The preceding lemma implies that $\{D^\alpha f_\mu\}_{\mu=1}^\infty$ is a Cauchy sequence in $C_0(\mathbb{R}^n)$ for $|\alpha| \leq k$. The result follows.   ■

5.9. Proof of Theorem 5.3. The theorem follows by letting $k \to \infty$ in Lemma 5.8.   ■

5.10. REMARK. In Example 4.3, $T(t)$ is not a compact operator for any $t > 0$, but the corresponding operators are compact when $L^2(\mathbb{R}^n)$ is replaced by $L^2(\Omega)$ where Ω is a nice bounded set in $\mathbb{R}^n$. One is led to ask whether there is a connection between analyticity of a semigroup T and compactness of $T(t)$ for $t > 0$.

Let $\mathscr{X}$ be infinite dimensional, and let $A \in \mathscr{B}(\mathscr{X})$. Then $T = \{T(t) = e^{tA} : t \in \mathbb{C}\}$ is an analytic group, none of whose elements are compact operators.

Let $\{x_n : n = 1,2,\ldots\}$ be an orthonormal basis for a separable Hilbert space $\mathscr{H}$. Define $T(t)$ for $t \geq 0$ by

$$T(t)x_n = \exp(-nt + ie^{n^2}t)x_n,$$

and extend $T(t)$ to all of $\mathscr{H}$ by linearity and continuity. $T = \{T(t) : t \in R^+\}$ is a (C_0) contraction semigroup, $T(t)$ is compact for each $t > 0$, $T(t)(\mathscr{H}) \not\subset \mathscr{D}(A)$ where A is the generator of T, and T does not have an analytic continuation into any sector in $\mathbb{C}$ containing $(0,\infty)$ in its interior. The details of the proof are straightforward and are left to the reader.

5.11. EXERCISES

1. Verify the details in Remark 5.10.
2. Show that if T is a (C_0) semigroup with generator A and if $T(t_0)(\mathcal{X}) \subset \mathcal{D}(A)$ for some $t_0 > 0$, then

$$\lim_{t \to s} \| T(t) - T(s)\| = 0$$

for all $s > t_0$.

6. The Spectral Theorem

6.1. DEFINITION.   A function $g: \mathbb{R} \to \mathbb{C}$ is called *nonnegative definite* if for each positive integer n, for each $t \in \mathbb{R}^n$, and for each $\lambda \in \mathbb{C}^n$,

$$\sum_{j,k=1}^{n} g(t_j - t_k)\lambda_j \bar{\lambda}_k \geq 0. \tag{6.1}$$

6.2. BOCHNER'S THEOREM. *A function $g: \mathbb{R} \to \mathbb{C}$ which is continuous at the origin is nonnegative definite iff there exists a (unique) nonnegative finite (Radon) measure μ on $\mathbb{R}$ such that*

$$g(t) = \int_{\mathbb{R}} e^{itx} \mu(dx) \qquad t \in \mathbb{R}. \tag{6.2}$$

Proof. (Necessity). Define g by (6.2), where μ is a nonnegative finite (Radon) measure on $\mathbb{R}$. Then if $t \in \mathbb{R}^n$, $\lambda \in \mathbb{C}^n$,

$$\sum_{j,k=1}^{n} g(t_j - t_k)\lambda_j \bar{\lambda}_k = \int_{\mathbb{R}} \sum_{j,k=1}^{n} \exp[i(t_j - t_k)x]\lambda_j \bar{\lambda}_k \mu(dx)$$

$$= \int_{\mathbb{R}} \sum_{j,k=1}^{n} \lambda_j \exp(it_j x)[\overline{\lambda_k \exp(it_k x)}]\mu(dx)$$

$$= \int_{\mathbb{R}} \left| \sum_{l=1}^{n} \lambda_l \exp(it_l x)\right|^2 \mu(dx) \geq 0.$$

Also, by the dominated convergence theorem and (6.2), g is continuous on $\mathbb{R}$ (in fact, uniformly continuous on $\mathbb{R}$). Finally, we note that for all $t \in \mathbb{R}$,

$$|g(t)| \leq g(0) = \int_{\mathbb{R}} \mu(dx) = \mu(\mathbb{R}).$$

(Sufficiency). Here we briefly sketch a classical proof of the sufficiency part of Bochner's theorem. For an alternate modern proof involving Banach algebra notions, see Rudin [1, pp. 17–21].

First, nonnegative definiteness implies $|g(t)| \leq g(0)$ for all $t \in \mathbb{R}$, and g is (uniformly) continuous on $\mathbb{R}$. Let $\tau > 0$, $x \in \mathbb{R}$ and set

$$p_\tau(x) = (2\pi\tau)^{-1} \int_0^\tau \int_0^\tau g(u - v)e^{-i(u-v)x}\, du\, dv. \tag{6.3}$$

Then $p_\tau(x) \geq 0$ since the integral is a limit of nonnegative Riemann sums. Let

$$g_\tau(t) = \begin{cases} \left(1 - \dfrac{|t|}{\tau}\right) g(t) & \text{if } |t| \leq \tau \\ \\ 0 & \text{if } |t| > \tau; \end{cases}$$

making the substitution $s = v, t = u - v$ in (6.3) and integrating with respect to s yields

$$0 \leq p_\tau(x) = (2\pi)^{-1} \int_{\mathbb{R}} g_\tau(t) e^{-itx}\, dt.$$

By the Fourier transform inversion formula (Theorem 3.10),

$$g_\tau(t) = \int_{\mathbb{R}} e^{itx} p_\tau(x)\, dx.$$

Set $P_\tau(x) = \int_{-\infty}^{x} p_\tau(y)\, dy$. Then P_τ is monotone nondecreasing on $\mathbb{R}$ and bounded independent of τ since g is nonnegative definite. Let $\{\tau_n\}$ be a sequence of positive reals tending to ∞. Then a diagonalization argument shows that $\{P_{\tau_n}\}$ has a subsequence which we again denote by $\{P_{\tau_n}\}$ which converges to a monotone nondecreasing bounded function P at all continuity points of P. (We are using the Helly selection principle; cf. Loève [1, p. 179].)[†] It follows that

$$g_{\tau_n}(x) = \int_{\mathbb{R}} e^{itx}\, dP_n(t) \rightarrow \int_{\mathbb{R}} e^{itx}\, dP(t)$$

by the Helly-Bray theorem (Loève [1, p. 180]) as $k \rightarrow \infty$. Thus

$$g(x) = \int_{\mathbb{R}} e^{itx}\, dP(t)$$

for all $x \in \mathbb{R}$. Setting $\mu(E) = \int_{\mathbb{R}} \chi_E(t) dP(t)$ where $\chi_E(t) = 1$ or 0 according as $t \in E$ or $t \notin E$ (where E is a Borel set in $\mathbb{R}$), the result follows. The uniqueness of μ follows from a uniqueness result for Fourier transforms (Loève [1, p. 186]). ∎

6.3. DEFINITION. Let (Ω, Σ, μ) be a measure space. Let $q : \Omega \rightarrow \mathbb{C}$ be Σ-measurable. The *multiplication operator* M_q on $\mathscr{H} = L^2(\Omega, \Sigma, \mu)$ is defined by $(M_q f)(x) = q(x) f(x)$ for $f \in \mathscr{D}(M_q) = \{g \in \mathscr{H} : \int_\Omega |q(x) g(x)|^2\, \mu(dx) < \infty\}$.

Note that $M_{q_1} = M_{q_2}$ iff $q_1 = q_2$ a.e. $[\mu]$. In particular, q need only be defined on Ω / N where $\mu(N) = 0$ for M_q to make sense.

6.4. LEMMA. *The multiplication operator M_q is self-adjoint if q is real-valued.*

[†] We briefly indicate an alternative proof of the Helly selection principle using functional analysis rather than the diagonalization process. Identify a function F of bounded variation on $\mathbb{R}$ with its "indefinite integral," the (complex) signed measure v defined by $v(E) = \int_{\mathbb{R}} \chi_E(x) dF(x)$. The space $\mathscr{M}(\mathbb{R})$ of all such functions of bounded variation is the dual space of $C_0(\mathbb{R})$ by the Riesz representation theorem. Hence bounded sets in $\mathscr{M}(\mathbb{R})$ are relatively compact in the weak* topology. This last fact implies the Helly selection theorem.

Proof. First, $\mathscr{D}(M_q)$ is dense, *for*, set $E_n = \{x \in \Omega : |q(x)| \le n\}$, $n = 1, 2, \ldots$. $\bigcup_{n=1}^{\infty} E_n = \Omega \backslash N$ where $\mu(N) = 0$. For each $f \in \mathscr{H}$, $f\chi_{E_n} \in \mathscr{D}(M_q)$ since

$$\int_{\Omega} |qf\chi_{E_n}|^2 \, d\mu \le n^2 \int_{\Omega} |f|^2 \, d\mu < \infty. \, \|f\chi_{E_n} - f\| \to 0 \text{ as } n \to \infty$$

by the dominated convergence theorem; hence $\overline{\mathscr{D}(M_q)} = \mathscr{H}$. Next, M_q is symmetric, *for*, if $f, g \in \mathscr{D}(M_q)$,

$$\langle M_q f, g \rangle = \int_{\Omega} q f \bar{g} \, d\mu = \int_{\Omega} f \overline{(qg)} \, d\mu = \langle f, M_q g \rangle$$

since q is real-valued. $\pm iI + M_q : \mathscr{D}(M_q) \to \mathscr{H}$ is clearly a bijection and $(\pm i + M_q)^{-1} h(x) = [\pm i + q(x)]^{-1} h(x)$. Assume $M_q^* \supsetneq M_q$. Then $iI + M_q^*$ is not injective. Choose $h \ne 0$ such that $(iI + M_q^*)h = 0$. Then for all $f \in \mathscr{D}(M_q)$,

$$\langle M_q f, h \rangle = \langle f, M_q^* h \rangle = \langle f, -ih \rangle = \langle if, h \rangle,$$

whence

$$\int_{\Omega} [q(x) - i] f(x) \overline{h(x)} \, \mu(dx) = 0$$

for all $f \in \mathscr{D}(M_q)$. Thus $h \perp (M_q - iI)[\mathscr{D}(M_q)] = \mathscr{H}$, whence $h = 0$. This contradiction shows that M_q is self-adjoint. ∎

6.5. COROLLARY. *Let q be real-valued. Then the (C_0) unitary group generated by iM_q is $\{M_{e^{itq}} : t \in \mathbb{R}\}$.*

Proof. iM_q generates a unitary group by Lemma 6.4 and Stone's Theorem I.4.7. The rest is obvious. ∎

6.6. DEFINITION. Let J be a nonempty index set. Let $\{(\Omega_\alpha, \mu_\alpha) : \alpha \in J\}$ be a family of measure spaces. (We omit reference to the underlying σ-algebras, which should cause no confusion.) Let the Hilbert space $\mathscr{H}$ consist of all $f : J \to \bigcup_{\alpha \in J} L^2(\Omega_\alpha, \mu_\alpha)$ such that $f(\alpha) \in L^2(\Omega_\alpha, \mu_\alpha)$ for each $\alpha \in J$ and

$$\|f\|^2 = \sum_{a \in J} \|f(\alpha)\|_{L^2(\Omega_\alpha, \mu_\alpha)}^2 < \infty.$$

Let $q_\alpha : \Omega_\alpha \to \mathbb{C}$ be measurable for each $\alpha \in J$, and let $\underline{q} = \{q_\alpha : \alpha \in J\}$. The *generalized multiplication operator* $\mathscr{M}_{\underline{q}}$ is defined by

$$(\mathscr{M}_{\underline{q}} f)(\alpha) = M_{q_\alpha} f(\alpha), \qquad \alpha \in J,$$

$$\mathscr{D}(\mathscr{M}_{\underline{q}}) = \left\{ f \in \mathscr{H} : \sum_{a \in J} \|M_{q_\alpha} f(\alpha)\|_{L^2(\Omega_\alpha, \mu_\alpha)}^2 < \infty \right\}.$$

Of course, $\mathscr{M}_{\underline{q}} = \mathscr{M}_{\underline{p}}$ iff $p_\alpha = q_\alpha$ a.e. $[\mu_\alpha]$ for each $\alpha \in J$. Note that if $f \in \mathscr{H}$, then $f(\alpha) = 0$ except for at most countably many values of α.

6.7. PROPOSITION. *If each q_α is real-valued, then the generalized multiplication operator $\mathscr{M}_{\underline{q}}$ is self-adjoint on $\mathscr{H}$.*

Proof. Let $h \perp \mathscr{D}(\mathscr{M}_q)$, and let $\beta \in J$, $f_\beta \in \mathscr{D}(M_{q\beta})$. Define $f \in \mathscr{K}$ by $f(\alpha) = f_\beta$ or 0 according as $\alpha = \beta$ or $\alpha \neq \beta$. Then $f \in \mathscr{D}(\mathscr{M}_q)$ and

$$0 = \langle f, h \rangle_{\mathscr{K}} = \langle f_\beta, h(\beta) \rangle_{L^2(\Omega_\beta, \mu_\beta)}.$$

Therefore $h(\beta) \perp \mathscr{D}(M_{q\beta})$, whence $h(\beta) = 0$. Since $\beta \in J$ is arbitrary, it follows that $h = 0$ and $\overline{\mathscr{D}(\mathscr{M}_q)} = \mathscr{K}$. We omit the rest of the proof, which essentially coincides with the corresponding part of the proof of Lemma 6.4. ∎

6.8. Suppose that $\Omega_\alpha = \Omega$ does not depend on α. (The measures μ_α can still vary with $\alpha \in J$.) Let $q: \Omega \to \mathbb{C}$. Let $q_\alpha = q$ for each $\alpha \in J$. We denote by $\mathscr{M}_q$ the resulting generalized multiplication operator $\mathscr{M}_q$. In particular, if $\Omega_\alpha = \mathbb{R}$ for all $\alpha \in J$, $\mathscr{M}_{id}$ denotes the generalized multiplication operator $\mathscr{M}_q$, where q is the identity function: $q(x) = x$ for all $x \in \mathbb{R}$.

6.9. SPECTRAL THEOREM. *Every self-adjoint operator is unitarily equivalent to a generalized multiplication operator of the form $\mathscr{M}_{id}$.*

Proof. Let S be self-adjoint on $\mathscr{H}$. Then by Stone's Theorem I.4.7, iS generates a (C_0) unitary group U. Let $f \in \mathscr{H}$ and let $g_f(t) = \langle U(t)f, f \rangle$. Then g_f is continuous and nonnegative definite since

$$\sum_{j,k=1}^{n} g_f(t_j - t_k) \lambda_j \bar{\lambda}_k = \sum_{j,k=1}^{n} \lambda_j \bar{\lambda}_k \langle U(t_j - t_k)f, f \rangle$$

$$= \sum_{j,k=1}^{n} \langle \lambda_j U(t_j)f, \lambda_k U(t_k)f \rangle$$

$$= \left\| \sum_{l=1}^{n} \lambda_l U(t_l)f \right\|^2 \geq 0.$$

Hence by Bochner's theorem 6.2 there is a finite (Radon) measure μ_f on $\mathbb{R}$ such that

$$\langle U(t)f, f \rangle = \int_{\mathbb{R}} e^{itx} \mu_f(dx) \qquad t \in \mathbb{R}, \tag{6.4}$$

and $\mu_f(\mathbb{R}) = \langle U(0)f, f \rangle = \|f\|^2$. Let $\mathscr{H}_f$ be the closure of the span of $\{U(t)f : t \in \mathbb{R}\}$. Then $U(t)(\mathscr{H}_f) \subset \mathscr{H}_f$ for each $t \in \mathbb{R}$, so that $U|_{\mathscr{H}_f}$ is a (C_0) unitary group on $\mathscr{H}_f$. Define V_f: span $\{U(t)f : t \in \mathbb{R}\} \to L^2(\mathbb{R}, \mu_f)$ by $V_f(\sum_{j=1}^{n} a_j U(t_j)f)(x) = \sum_{j=1}^{n} a_j e^{itjx}$. Then

$$\left\langle V_f\left(\sum_{j=1}^{n} a_j U(t_j)f \right), V_f\left(\sum_{k=1}^{m} b_k U(s_k)f \right) \right\rangle_{L^2(\mathbb{R}, \mu_f)}$$

$$= \sum_{j=1}^{n} \sum_{k=1}^{m} a_j \bar{b}_k \int_{\mathbb{R}} \exp[i(t_j - s_k)x] \mu_f(dx)$$

$$= \left\langle \sum_{j=1}^{n} a_j U(t_j)f, \sum_{k=1}^{m} b_k U(s_k)f \right\rangle_{\mathscr{H}_f}$$

by (6.4). V_f is thus isometric and extends by continuity to a unitary map, which we also denote by V_f, from $\mathscr{H}_f$ to $L^2(\mathbb{R}, \mu_f)$; the surjectivity of this extension follows from the finiteness of μ_f and the Weierstrass approximation theorem. We thus have shown that $U(t)|_{\mathcal{H}_f} = V_f^{-1} M_{e^{it}} V_f$ for each $t \in \mathbb{R}$. ($e^{it \cdot}$ is the function $x \to e^{itx}$.) It follows from Lemma 3.17 and by Corollary 6.5 that $S|_{\mathcal{H}_f} = V_f^{-1} M_{id} F_f$ since $iS|_{\mathcal{H}_f}$ is the generator of $U|_{\mathcal{H}_f}$. In particular, $S|_{\mathcal{H}_f}$ is unitarily equivalent to M_{id} on $L^2(\mathbb{R},\mu_f)$.

By Zorn's Lemma, there is a family of vectors $\{f_\alpha : \alpha \in J\} \subset \mathcal{H}$ such that $\mathcal{H}_{f_\alpha} \perp \mathcal{H}_{f_\beta}$ if $\alpha \neq \beta$ and $\mathcal{H} = \bigoplus_{\alpha \in J} \mathcal{H}_{f_\alpha}$, the direct sum of the $\mathcal{H}_{f_\alpha}$, $\alpha \in J$; i.e. for each $h \in \mathcal{H}$, there exists a unique $h_\alpha \in \mathcal{H}_{f_\alpha}$ (at most countably many of which are nonzero) such that $h = \sum_{\alpha \in J} h_\alpha$ (and $\|h\|^2 = \sum_{\alpha \in J} \|h_\alpha\|^2$). Let

$$\mathcal{K} = \{g : J \to \bigcup_{\alpha \in J} L^2(\mathbb{R},\mu_{f_\alpha}) : g(\alpha) \in L^2(\mathbb{R},\mu_{f_\alpha}) \text{ for each } \alpha \in J$$

$$\text{and } \|g\|^2 = \sum_{\alpha \in J} \|g(\alpha)\|^2 L^2(\Omega,\mu_{f_\alpha}) < \infty\}.$$

Define $V : \mathcal{H} \to \mathcal{K}$ by $V|_{\mathcal{H}_{f_\alpha}} = V_{f_\alpha}$, $\alpha \in J$, and extend V by linearity and continuity to all of $\mathcal{H}$. Then $V \in \mathcal{B}(\mathcal{H},\mathcal{K})$ is unitary. Moreover, $U(t) = V^{-1} M_{e^{it}} V$ for each $t \in \mathbb{R}$, whence by Lemma 3.17,

$$S = V^{-1} M_{id} V.$$

The proof is complete.   ∎

6.10. Further Developments of the Spectral Theorem (sketch). Let $F : \mathbb{R} \to \mathbb{C}$ be Borel measurable. Let S be a self-adjoint operator on $\mathcal{H}$. Write $S = V^{-1} M_{id} V$ by the spectral theorem. Define

$$F(S) = V^{-1} M_F V.$$

$F(S)$ is a well-defined operator on $\mathcal{H}$ which is independent of the spectral representation of S; i.e. if also $S = W^{-1} M_{id} W$ where $W \in \mathcal{B}(\mathcal{H},\mathcal{K}_1)$ is unitary, then

$$V^{-1} M_F V = W^{-1} M_F W.$$

$F(S) \in \mathcal{B}(\mathcal{H})$ if the function F is bounded. $F(S)G(S) = G(S)F(S) = (F \cdot G)(S)$ for all Borel functions $F,G : \mathbb{R} \to \mathbb{C}$. $F(S)^* = \bar{F}(S)$, and $F(S)$ is self-adjoint if F is real-valued.

As before, let $\chi_B(x) = 1$ or 0 according as $x \in B$ or $x \notin B$, where B is a Borel set in $\mathbb{R}$. $E_B = \chi_B(S)$ is a self-adjoint (orthogonal contraction) projection, called a *spectral projection* for S. For $\lambda \in \mathbb{R}$ let $E_\lambda = E_{(-\infty,\lambda)}$. Then $\{E_\lambda : \lambda \in \mathbb{R}\}$ is a (left continuous) *resolution of the identity*:

   (i) $\lim_{\lambda \to -\infty} E_\lambda f = 0$, $\lim_{\lambda \to +\infty} E_\lambda f = f$ for all $f \in \mathcal{H}$,
   (ii) $E_\lambda E_\mu = E_\lambda E_\mu = E_\mu$ if $\mu \leq \lambda$,
   (iii) $\lim_{\lambda \to \mu-} E_\lambda f = E_\mu f$ for each $f \in \mathcal{H}$, $\mu \in \mathbb{R}$.

$$M_{id} \text{ on } L^2(\mathbb{R},\mu_f) \text{ is given by}$$

$$M_{id} f(x) = x f(x) = \int_{-\infty}^{\infty} \lambda \, d\chi_{(-\infty,\lambda)}(x) f(x)$$

or

$$M_{id}f = \int_{-\infty}^{\infty} \lambda \, d\chi_{(-\infty,\lambda)}f$$

(the integral being an improper Riemann-Stieltjes integral). This observation together with the spectral theorem implies that for all $f \in \mathscr{D}(S)$,

$$Sf = \int_{-\infty}^{\infty} \lambda \, dE_\lambda f, \tag{6.5}$$

and

$$\mathscr{D}(S) = \left\{ f \in \mathscr{H} : \int_{-\infty}^{\infty} |\lambda|^2 \, d(\|E_\lambda f\|^2) < \infty \right\}.$$

The integral in (6.5) is the improper Riemann-Stieltjes integral

$$\lim_{\substack{a \to -\infty \\ b \to \infty}} \lim_{\text{mesh}\,\pi \to 0} \sum_{k=1}^{n} \lambda_k' (E_{\lambda_k} - E_{\lambda_{k-1}})f;$$

here $\pi : a = \lambda_0 < \cdots < \lambda_n = b$ is a partition of $[a,b]$, mesh $\pi = \max(\lambda_k - \lambda_{k-1})$, $\lambda_{k-1} \le \lambda_k' \le \lambda_k$. One also has

$$F(S) = \int_{\mathbb{R}} F(\lambda) \, dE_\lambda,$$

$$\mathscr{D}[F(S)] = \left\{ f \in \mathscr{H} : \int_{\mathbb{R}} |F(\lambda)|^2 \, d(\|E_\lambda f\|^2) < \infty \right\}$$

for all Borel functions $F : \mathbb{R} \to \mathbb{C}$.

6.11. A self-adjoint operator S is called *nonnegative* $(S \ge 0)$ if $\langle Sf, f \rangle \ge 0$ for all $f \in \mathscr{D}(S)$. $S \ge 0$ iff $\mu_f(-\infty,0) = 0$ for all $f \in \mathscr{H}$, where μ_f is as in the spectral theorem: $\langle e^{itS}f, f \rangle = \int_{\mathbb{R}} e^{itx} \mu_f(dx), f \in \mathscr{H}$. Every nonnegative self-adjoint square operator S has a nonnegative self-adjoint square root $S^{1/2}$. In fact, if $S = V^{-1}\mathscr{M}_{id}V$, then $S^{1/2} = V^{-1}\mathscr{M}_q V$ where $q(x) = \sqrt{x}$, $x \in \mathbb{R}^+$. (We need not bother to define q on $(-\infty,0)$ since $(-\infty,0)$ is μ_f-null for each $f \in \mathscr{H}$.) Alternatively, $S = \int_0^\infty \lambda \, dE_\lambda$ and $S^{1/2} = \int_0^\infty \lambda^{1/2} \, dE_\lambda$.

6.12. THEOREM. *Let $S \ge 0$ be self-adjoint. Then $-S \in \mathscr{GA}_b(\pi/2)$ so that $-S$ generates an analytic semigroup of type $(\pi/2)$.*

This is a generalization of Theorem 5.1 and can be proven by adapting the proof of Theorem 5.1 to the present situation. The analytic semigroup generated by $-S$ is given by

$$e^{-tS} = V^{-1} \mathscr{M}_{e^{-t\cdot}} V = \int_0^\infty e^{-\lambda t} \, dE_\lambda$$

for Re $t > 0$, where

$$S = V^{-1}\mathscr{M}_{id}V = \int_0^\infty \lambda \, dE_\lambda. \qquad \blacksquare$$

6.13. We shall apply some of the techniques developed thus far to prove a fundamental result of Kato in the theory of potential scattering. Let $v > 0$ and let $P : \mathbb{R}^n \to \mathbb{R}$ be measurable and satisfy

$$c|\xi|^v \le P(\xi) \le c^{-1}|\xi|^v$$

for all $\xi \in \mathbb{R}^n$ and some $c > 0$ (independent of ξ). Let S be the self-adjoint operator $S = \mathscr{F}^{-1} M_P \mathscr{F}$. (For example, $-S = -P(D)$ can be a strictly elliptic constant coefficient partial differential operator of order $v = 2m$.)

THEOREM. *Let S be as above and suppose $2v > n$. Let $q = q_1 + q_2$ where $q_1 \in L^2(\mathbb{R}^n)$, $q_2 \in L^\infty(\mathbb{R}^n)$, and q_1, q_2 are real-valued. Then $S + M_q$ is self-adjoint on $\mathscr{H} = \mathscr{L}^2(\mathbb{R}^n)$ and $\mathscr{D}(S + M_q) = \mathscr{D}(S) = H^v(\mathbb{R}^n)$.*

In the applications this result is most frequently used when $-S = \Delta$ is the Laplacian on $L^2(\mathbb{R}^3)$ (so that $2v = 4 > 3 = n$).

Proof. M_{q_2} is a bounded self-adjoint operator and hence a symmetric Kato perturbation of S. It suffices to show that M_{q_1} (which is self-adjoint) is a Kato perturbation of S; the result then follows from I.6.11 and I.6.12.

Let

$$C = \int_{\mathbb{R}^n} (1 + |\eta|^v)^{-2} \, d\eta. \ C < \infty \text{ since } 2v > n.$$

For

$$f \in \mathscr{D}(S) = \left\{ g \in L^2(\mathbb{R}^n) : \int_{\mathbb{R}^n} \left| |\xi|^v \hat{g}(\xi) \right|^2 d\xi < \infty \right\} = H^v(\mathbb{R}^n)$$

and $\alpha > 0$,

$$\left(\int_{\mathbb{R}^n} |\hat{f}(\xi)| \, d\xi \right)^2$$

$$\le \left\{ \int_{\mathbb{R}^n} (\alpha + |\xi|^v)^2 |\hat{f}(\xi)|^2 \, d\xi \right\} \left\{ \int_{\mathbb{R}^n} (\alpha + |\xi|^v)^{-2} \, d\xi \right\}$$

by the Schwarz inequality; let $\eta = \alpha^{-1/v} \xi$

$$= \left\{ \int_{\mathbb{R}^n} (\alpha + |\xi|^v)^2 |\hat{f}(\xi)|^2 \, d\xi \right\} \cdot C\alpha^{n/v - 2}$$

$$\le Cc^{-1} \alpha^{n/v - 2} \|(\alpha + S)f\|^2.$$

Hence, letting $K = \sqrt{Cc^{-1}}$,

$$\int_{\mathbb{R}^n} |\hat{f}(\xi)| \, d\xi \le K \{ \alpha^{n/2v - 1} \|Sf\| + \alpha^{n/2v} \|f\| \};$$

in particular, $\hat{f} \in L^1(\mathbb{R}^n)$. It follows that f is (after correction on a null set) a bounded (and uniformly continuous) function on $\mathbb{R}^n$; in fact, for all $x \in \mathbb{R}^n$,

$$|f(x)| \le (2\pi)^{-n/2} \int_{\mathbb{R}^n} |\hat{f}(\xi)| \, d\xi$$

$$\le K' \{ \alpha^{n/2v - 1} \|Sf\| + \alpha^{n/2v} \|f\| \} \tag{6.6}$$

where $K' = (2\pi)^{-n/2} K$. Hence for all $f \in \mathcal{D}(S)$,

$$\|M_{q_1} f\| \leq \|f\|_\infty \|q_1\|$$

$$\leq K' \|q_1\| \{\alpha^{n/2\nu - 1} \|Sf\| + \alpha^{n/2\nu} \|f\|\}$$

by (6.6). Since α can be taken arbitrarily large and $2\nu > n$, it follows that M_{q_1} is a Kato perturbation of S. The proof is complete. ∎

6.14. EXERCISES

1. Give the details in the following outline of the proof that the spectral theorem (Theorem 6.9) implies Bochner's theorem (Theorem 6.2). First, $g : \mathbb{R} \to \mathbb{C}$ is nonnegative definite iff $\int_{-\infty}^\infty \int_{-\infty}^\infty g(t - s)\phi(t)\overline{\phi(s)}dt\, ds > 0$ for every continuous $\phi : \mathbb{R} \to \mathbb{C}$ having compact support. Let $\mathcal{F}$ consist of all functions $f : \mathbb{R} \to \mathbb{C}$ which vanish except at finitely many points of $\mathbb{R}$. Define $\langle \cdot , \cdot \rangle : \mathcal{F} \times \mathcal{F} \to \mathbb{C}$ by $\langle f_1, f_2 \rangle = \Sigma g(t - s) f_1(t) \overline{f_2(s)}$, the sum being over all real t and s. Let $\mathcal{N} = \{f \in \mathcal{F} : \langle f, f \rangle = 0\}$. Complete $\mathcal{F}/\mathcal{N}$ into a Hilbert space $\mathcal{H}$, and define $U = \{U(t) : t \in \mathbb{R}\}$ on $\mathcal{H}$ by

    $$U(t)f(s) = f(s + t)$$

    for $f \in \mathcal{H}$ and $s,t \in \mathbb{R}$. U is a (C_0) unitary group; hence by Stone's theorem it has the form

    $$U(t)f = \int_{-\infty}^\infty e^{it\lambda}\, dE_\lambda f$$

    for a suitable resolution of the identity $\{E_\lambda : \lambda \in \mathbb{R}\}$. Given $t \in \mathbb{R}$ let $f_0 = \chi_{\{0\}}$ and conclude that

    $$g(t) = \langle U(t - s)f_0, U(-s)f_0 \rangle = \langle U(t)f_0, f_0 \rangle$$

    $$= \int_{-\infty}^\infty e^{it\lambda}\, d\mu(\lambda)$$

    where μ is the monotone nondecreasing function whose value at λ is $\|E_\lambda(f_0 + \mathcal{N})\|^2$.

2. Let $\mathcal{H} = \mathbb{C}^n$. Deduce from the spectral theorem that every self-adjoint A on $\mathcal{H}$ has an orthonormal basis of eigenvectors.

3. Let $S \geq 0$ be self-adjoint. Show that the definition of S^α (for $0 < \alpha < 1$) given in 6.11 agrees with the definition in I.9.2.

4. Let $P(\xi) = -|\xi|^2$ for $\xi \in \mathbb{R}^n$. Let $S = \mathcal{F}^{-1} M_P \mathcal{F}$ be the Laplacian on $\mathcal{H} = L^2(\mathbb{R}^n)$. Let $q = q_1 + q_2$ be real-valued, where $q_1 \in L^p(\mathbb{R}^n)$, $q_2 \in L^\infty(\mathbb{R}^n)$, and where $p > n/2$, and $p \geq 2$ if $n \leq 3$. Show that M_q is a Kato perturbation of S; thus $S + M_q$ is self-adjoint on $\mathcal{D}(S + M_q) = H^2(\mathbb{R}^n)$. Hint. Use the Hausdorff-Young-Titchmarsh Theorem:

    $$\|\mathcal{F}f\|_p \leq \|f\|_{p'} \text{ for } f \in L^p(\mathbb{R}^n),$$

    $$2 \leq p \leq \infty, \text{ and } 1/p + 1/p' = 1.$$

    This is trivial for $p = \infty$ [and $f \in \mathcal{S}(\mathbb{R}^n)$] and the $p = 2$ case follows from the unitarity of $\mathcal{F}$. To get the general result use the Riesz-Thorin interpolation theorem (see, for example, Dunford-Schwartz [1, p. 525]). Alternatively, use the method of 6.13. (See also Exercise 11 below.)

5. This problem sketches another proof of Bochner's theorem. Let f be nonnegative definite. Assume that $f \in \mathscr{S}(\mathbb{R})$. For $h \in \mathscr{S}(\mathbb{R})$,

$$0 \le \langle f * h, h \rangle_{L^2(\mathbb{R})} = \sqrt{2\pi} \langle \hat{f}\hat{h}, \hat{h} \rangle_{L^2(\mathbb{R})};$$

the inequality follows from the nonnegativity of the associated Reimann sums. Since h is arbitrary, argue that $\hat{f}(\xi) \ge 0$ for all ξ. Conclude that

$$f(x) = \int_{-\infty}^{\infty} e^{ix\xi} \hat{f}(\xi) \, d\xi.$$

For the case when $f \notin \mathscr{S}(\mathbb{R})$, approximate f by a sequence of nonnegative definite functions in $\mathscr{S}(\mathbb{R})$.

6. Let S be a self-adjoint operator on $\mathscr{H}$. Show that S is unitarily equivalent to a multiplication operator. {Hints. We know that for a certain $\mathscr{K} = \bigoplus_{\alpha \in J} L^2(\mathbb{R}, \mu_\alpha)$ and for a certain unitary operator $V \in \mathscr{B}(\mathscr{H}, \mathscr{K})$, $VSV^{-1} = \mathscr{M}_{id}$. Let Ω be the disjoint union of $\mathbb{R}_\alpha$, $\alpha \in J$, where $\mathbb{R}_\alpha = \mathbb{R}$, and define the measure μ on the obvious subsets of Ω by $\mu(\bigcup_{\alpha \in J} E_\alpha) = \sum_{\alpha \in J} \mu_\alpha(E_\alpha)$. Define $q: \Omega \to \mathbb{R}$ as follows: if $w \in \Omega$, then $w \in \bigcup \mathbb{R}_\alpha$, and consequently w is identified with some real number x (in a particular copy of $\mathbb{R}$); set $q(w) = x$. Then S is unitarily equivalent to M_q on $L^2(\Omega, \mu)$.}

*7. $N \in \mathscr{B}(\mathscr{H})$ is called *normal* if $NN^* = N^*N$. Show that N is normal iff N is unitarily equivalent to a multiplication operator M_q where q is complex-valued and bounded. {Hints. Let $S_1 = 1/2(N + N^*), S_2 = 1/2i(N - N^*)$, so that S_1, S_2 are bounded commuting self-adjoint operators and $N = S_1 + iS_2$. Write $S_j = \int_{-\infty}^{\infty} \lambda dE^j$ for $j = 1,2$. If M_j is a Borel set in $\mathbb{R}$, define $E^j(M_j) = \int_{M_j} dE^j_\lambda$ and define $E(M_1 \times M_2) = E^1(M_1)E^2(M_2)$. E extends to the Borel sets of $\mathbb{R}^2$ and is a projection valued measure. The integral

$$\int_{\mathbb{R}^2} f(\lambda) E(d\lambda)$$

is defined for $f = \chi_M$, M a Borel set in $\mathbb{R}^2$, and by the usual linearity and limiting arguments it extends to measurable functions f on $\mathbb{R}^2$. Note that $g \in \mathscr{D}[\int_{\mathbb{R}^2} f(\lambda) \, dE(d\lambda)]$ iff $\int_{\mathbb{R}^2} |f(\lambda)|^2 \, d[\|E(d\lambda)g\|^2] < \infty$. Check that

$$N = \int_{\mathbb{R}^2} (\lambda_1 + i\lambda_2) E(d\lambda).$$

Note that this construction works for $N = S_1 + iS_2$ where S_1, S_2 are self-adjoint commuting operators; they need not be bounded. Now translate from the spectral measure picture to the multiplication operator picture.}

8. Let $A_1, \ldots, A_n$ generate (C_0) semigroups on $\mathscr{X}$. Then $(\lambda_1 - A_1)^{-1}, \ldots, (\lambda_n - A_n)^{-1}$ all commute, for all $\lambda_j \in \rho(A_j)$ iff $T_1(t_1), \ldots, T_n(t_n)$ all commute for $t_j > 0$, where T_j is the semigroup generated by A_j. Prove this and use it to define the notion of commuting self-adjoint operators.

*9. If $U \in \mathscr{B}(\mathscr{H})$ is unitary, then U is normal and it follows from Exercise 7 above that U is unitarily equivalent to M_q when q takes values in $\{z \in \mathbb{C}: |z| = 1\}$. Give an alternate proof of this based on the Cayley transform. (U is unitary with $I + U$ being injective and having dense range iff S is self-adjoint, where S and U are in one-to-one correspondence via the formulas

$$U = (I + iS)(I - iS)^{-1}, \ S = i(I - U)(I + U)^{-1}.)$$

10. If $S_1, \ldots, S_n$ are commuting self-adjoint operators on $\mathcal{H}$, then there is a measure space (Ω, μ), a unitary operator $U \in \mathcal{B}(\mathcal{H}, L^2(\Omega, \mu))$ and measurable functions $q_1, \ldots, q_n : \Omega \to \mathbb{R}$ such that $S_j = U^{-1} M_{q_j} U$ for $j = 1, \ldots, n$.

11. A self-adjoint operator A is *bounded from below* if for some $c \in \mathbb{R}$, $A + cI$ is nonnegative. Let A be self-adjoint and B symmetric with $\mathscr{D}(B) \supset \mathscr{D}(A)$. Suppose there are numbers $a < 1$ and $b \geq 0$ such that

$$\|Bf\| \leq a \|Af\| + b \|f\|$$

for all $f \in \mathscr{D}(A)$. Then $A + B$ is (self-adjoint and) bounded from below if A is.

12. Establish the first assertion of 6.10. That is, if $F : \mathbb{R} \to \mathbb{C}$ is Borel measurable and $S = S^*$, then $F(S)$ is well-defined (and independent of the spectral representation of S used to define it).

7.   Second-Order Equations

7.1.   Consider the abstract Cauchy problem

$$u''(t) = Au(t) \qquad (t \in \mathbb{R}) \tag{7.1}$$

$$u(0) = f_1, u'(0) = f_2. \tag{7.2}$$

Here A is a linear operator on $\mathscr{X}$, and we seek a unique solution $u \in C^2(\mathbb{R}, \mathscr{X})$ of (7.1), (7.2). Let $U(t) = \begin{pmatrix} u(t) \\ u'(t) \end{pmatrix} \in \mathscr{X} \times \mathscr{X}$. Then the problem (7.1), (7.2) becomes (formally)

$$U'(t) = MU(t) \qquad (t \in \mathbb{R}), U(0) = f \tag{7.3}$$

where $M = \begin{pmatrix} 0 & I \\ A & 0 \end{pmatrix}$, $f = \begin{pmatrix} f_1 \\ f_2 \end{pmatrix}$. Formally, the solution of (7.3) is given by

$$U(t) = \exp\left\{ t \begin{pmatrix} 0 & I \\ A & 0 \end{pmatrix} \right\} f.$$

Consider the special case when $\mathscr{X} = \mathbb{K}$. Let $a \in \mathbb{K}$. Then $\begin{pmatrix} 0 & 1 \\ a & 0 \end{pmatrix}^2 = a \begin{pmatrix} 1 & 0 \\ 0 & 1 \end{pmatrix}$. Hence

$$\exp t \begin{pmatrix} 0 & 1 \\ a & 1 \end{pmatrix} = \sum_{n=0}^{\infty} t^n \begin{pmatrix} 0 & 1 \\ a & 0 \end{pmatrix}^n \Big/ n!$$

$$= \sum_{n \text{ even}} + \sum_{n \text{ odd}}$$

$$= \cosh(t\sqrt{a}) \begin{pmatrix} 1 & 0 \\ 0 & 1 \end{pmatrix} + \frac{1}{\sqrt{a}} \sinh(t\sqrt{a}) \begin{pmatrix} 0 & 1 \\ a & 0 \end{pmatrix}$$

if $a \neq 0$. It is therefore plausible to expect that: $M = \begin{pmatrix} 0 & I \\ A & 0 \end{pmatrix}$ *generates a* (C_0)

group T if A has a square root B which generates a (C_0) group S on $\mathscr{X}$ and if $0 \in \rho(B)$;

$$T(t) = \cosh(tB)\begin{pmatrix} I & 0 \\ 0 & I \end{pmatrix} + B^{-1}\sinh(tB)\begin{pmatrix} 0 & I \\ B^2 & 0 \end{pmatrix} \qquad (7.4)$$

where

$$\cosh(tB) = 2^{-1}[S(t) + S(-t)], \sinh(tB) = 2^{-1}[S(t) - S(-t)].$$

If $a < 0$, then

$$\frac{1}{\sqrt{a}}\sinh(t\sqrt{a}) = -t\frac{\sin t\sqrt{-a}}{t\sqrt{-a}},$$

which makes sense even if $a = 0$, since we can interpret $\left.\dfrac{\sin x}{x}\right|_{x=0}$ to be 1. This

suggests that $0 \in \rho(B)$ is not really crucial.

Before precisely stating the correct version of the above conjecture, we shall require some preliminaries. If B is a closed operator on $\mathscr{X}$, let $[\mathscr{D}(B)]$ be $\mathscr{D}(B)$ equipped with the graph norm: $|f| = \|Bf\| + \|f\|$ [or $\|f\| = (\|Bf\|^2 + \|f\|^2)^{1/2}$ if $\mathscr{X}$ is a Hilbert space]. Then $[\mathscr{D}(B)]$ is a Banach [or Hilbert] space. If $\mathscr{Z}, \mathscr{X}$ are Banach [or Hilbert] spaces, then so is $\mathscr{Y} = \mathscr{Z} \times \mathscr{X}$ if $\mathscr{Y}$ is given the norm

$$\left\|\begin{pmatrix} f_1 \\ f_2 \end{pmatrix}\right\|_{\mathscr{Y}} = \|f_1\|_{\mathscr{Z}} + \|f_2\|_{\mathscr{X}} \left[\text{or } \left\|\begin{pmatrix} f_1 \\ f_2 \end{pmatrix}\right\|_{\mathscr{Y}} = (\|f_1\|_{\mathscr{Z}}^2 + \|f_2\|_{\mathscr{X}}^2)^{1/2}\right].$$ Note that

the two ways of defining $\|\cdot\|_{\mathscr{Y}}$ yield equivalent norms.

7.2.   THEOREM.   *Suppose $0 \in \rho(B)$ and B generates a (C_0) group S on $\mathscr{X}$. Then*

$M = \begin{pmatrix} 0 & I \\ B^2 & 0 \end{pmatrix}$ *with domain $\mathscr{D}(M) = \mathscr{D}(B^2) \times \mathscr{D}(B)$ generates a (C_0) group T on*

$\mathscr{Y} = [\mathscr{D}(B)] \times \mathscr{X}$ *given by (7.4). Moreover, $0 \in \rho(M)$. The Cauchy problem (7.1), (7.2) is well-posed if $A = B^2$, $f_1 \in \mathscr{D}(B^2)$, $f_2 \in \mathscr{D}(B)$. Denoting the solution by u, we have $u: \mathbb{R} \to \mathscr{D}(B^2)$, $u': \mathbb{R} \to \mathscr{D}(B)$, and there are constants K, ω such that*

$$\|Bu(t)\| + \|u(t)\| + \|u'(t)\| \le Ke^{\omega|t|}(\|Bf_1\| + \|f_1\| + \|f_2\|) \qquad (7.5)$$

for all $t \in \mathbb{R}$, $f_1 \in \mathscr{D}(B^2)$, $f_2 \in \mathscr{D}(B)$.

The theorem is false if $\mathscr{Y}$ is replaced by $\mathscr{X} \times \mathscr{X}$; cf. Example 7.3. The norm $\|\cdot\|_{\mathscr{Y}}$ is often called the *energy norm*.

Proof of Theorem 7.2.   The notation is as above; the norm in $[\mathscr{D}(B)]$ is taken to

be $|g_1| = \|Bg_1\| + \|g_1\|$; the norm in $\mathscr{Y}$ is $\|g\|_{\mathscr{Y}} = |g_1| + \|g_2\|$ where $g = \begin{pmatrix} g_1 \\ g_2 \end{pmatrix}$.

Define T by (7.4), i.e. for $g \in \mathscr{Y}$,

$$T(t)g = \begin{pmatrix} \cosh(tB)g_1 + B^{-1}\sinh(tB)g_2 \\ \cosh(tB)g_2 + B \ \sinh(tB)g_1 \end{pmatrix}.$$

Note that $T(t)(\mathscr{Y}) \subset \mathscr{Y}$ by the definition of $\mathscr{Y}$. Choose L, ω such that

$\|S(t)\| \le Le^{\omega|t|}$, $t \in \mathbb{R}$ (cf. I.2.17). Then for all $g \in \mathscr{Y}$, $t \in \mathbb{R}$,

$$\|T(t)g\|_{\mathscr{Y}} = \|\cosh(tB)Bg_1 + \sinh(tB)g_2\|$$
$$+ \|\cosh(tB)g_1 + \sinh(tB)B^{-1}g_2\| + \|\cosh(tB)g_2 + \sinh(tB)Bg_1\|$$
$$\le Le^{\omega|t|}\{\|Bg_1\| + \|g_2\| + \|g_1\| + \|B^{-1}g_2\| + \|g_2\| + \|Bg_1\|\}$$
$$\le (2L + \|B^{-1}\|)\,e^{\omega|t|}\|g\|_{\mathscr{Y}}, \tag{7.6}$$

whence $T: \mathbb{R} \to \mathscr{B}(\mathscr{Y})$ and $\|T(\cdot)\|_{\mathscr{Y}}$ is locally bounded. The group property of T follows easily from the group property of S, and $T(0)$ is the identity.

Let $\mathscr{D}_0 = \mathscr{D}(B^2) \times \mathscr{D}(B) \subset \mathscr{Y}$. $\mathscr{D}_0$ is dense in $\mathscr{Y}$ because $\mathscr{D}(B)$ is dense in $\mathscr{X}$ and $\mathscr{D}(B^2)$ is dense in $[\mathscr{D}(B)]$. To see the last statement, let $f \in \mathscr{D}(B)$ and note that $B[\mathscr{D}(B^2)] = \mathscr{D}(B)$ is dense in $\mathscr{X}$ [since $0 \in \rho(B)$]; choose $f_n \in \mathscr{D}(B^2)$ such that $\|Bf_n - Bf\| \to 0$ as $n \to \infty$. Then

$$|f_n - f| = \|Bf_n - Bf\| + \|f_n - f\|$$
$$\le (1 + \|B^{-1}\|)\|Bf_n - Bf\| \to 0 \text{ as } n \to \infty.$$

Next, let $M = \begin{pmatrix} 0 & I \\ B^2 & 0 \end{pmatrix}$ with $\mathscr{D}(M) = \mathscr{D}_0$ and let $N = \begin{pmatrix} 0 & B^{-2} \\ I & 0 \end{pmatrix}$. For $g \in \mathscr{Y}$,

$$\|Ng\|_{\mathscr{Y}} = \|B^{-1}g_2\| + \|B^{-2}g_2\| + \|g_1\|$$
$$\le (1 + \|B^{-1}\|)^2\|g\|_{\mathscr{Y}},$$

so that $N \in \mathscr{B}(\mathscr{Y})$. Also, clearly

$$NMf = f \text{ for all } f \in \mathscr{D}(M),$$
$$MNf = f \text{ for all } f \in \mathscr{Y}.$$

It follows that $0 \in \rho(M)$ and $M^{-1} = N$. Furthermore, if G denotes the generator of T, then a simple straightforward calculation shows that $G \supset M$. If $\mathscr{D}(G) \ne \mathscr{D}(M)$, then there is an $h \in \mathscr{D}(G)$ such that $h \ne 0$ and $Gh = 0$ [since $0 \in \rho(M)$]. Then $T(t)h \equiv h$ independent of t, i.e.

$$\cosh(tB)h_1 + B^{-1}\sinh(tB)h_2 = h_1, \tag{7.7}$$

$$\sinh(tB)Bh_1 + \cosh(tB)h_2 = h_2. \tag{7.8}$$

Differentiating (7.7) yields

$$\sinh(tB)Bh_1 + \cosh(tB)h_2 = 0. \tag{7.9}$$

Equations (7.8) and (7.9) imply $h_2 = 0$. Therefore by (7.9), $B\sinh(tB)h_1 = 0$; hence $\sinh(tB)h_1 = 0$. Differentiation at $t = 0$ yields $Bh_1 = 0$, and so $h_1 = 0$. Therefore $h = 0$; this contradiction shows that $G = M$.

The well-posedness of (7.3) with $f \in \mathscr{D}(M)$ implies the well-posedness of (7.1), (7.2) if $f_1 \in \mathscr{D}(B^2)$, $f_2 \in \mathscr{D}(B)$. The last assertion of Theorem 7.2 [i.e. (7.5)] follows from the fact that $T(t)$ leaves $\mathscr{D}(M)$ invariant and from (7.6). This completes the proof. ∎

7.3. EXAMPLE. Theorem 7.2 is false if $\mathcal{Y} = [\mathcal{D}(B)] \times \mathcal{X}$ is replaced by $\mathcal{X} \times \mathcal{X}$. The reason is that the operator $T(t)$ defined by (7.4) is not necessarily bounded on $\mathcal{X} \times \mathcal{X}$. For a specific instance in which this happens, let $\mathcal{X} = L^2(\mathbb{R})$ and $B = iM_q$ where $q(x) = 1 + x^2$. The verification that $T(t) \notin \mathcal{B}(\mathcal{X} \times \mathcal{X})$ for each $t \neq 0$ is left as an exercise.

7.4. THEOREM.   *Let C be self-adjoint on $\mathcal{H}$. Then the Cauchy problem*

$$u''(t) + C^2 u(t) = 0 \qquad (t \in \mathbb{R}),$$

$$u(0) = f_1 \in \mathcal{D}(C^2),\, u'(0) = f_2 \in \mathcal{D}(C)$$

is well-posed.

Proof. This is an immediate consequence of Stone's theorem I.4.7 and Theorem 7.2 with $B = iC$ if, in addition, $0 \in \rho(C)$. The hypothesis $0 \in \rho(B)$ in Theorem 7.2 was used in order to make sense out of $B^{-1} \sinh(tB)$. But if iB is self-adjoint, then $B^{-1} \sinh(tB) = F(iB)$ where $F(x) = (\sin tx)/x$ if $x \neq 0$, $F(0) = t$; F is bounded and so $F(iB) \in \mathcal{B}(\mathcal{H})$ [even if $0 \notin \rho(B)$]. Therefore, the theorem is valid whether or not $0 \in \rho(B)$.

The proof of the above theorem need not be based on Theorem 7.2; an alternative proof can be based directly on the operational calculus associated with the spectral theorem (cf. 6.10). Formula (7.4) takes the form

$$T(t) = \cos(tC)\begin{pmatrix} I & 0 \\ 0 & I \end{pmatrix} + C^{-1}\sin(tC)\begin{pmatrix} 0 & I \\ -C^2 & 0 \end{pmatrix}$$

where $C = -iB$ and $C^{-1}\sin(tC) = F(C)$ as above. ∎

7.5. REMARK.   In Theorem 7.4, C^2 can be replaced by any nonnegative self-adjoint operator D. One can let C be any self-adjoint square root of D; in particular, D has a nonnegative self-adjoint square root by the operational calculus associated with the spectral theorem (cf. 6.11).

7.6. REMARK.   Theorem 7.4 takes place in the context of the space $\mathcal{Y} = [\mathcal{D}(B)] \times \mathcal{X}$ with the norm, say

$$\left\| \begin{pmatrix} f \\ g \end{pmatrix} \right\|_{\mathcal{Y}} = (\|Bf\|_{\mathcal{X}}^2 + \|f\|_{\mathcal{X}}^2 + \|g\|_{\mathcal{X}}^2)^{1/2}.$$

The $\|f\|_{\mathcal{X}}$ term can be eliminated as follows. Let $\mathcal{Z}$ be the completion of the inner product space $[\mathcal{D}(B)/\mathcal{N}(B)] \times \mathcal{X}$ in the norm $\left\| \begin{pmatrix} f \\ g \end{pmatrix} \right\|_{\mathcal{Z}} = (\|Bf\|_{\mathcal{X}}^2 + \|g\|_{\mathcal{X}}^2)^{1/2}$; here $\mathcal{N}(B)$ is the (closed) null space (or kernel) of B and $\|Bf\|_{\mathcal{X}}$ stands for the norm of $Bf + \mathcal{N}(B)$ in $\mathcal{X}/\mathcal{N}(B)$. Note that $(\mathcal{D}(B)/\mathcal{N}(B))$ is complete iff 0 is in the resolvent set of B, viewed as an operator on $\mathcal{X}/\mathcal{N}(B)$; in this case $\mathcal{Z} = \mathcal{Y}$. Clearly $\mathcal{Z} \supset \mathcal{Y}$. $\mathcal{Z} \neq \mathcal{Y}$ can actually happen; see Exercise 7.14.2.

7.7.   Suppose we are in the situation of Theorem 7.4 (or 7.6). If $B = iC$ is skew-adjoint, so that the group S generated by B is a (C_0) unitary group, is it possible to choose the norm in $\mathcal{Y}$ (or in $\mathcal{Z}$) so that T, the group generated by

$M = \begin{pmatrix} 0 & I \\ -C^2 & 0 \end{pmatrix}$, becomes a (C_0) unitary group, i.e. so that M becomes skew-adjoint? The affirmative answer is given in the following lemma.

LEMMA.   *Let B be skew-adjoint on $\mathcal{H}$. Then so is the closure of $M = \begin{pmatrix} 0 & I \\ B^2 & 0 \end{pmatrix}$ with domain $\mathcal{D}(M) = [\mathcal{D}(B^2)/\mathcal{N}(B)] \times \mathcal{D}(B)$ on the completion $\mathcal{Z}$ of $\mathcal{W} = (\mathcal{D}(B)) \times \mathcal{X}$ if $\mathcal{Z}$ is given the norm*

$$\left\| \begin{pmatrix} g_1 \\ g_2 \end{pmatrix} \right\|_{\mathcal{Z}} = \{\|Bg_1\|^2 + \|g_2\|^2\}^{1/2}.$$

Proof.   We identify $g_1 \in \mathcal{D}(B)$ with the coset $g_1 + \mathcal{N}(B)$; this should cause no confusion. For all $g = \begin{pmatrix} g_1 \\ g_2 \end{pmatrix} \in \mathcal{W}, t \in \mathbb{R}$, using (7.4),

$$\|T(t)g\|_{\mathcal{Z}}^2 = \|\cosh(tB)Bg_1 + \sinh(tB)g_2\|^2$$
$$+ \|\cosh(tB)g_2 + \sinh(tB)Bg_1\|^2$$
$$= 4^{-1}\{\|S(t)(Bg_1 + g_2) + S(-t)(Bg_1 - g_2)\|^2$$
$$+ \|S(t)(Bg_1 + g_2) - S(-t)(Bg_1 - g_2)\|^2\}$$
$$= 2^{-1}\{\|S(t)(Bg_1 + g_2)\|^2 + \|S(-t)(Bg_1 - g_2)\|^2\}$$

by the parallelogram law

$$= 2^{-1}\{\|Bg_1 + g_2\|^2 + \|Bg_1 - g_2\|^2\}$$

since $S(t), S(-t)$ are unitary

$$= \|Bg_1\|^2 + \|g_2\|^2 = \|g\|_{\mathcal{Z}}^2$$

by the parallelogram law again. T is thus a (C_0) isometric group on $\mathcal{W}$ and therefore a (C_0) unitary group on $\mathcal{Z}$. An appeal to Stone's theorem I.4.7 completes the proof.   ∎

7.8.   PERTURBATION THEOREM.   *Let the hypotheses of Theorem 7.2 (or Theorem 7.4 with $B = iC$) hold. Let E be a closed operator on $\mathcal{X}$ such that $\mathcal{D}(E) \supset \mathcal{D}(B)$. Then the Cauchy problem*

$$u''(t) = (B^2 + E)u(t)   (t \in \mathbb{R}), \tag{7.10}$$
$$u(0) = f_1 \in \mathcal{D}(B^2), u'(0) = f_2 \in \mathcal{D}(B)$$

is well-posed.

Proof. $E \in \mathcal{B}([\mathcal{D}(B)], \mathcal{X})$ by the closed graph theorem. Hence

$$\|Eg_1\| \leq K(\|Bg_1\| + \|g_1\|)$$

for some constant K and all $g_1 \in \mathcal{D}(B)$. $M = \begin{pmatrix} 0 & I \\ B^2 & 0 \end{pmatrix}$ generates a (C_0) group on

$\mathcal{Y}$. Let $N = \begin{pmatrix} 0 & 0 \\ E & 0 \end{pmatrix}$. Then

$$\|Ng\|_{\mathcal{Y}} = \left\|\begin{pmatrix} 0 \\ Eg_1 \end{pmatrix}\right\|_{\mathcal{Y}} = \|Eg_1\|$$
$$\leq K(\|Bg_1\| + \|g_2\|) \leq K\|g\|_{\mathcal{Y}}$$

for all $g \in \mathcal{Y}$. Hence $N \in \mathcal{B}(\mathcal{Y})$. Therefore $M + N = \begin{pmatrix} 0 & I \\ B^2 + E & 0 \end{pmatrix}$ generates a (C_0) group on $\mathcal{Y}$ by Theorem I.6.5. The desired result follows.   ∎

7.9.   COROLLARY. *If B generates a (C_0) group on $\mathcal{X}$, then $M = \begin{pmatrix} 0 & I \\ B^2 & 0 \end{pmatrix}$ with domain $\mathcal{D}(M) = \mathcal{D}(B^2) \times \mathcal{D}(B)$ generates a (C_0) group on $\mathcal{Y} = [\mathcal{D}(B)] \times \mathcal{X}$.*

Proof. $B - \lambda I$ satisfies the hypotheses of Theorem 7.2 for any $\lambda \in \rho(B)$ $(\neq \emptyset)$. Theorem 7.8 and the observation that

$$B^2 = (B - \lambda I)^2 + (2\lambda B - \lambda^2 I)$$

yields the desired conclusion.   ∎

7.10.   THEOREM.   *Let $\mathcal{H} = L^2(\mathbb{R}^n)$. Let P be a strictly elliptic real polynomial of degree $2m(m = 1,2,\ldots)$. Let $Q(\cdot,D) = \sum_{|\alpha| \leq m} M_{b_\alpha} D^\alpha$ be a (variable coefficient) partial differential operator of order at most m, where each $b_\alpha \in L^\infty(\mathbb{R}^n)$. Then the Cauchy problem*

$$u''(t) = P(D)u + Q(\cdot,D)u \qquad (7.11)$$
$$u(0) = f_1 \in H^{2m}(\mathbb{R}^n), u'(0) = f_2 \in H^m(\mathbb{R}^n)$$

is well-posed.

Of course, (7.11) is the Banach space version of the partial differential equation

$$\frac{\partial^2 u}{\partial t^2} = P(D)u + Q(x,D)u.$$

A special case of this is the second-order *hyperbolic equation*

$$\frac{\partial^2 u}{\partial t^2} = \Delta u + \sum_{j=1}^{n} b_j(x) \frac{\partial u}{\partial x_j} + c(x)u$$

where $\Delta = \sum_{j=1}^{n} \partial^2/\partial x_j 2$ is the Laplacian. This is the *n*-dimensional *wave equation* when $b_1 = \ldots = b_n = c \equiv 0$.

Proof of Theorem 7.10.   Let $R = \sqrt{-P}$, so that $(iR)^2 = P$. iM_R is skew-adjoint; therefore so is $B = i \mathcal{F}^{-1} M_R \mathcal{F}$. Also, $B^2 = P(D)$,

$$\mathcal{D}(B) = \left\{ f \in \mathcal{H} : \int_{\mathbb{R}^n} (|\xi|^m |\hat{f}(\xi)|)^2 \, d\xi < \infty \right\} = H^m(\mathbb{R}^n).$$

Set $E = Q(\cdot,D)$. The theorem now follows immediately from Theorem 7.8.   ∎

7.11. DEFINITIONS.   Let C be a self-adjoint operator on a Hilbert space $\mathscr{H}$. Let $u:\mathbb{R} \to \mathscr{H}$ be a solution of $u''(t) + C^2 u(t) = 0$, $t \in \mathbb{R}$. We call

$$K(t) = \|u'(t)\|^2 \quad \text{and} \quad P(t) = \|Cu(t)\|^2$$

the *kinetic* and *potential energies* of the solution at time t. By Lemma 7.7, the *total energy* E is conserved, i.e. $E(t) = K(t) + P(t) = K(0) + P(0) = E(0)$ for all $t \in \mathbb{R}$. E thus depends only on the initial data. The equation

$$u'' + C^2 u = 0$$

is said to admit *(asymptotic) equipartition of energy* if

$$\lim_{t \to \pm \infty} K(t) = \lim_{t \to \pm \infty} P(t) = \frac{1}{2} E(0)$$

for all choices of initial data [in $\mathscr{D}(C^2) \times \mathscr{D}(C)$].

7.12. THEOREM.   *Let* $C = \int_{-\infty}^{\infty} \lambda \, dE_\lambda$ *be a self-adjoint operator on* $\mathscr{H}$. *The equation* $u'' + C^2 u = 0$ *admits equipartition of energy iff for all* $h \in \mathscr{H}$,

$$\lim_{t \to \pm \infty} \int_{-\infty}^{\infty} e^{it\lambda} \, d(\|E_\lambda h\|^2) = 0.$$

Proof.   First, if $u'' + C^2 u = 0$ admits equipartition of energy, then C is injective, for, if $Cf = 0$, then taking the initial data to be $u(0) = f$, $u'(0) = f$, the solution becomes $u(t) = f + tf$, whence $K(t) = \|f\|^2$, $P(t) = 0$. Equipartition of energy cannot hold if $f \neq 0$. So we may assume that C is injective.

Let $\mathscr{Y} = \mathscr{D}(C) \times \mathscr{H}$ with inner product

$$\left\langle \begin{pmatrix} f_1 \\ f_2 \end{pmatrix}, \begin{pmatrix} g_1 \\ g_2 \end{pmatrix} \right\rangle_{\mathscr{Y}} = \langle Cf_1, Cg_1 \rangle + \langle f_2, g_2 \rangle.$$

According to Lemma 7.7, $u'' + C^2 u = 0$ is governed by a (C_0) unitary group on the completion of $\mathscr{Y}$. Formula (7.4) gives the solution of the Cauchy problem (7.1), (7.2) with $A = -C^2$, $B = iC$. For initial data take $f_1 \in \mathscr{D}(C^2)$, $f_2 \in \mathscr{D}(C)$. Let $t \in \mathbb{R}$ and let $V = \exp(-2tB)$. Then by (7.4),

$$4\|u'(t)\|^2 = \|\exp(tB)\{(I + V)f_2 - (I - V)Bf_1\}\|^2$$
$$= \|f_2 + Bf_1 + V(f_2 - Bf_1)\|^2$$
$$= \|f_2 + Bf_1\|^2 + \|V(f_2 - Bf_1)\|^2$$
$$\quad + 2\,\mathrm{Re}\langle f_2 + Bf_1, V(f_2 - Bf_1)\rangle$$
$$= 2(\|f_2\|^2 + \|Bf_1\|^2) + 2\mathrm{Re}\langle f_2 + Bf_1, V(f_2 - Bf_1)\rangle,$$

by the parallelogram law and the unitarity of V. It follows easily that equipartition of energy holds iff

$$\mathrm{Re}\langle h_1, Vh_2 \rangle \to 0$$

as $t \to \pm \infty$ for all $h_1, h_2 \in \mathscr{D}(C)$, hence for all $h_1, h_2 \in \mathscr{H}$. Polarization shows that this is equivalent to

$$\lim_{t \to \pm\infty} \text{Re}\langle e^{itC}h,h\rangle = 0$$

for all $h \in \mathcal{H}$. Since

$$\langle e^{itC}h,h\rangle = \int_{-\infty}^{\infty} e^{it\lambda}\, d(\|E_\lambda h\|^2),$$

the result follows.   ∎

7.13.   THEOREM.   *Let C be self-adjoint on $\mathcal{H}$. Let $K(t)$, $P(t)$ denote the kinetic and potential energy at time t of a solution of $u'' + C^2u = 0$. Then*

$$\lim_{T \to \pm\infty} \frac{1}{T}\int_0^T K(t)\, dt = \lim_{T \to \pm\infty} \frac{1}{T}\int_0^T P(t)\, dt = \frac{1}{2} E(0)$$

for all choices of initial data iff C is injective.

The proof is left as an exercise.

7.14.   EXERCISES

1. Verify that $T(t) \notin \mathcal{B}(\mathcal{X} \times \mathcal{X})$ for each $t \neq 0$ in Example 7.3.
2. Let A be the Laplacian Δ acting as $\mathcal{X} = L^2(\mathbb{R}^n)$. Show that $[\mathcal{D}(B)] = H^1(\mathbb{R}^n)$ where $B^2 = A$. Let $\mathcal{H}$ be the completion of $\mathcal{D}(B)$ in the norm $|f| = \|Bf\|$. Show that $\mathcal{H} \supsetneq H^1(\mathbb{R}^n)$. (Hints: When $n = 1$, $\mathcal{H} = \{f: \mathbb{R} \to \mathbb{C}: f$ absolutely continuous, $f' \in \mathcal{X}\}$. When $n \geq 1$, $\mathcal{H} = \{f: \mathbb{R} \to \mathbb{C}:$ the distributional derivatives of f satisfy $\partial f/\partial x_j \in \mathcal{X}, j = 1, \dots, n\}$. For $n = 1$ and $f(x) = x^\alpha$ for $x > 1$, one can choose α so that $f' \in L^2([1,\infty[)$ but $f \notin L^2([1,\infty[).)$
3. Give an estimate analogous to (7.5) for solutions of (7.10).
4. Let $A = B^2 + E$ where B generates a (C_0) group on $\mathcal{X}$, $0 \in \rho(B)$, E is closed, and $\mathcal{D}(E) \supset \mathcal{D}(B)$. Using the results of this section and of Section 2, find a sufficient condition in order that the nonlinear Cauchy problem

$$u''(t) = Au(t) + g[t, u(t), u'(t)]$$

$$u(0) = f_1 \in \mathcal{D}(A), u'(0) = f_2 \in \mathcal{D}(B)$$

   has a unique mild solution.
5. Use the idea of Theorem 7.8 to obtain an existence and uniqueness theorem for the Cauchy problem

$$u''(t) + Du'(t) = (B^2 + E)u(t) \qquad (t \in \mathbb{R}),$$

$$u(0) = f_1 \in \mathcal{D}(B^2), u'(0) = f_2 \in \mathcal{D}(B),$$

   where $\mathcal{D} \in \mathcal{B}(\mathcal{X})$. What if D is unbounded and $t \in \mathbb{R}$ is replaced by $t \in \mathbb{R}^+$? Note that if $D = d/dx$, then $Du'(t)$ becomes $\dfrac{\partial^2 u}{\partial t \partial x}(t, \cdot)$.

*6. Prove Theorem 7.13.
*7. (i) Let $C^2 = -\Delta$ on $L^2(\mathbb{R}^n)$. Show that the wave equation $u'' + C^2u = 0$ admits equipartition of energy.
   (ii) Let C^2 be as above and let $u'' + C^2u = 0$, $u(0) = f_1$, $u'(0) = f_2$. If f_1, f_2 are smooth functions having compact support, does there exist a $T = T(f_1, f_2)$ such that $K(t) = P(t) = E(0)/2$ for $|t| \geq T$? The answer is yes for all such f_1, f_2 iff n is odd.

*8. Let S be a self-adjoint operator on a Hilbert space $\mathcal{H}$. For $w \colon \mathbb{R}^+ \to \mathcal{H}$ let

$$h(w) = \limsup_{t \to \infty} t^{-1} \log \|w(t)\|.$$

Let $\beta > 0$ and suppose $S \le \beta^2 I$ but $S \nleq (\beta^2 - \epsilon)I$ for each $\epsilon > 0$; i.e. $\beta^2 = \sup \sigma(S)$. Then all solutions of $u'' = Su$ satisfy

$$h(u) \le \beta, h(u') \le \beta;$$

and there exists a solution u which satisfies

$$h(u) = \beta, h(u') = \beta.$$

8. Cosine Functions

Cosine functions bear the same relationship to the (second-order) Cauchy problem for $u'' = Au$ as do (C_0) semigroups to the (first-order) Cauchy problem for $u' = Au$. This section consists of discussion only. No proofs are given. References are given in the last section of this chapter.

8.1. DEFINITIONS. A (strongly continuous) *cosine function* is a family of operators $C = \{C(t) \colon t \in \mathbb{R}\} \subset \mathcal{B}(\mathcal{X})$ satisfying

(i) $C(t + s) + C(t - s) = 2C(t)C(s)$ for all $t,s \in \mathbb{R}$,
(ii) $C(0) = I$,
(iii) $C(\cdot)f \in C(\mathbb{R},\mathcal{X})$ for each $f \in \mathcal{X}$.

The *(infinitesimal) generator* A of a cosine function C is $A = C''(0)$. The domain of A is the set of $f \in \mathcal{X}$ for which $C(t)f$ is twice differentiable at $t = 0$; equivalently,

$$\mathcal{D}(A) = \{f \in \mathcal{X} \colon C(\cdot)f \in C^2(\mathbb{R},\mathcal{X})\}.$$

Let B be a linear operator on $\mathcal{D}(B) \subset \mathcal{X}$ to $\mathcal{X}$. The Cauchy problem for $u'' = Bu$ is said to be *well-posed of type* ω on $\mathbb{R}^+$ [or $\mathbb{R}$], where $\omega \in \mathbb{R}$, if the following three conditions are satisfied.

(a)   There is a dense subset $D \subset \mathcal{D}(B)$ such that for each $f,g \in D$, there is a unique solution u on $\mathbb{R}^+$ [or $\mathbb{R}$] of $u'' = Bu$, $u(0) = f$, $u'(0) = g$.
(b)   For each solution u as in (a),

$$\limsup_{t \to \infty} \frac{1}{t} \log\|u(t)\| \le \omega.$$

(c)   If $\{u_n\}$ is a sequence of solutions of $u'' = Bu$ with $\lim_{n \to \infty} u_n(0) = \lim_{n \to \infty} u_n'(0) = 0$, then $\lim_{n \to \infty} U_n(t) = 0$, uniformly for t in bounded subsets of $\mathbb{R}^+$ [or $\mathbb{R}$].

8.2. WELL-POSEDNESS THEOREM. *The Cauchy problem for $u'' = Au$ is well-posed of type ω on $\mathbb{R}^+$ for some ω iff it is well-posed of type ω on $\mathbb{R}$ for some ω iff A generates a cosine function iff $\rho(A) \neq \emptyset$ and for each $f \in \mathcal{D}(A)$, the problem $u'' = Au(t \ge 0)$, $u(0) = f$, $u'(0) = 0$ has a unique solution u in $C^2(\mathbb{R}^+,\mathcal{X})$.*

Compare this with Exercise 1.5.4 and Theorem 1.2.

8.3. GENERATION THEOREM. *A generates a cosine function C iff A is closed, densely defined, and there are constants $M \geq 1$, $\omega \geq 0$ such that for $\lambda > \omega$, $\lambda^2 \in \rho(A)$ and*

$$\left\| \frac{d^m}{d\lambda^m} \left[\lambda(\lambda^2 - A)^{-1} \right] \right\| \leq Mm!\,(\lambda - \omega)^{-m-1}$$

for all $m \in \mathbb{N}_0$. In this case,

$$\|C(t)\| \leq Me^{\omega|t|}$$

for all $t \in \mathbb{R}$, and

$$\lambda(\lambda^2 - A)^{-1}f = \int_0^\infty e^{-\lambda t}\, C(t)f \, dt$$

for all $\lambda > \omega$ and $f \in \mathscr{X}$.

Compare this with Chapter I, Theorem 2.13.

8.4. Let A generate a cosine function C. Let

$$\mathscr{D}_{1/2}(A) = \{ f \in \mathscr{X} : C(\cdot)f \in C^1(\mathbb{R},\mathscr{X}) \}.$$

Let

$$S(t)f = \int_0^t C(s)f \, ds$$

for $f \in \mathscr{X}$, $t \in \mathbb{R}$. $S = \{ S(t): t \in \mathbb{R} \}$ is the *sine function* associated with C.

THEOREM. *Let A generate a cosine function C. Let $h \in C^1(\mathbb{R},\mathscr{X})$. The unique solution of*

$$u'' = Au + h \ (t \in \mathbb{R}),\ u(0) = f \in \mathscr{D}(A),\ u'(0) = g \in \mathscr{D}_{1/2}(A)$$

is given by

$$u(t) = C(t)f + S(t)g + \int_0^t S(t-s)h(s) \, ds$$

for all $t \in \mathbb{R}$.

Compare this with Theorem 1.3.

8.5. PERTURBATION THEOREM. *For $n \in \mathbb{N}_0$ let A generate a cosine function and let $B \in \mathscr{B}(\mathscr{X})$. Then $A + B$ generates a cosine function.*

Compare this with Chapter I, Theorem 6.5.

8.6. APPROXIMATION THEOREM. *For $n \in \mathbb{N}_0$ let A_n generate a cosine function C_n satisfying the stability condition: There are constants $M \geq 1$, $\omega \geq 0$ such that $\|C_n(t)\| \leq Me^{\omega|t|}$ for all $t \in \mathbb{R}$ and all $n \in \mathbb{N}$. Then*

$$\lim_{n \to \infty} (\lambda^2 - A_n)^{-1}f = (\lambda^2 - A_0)^{-1}f$$

holds for all $f \in \mathscr{X}$ and all $\lambda > \omega$ iff

$$\lim_{n \to \infty} C_n(t)f = C_0(t)f$$

holds for all $f \in \mathscr{X}$ and all $t \in \mathbb{R}$,

Compare this with Chapter I Theorem 7.3.

8.7. We now turn to the connections between cosine functions and (C_0) semigroups.

THEOREM. *Let A generate a cosine function C. Then A generate a (C_0) semigroup T given by*

$$T(t)f = \frac{1}{\sqrt{\pi t}} \int_0^\infty e^{-y^2/4t} C(y)f\, dy$$

When $\mathscr{H}$ is complex, T (defined by the above formula) is an analytic semigroup in the right half plane; i.e. for some $\alpha > 0$, $A - \alpha \in \mathscr{GA}_b(\pi/2)$. When $\mathscr{H}$ is real, $T(t)(\mathscr{X}) \subset \mathscr{D}(A)$ holds for $t > 0$.

   The converse is false. That is, there are operators in $\mathscr{GA}_b(\pi/2)$ which do not generate cosine functions.

8.8. Let A generate a cosine function C. Then so does $A - \alpha I$. By choosing α appropriately, one can construct an operator B such that $B^2 = A - \alpha I$ and $0 \in \rho(B)$. Moreover, B commutes with all bounded operators commuting with A. We can assume without loss of generality that $\alpha = 0$.

THEOREM. *Let A generate a cosine function C_0 in $\mathscr{X}$. Let B satisfy $B^2 = A$, $0 \in \rho(B)$, and B commutes with every bounded operator which commutes with A. Then the following four conditions are equivalent.*

   (I)   *If S is the sine function associated with C, then $S(t)(\mathscr{X}) \subset \mathscr{D}(B)$ and $BS(t) \in \mathscr{B}(\mathscr{X})$ for each $t \in \mathbb{R}$, and $BS(\cdot)f \in C(\mathbb{R},\mathscr{X})$ for each $f \in \mathscr{X}$.*
   (II) *B generates a (C_0) group on $\mathscr{X}$.*
   (III) $\begin{pmatrix} 0 & B \\ B & 0 \end{pmatrix}$, *with domain $\mathscr{D}(B) \times \mathscr{D}(B)$, generates a (C_0) group in $\mathscr{X} \times \mathscr{X}$.*
   (IV) $\begin{pmatrix} 0 & I \\ A & 0 \end{pmatrix}$, *with domain $\mathscr{D}(A) \times \mathscr{D}(B)$, generates a (C_0) group in*

$$\mathscr{Y} = [\mathscr{D}(B)] \times \mathscr{X} \left(\text{with norm} \left\| \begin{pmatrix} f \\ g \end{pmatrix} \right\|_{\mathscr{Y}} = \|Bf\| + \|f\| + \|g\| \right).$$

8.9. REMARKS. Thus the equation $u'' = Au$ is equivalent to a first-order system (in the sense of Section 7) precisely when the conditions of the above theorem hold. These conditions automatically hold whenever $\mathscr{X}$ is a Hilbert space, or, more generally, when $\mathscr{X} = L^P(\Omega,\Sigma,\mu)$, $1 < p < \infty$. Moreover, when B is as in the first sentence of the above theorem, then $\mathscr{D}(B) = \mathscr{D}_{1/2}(A)$ iff conditions (I)–(IV) hold. But these conditions do not always hold.

8.10. EXAMPLE. LET $\mathscr{X}$ be the odd, 2π-periodic real functions in $BUC(\mathbb{R})$. Let $A = d^2/dx^2$ with $\mathscr{D}(A) = \{f \in \mathscr{X}: f'' \in \mathscr{X}\}$. A generates a cosine function C given by

$$C(t)f(x) = \frac{1}{2}[f(x + t) + f(x - t)].$$

Moreover 0 is in the resolvent set of A (and hence of B also, if $B^2 = A$), and

$$A^{-1}f(x) = \int_0^x (x - y)f(y)\,dy$$

for $f \in \mathscr{X}$, $x \in \mathbb{R}$. However, conditions (I)–(IV) of Theorem 8.8 do not hold.

8.11. REMARK. Let B generate a (C_0) group T. Then $A_a = aI + B^2$ generates a cosine function C_a. We have the d'Alembert formula

$$C_0(t) = \frac{1}{2}[T(t) + T(-t)],$$

while for $a > 0$ we have the representation formula

$$C_a(t)f = C_0(t)f + at \int_0^t (t^2 - s^2)^{-1/2} I_1[a(t^2 - s^2)^{1/2}] C_0(s)f\,ds$$

for $f \in \mathscr{X}$ and $t \in \mathbb{R}$, where $C_0(t)$ is as above and I_1 is the modified Bessel function of order 1.

8.12. REMARK. Let A generate a uniformly bounded cosine function C on a complex Hilbert space $\mathscr{H}$. Then there is a $U \in \mathscr{B}(\mathscr{H})$ with $U^{-1} \in \mathscr{B}(\mathscr{H})$ and there is a self-adjoint operator L on $\mathscr{H}$ such that

$$C(t) = U \cos(tL)U^{-1}$$

holds for each $t \in \mathbb{R}$. Compare this with Chapter I, Remark 2.19(3).

9. Symmetric Hyperbolic Systems

9.1. We shall consider the Cauchy problem for the *symmetric hyperbolic system*

$$E(x)\frac{\partial u}{\partial t}(t,x) = \sum_{j=1}^n A^j \frac{\partial u}{\partial x_j}(t,x) \qquad u(0,x) = f(x). \tag{9.1}$$

Here $t \in \mathbb{R}$, $x \in \mathbb{R}^n$, $u: \mathbb{R} \times \mathbb{R}^n \to \mathbb{R}^m$ so that $u(t,x)$ is an $m \times 1$ (complex) matrix (i.e. a column vector), $A^1,\ldots,A^n$ are real symmetric (constant) $m \times m$ matrices, and $E(x)$ is a real symmetric positive definite $m \times m$ matrix for each $x \in \mathbb{R}^n$.

If b is a matrix, then b^* will denote the conjugate transpose of b; b^* also equals the adjoint of b when we view b as a linear operator from one finite dimensional (complex) Hilbert space to another in the usual way.

9.2.   EXAMPLE.   The equation for acoustic waves in an inhomogeneous fluid at rest is

$$\frac{1}{c^2(x)}\frac{\partial^2 p}{\partial t^2} = \rho(x)\nabla \cdot \left(\frac{1}{\rho(x)}\nabla p\right) \qquad (t,x) \in \mathbb{R} \times \mathbb{R}^3, \qquad (9.2)$$

$$p(0,x) = g(x), \frac{\partial p}{\partial t}(0,x) = h(x). \qquad (9.3)$$

Here $p(t,x)$ represents the difference between the instantaneous pressure and the equilibrium pressure at the point x at time t, $\rho(x)$ is the equilibrium density at x, and $c(x)$ is the speed of sound at x. Equations (9.2), (9.3) can be put into the form (9.1) if we set

$$u = (\rho^{-1}\partial p/\partial x_1, \rho^{-1}\partial p/\partial x_2, \rho^{-1}\partial p/\partial x_3, \partial p/\partial t)^*,$$

$$E = \begin{pmatrix} \rho & 0 & 0 & 0 \\ 0 & \rho & 0 & 0 \\ 0 & 0 & \rho & 0 \\ 0 & 0 & 0 & (\rho c^2)^{-1} \end{pmatrix}$$

$$\sum_{j=1}^{3} A^j \partial/\partial x_j = \begin{pmatrix} 0 & 0 & 0 & \partial/\partial x_1 \\ 0 & 0 & 0 & \partial/\partial x_2 \\ 0 & 0 & 0 & \partial/\partial x_3 \\ \partial/\partial x_1 & \partial/\partial x_2 & \partial/\partial x_3 & 0 \end{pmatrix},$$

$$f = (\rho^{-1}\partial g/\partial x_1, \rho^{-1}\partial g/\partial x_2, \rho^{-1}\partial g/\partial x_3, h)^*.$$

Equation (9.2) becomes the three-dimensional wave equation when $\rho \equiv c \equiv 1$. For more examples (including Maxwell's equations, the equations of elasticity, etc.) see Wilcox [1].

9.3.   For simplicity we shall assume that $E(x) \equiv E$ does not depend on x; for the more general case see Wilcox [1]. Let $\mathscr{H}$ be the Hilbert space $\mathscr{H} = [L^2(\mathbb{R}^n)]^m = L^2(\mathbb{R}^n,\mathbb{C}^m) = \{f = (f_1,\ldots,f_m)^* : f_j \in L^2(\mathbb{R}^n), \quad j = 1,\ldots,m\}$. The inner product and norm in $\mathscr{H}$ will be taken to be

$$\langle f,g \rangle = \int_{\mathbb{R}^n} g(x)^* E f(x)\, dx,$$

$$\|f\| = \left\{\int_{\mathbb{R}^n} f(x)^* E f(x)\, dx\right\}^{1/2}.$$

If $E = (e_{jk})$, then we have

$$\langle f,g \rangle = \int_{\mathbb{R}^n} \sum_{j,k=1}^{m} e_{jk} f_j(x)\overline{g_k(x)}\, dx.$$

9.4.   THEOREM. $G = E^{-1}\sum_{j=1}^{n} A_j \partial/\partial x_j$ (with $\mathscr{D}(G)$ suitably defined) is a skew-adjoint operator on $\mathscr{H}$; hence the Cauchy problem for

$$E\, \partial u/\partial t = \sum_{j=1}^{n} A^j\, \partial u/\partial x_j$$

is well-posed and is governed by a (C_0) unitary group on $\mathscr{H}$.

Proof. First define $\tilde{\mathscr{F}} : \mathscr{H} \to \mathscr{H}$ by $(\tilde{\mathscr{F}}f)_j = \mathscr{F}f_j = \hat{f}_j, j = 1, \ldots, m$. We shall also write $\hat{f}$ for $\tilde{\mathscr{F}}f$. $\tilde{\mathscr{F}}$ commutes with multiplication by E (since E is a constant matrix). It follows that

$$\langle f,g \rangle = \int_{\mathbb{R}^n} g^*Ef \, dx = \int_{\mathbb{R}^n} \hat{g}^*(Ef)^\wedge \, dx \qquad \text{by Remark 3.12}$$

$$= \int_{\mathbb{R}^n} \hat{g}^*E\hat{f} \, dx = \langle \hat{f},\hat{g} \rangle,$$

and so $\tilde{\mathscr{F}} \in \mathscr{B}(\mathscr{H})$ is unitary. (The surjectivity of $\tilde{\mathscr{F}}$ follows from 3.12 and the definition of $\mathscr{H}$.) Let $P_j(x) = P_j(x_1, \ldots, x_n) = x_j$ and let

$$L = E^{-1} \sum_{j=1}^{n} A^j M_{P_j},$$

$$\mathscr{D}(L) = \left\{ f \in \mathscr{H} : \int_{\mathbb{R}^n} \left| \sum_{j=1}^{n} A^j x_j f(x) \right|^2 dx < \infty \right\}.$$

$[\mathscr{S}(\mathbb{R}^n)]^m \subset \mathscr{D}(L)$, hence $\overline{\mathscr{D}(L)} = \mathscr{H}$. L is symmetric since if $f, g \in \mathscr{D}(L)$,

$$\langle Lf,g \rangle = \left\langle \sum_{j=1}^{n} E^{-1}A_j M_{P_j}f,g \right\rangle = \sum_{j=1}^{n} \int_{\mathbb{R}^n} g^*A_j M_{P_j}f \, dx$$

$$= \sum_{j=1}^{n} \int_{\mathbb{R}^n} g^*M_{P_j}A_jf \, dx = \sum_{j=1}^{n} \int_{\mathbb{R}^n} A_j M_{P_j}g^*f \, dx$$

$$= \sum_{j=1}^{n} \int_{\mathbb{R}^n} E^{-1}A_j M_{P_j}g^*Ef \, dx = \langle f,Lg \rangle.$$

To show that L is self-adjoint, let $g \in \mathscr{D}(L^*)$. Then for all $f \in \mathscr{D}(L)$,

$$\langle Lf,g \rangle = \langle f,L^*g \rangle = \int_{\mathbb{R}^n} \sum_{j=1}^{n} g^*A^j M_{P_j}f \, dx$$

$$= \int_{\mathbb{R}^n} E^{-1} \sum_{j=1}^{n} A^j M_{P_j}g^*Ef \, dx;$$

whence

$$\int_{\mathbb{R}^n} \left(L^*g - E^{-1} \sum_{j=1}^{n} A^j M_{P_j}g \right)^* Ef \, dx = 0$$

for all $f \in \mathscr{D}(L)$. It follows that

$$L^*g = E^{-1} \sum_{j=1}^{n} A^j M_{P_j}g \text{ a.e. on } \mathbb{R}^n;$$

hence $g \in \mathscr{D}(L)$ and $L^*g = Lg$.

Therefore $\tilde{\mathscr{F}}^{-1}L\tilde{\mathscr{F}}$ is self-adjoint, being unitarily equivalent to L. But

$$\tilde{\mathscr{F}}^{-1}L\tilde{\mathscr{F}} = \sum_{j=1}^{n} \tilde{\mathscr{F}}^{-1}E^{-1}A^j M_{P_j}\tilde{\mathscr{F}} = \sum_{j=1}^{n} E^{-1}A^j\tilde{\mathscr{F}}^{-1}M_{P_j}\tilde{\mathscr{F}}$$

$$= \sum_{j=1}^{n} E^{-1}(i \, \partial/\partial x_j) = iG \qquad \text{by (5) of 3.6.}$$

Thus G is skew-adjoint, where

$$\mathscr{D}(G) = \left\{ f \in \mathscr{H} : \int_{\mathbb{R}^n} \left| \sum_{j=1}^{2} A^j x_j \hat{f}(x) \right|^2 dx < \infty \right\};$$

the final conclusion of Theorem 9.4 follows from Stone's theorem I.4.7. ∎

9.5. EXERCISES

1. Formulate and prove a version of Theorem 9.4 in which E depends on x.
2. Show that the wave equation

$$\frac{\partial^2 v}{\partial t^2} = \Delta v \qquad (t,x) \in \mathbb{R} \times \mathbb{R}^n$$

   can be written as a symmetric hyperbolic system to which Theorem 9.4 applies.

*3. Let E be the identity matrix, and let U denote the (C_0) unitary group generated by G, as in Theorem 9.4. U and G act on $[L^2(\mathbb{R}^n)]^m$. Let $U_+ = \{U(t) : t \in \mathbb{R}^+\}$. Let $\mathscr{X}_p$ be $[L^p(\mathbb{R}^n)]^m$ when $1 \le p < \infty$ and $\mathscr{X}_\infty = [BUC(\mathbb{R}^n)]^m$. For $n = 1$, U_+ is a (C_0) semigroup on $\mathscr{X}_p$, $1 \le p \le \infty$. For $n \ge 2$, and $p \ne 2$, U_+ is a (C_0) semigroup on $\mathscr{X}_p$ iff the matrices $A^1, \ldots, A^n$ all commute with one another.

4. Define the mapping P from $\mathbb{R}$ to the 4×4 complex matrices by

$$P(x_1,x_2,x_3) = c(x_1\alpha_1 + x_2\alpha_2 + x_3\alpha_3) + \mu c^2 \beta,$$

   where μ and c are positive constants,

$$\alpha_j = \begin{pmatrix} 0 & \sigma_j \\ \sigma_j & 0 \end{pmatrix} \text{ for } j = 1, 2, 3,$$

$$\sigma_1 = \begin{pmatrix} 0 & 1 \\ 1 & 0 \end{pmatrix}, \sigma_2 = \begin{pmatrix} 0 & -i \\ i & 0 \end{pmatrix}, \sigma_3 = \begin{pmatrix} 1 & 0 \\ 0 & -1 \end{pmatrix},$$

   and

$$\beta = \begin{pmatrix} 1 & 0 & 0 & 0 \\ 0 & 1 & 0 & 0 \\ 0 & 0 & -1 & 0 \\ 0 & 0 & 0 & -1 \end{pmatrix}. \text{ The operator } A = \mathscr{F} M_P \mathscr{F}^{-1}$$

   (where $\mathscr{F}$ is the Fourier transform, suitably interpreted) is self-adjoint on the Hilbert space $L^2[\mathbb{R}^3, \mathscr{B}(\mathbb{C}^4)]$. ($A$ is the *Dirac operator* for a free electron; here μ is the mass of the electron and c is the speed of light.) The eigenvalues of $P(x)$ are $h_j(x)$ where

$$h_1(x) = h_2(x) = -h_3(x) = -h_4(x) = \mu c^2 (1 + \mu^{-2} c^{-2} |x|^2)^{1/2}.$$

   A is unitarily equivalent to the operator of multiplication by

$$H(x) = \begin{pmatrix} h_1(x) & 0 & 0 & 0 \\ 0 & h_2(x) & 0 & 0 \\ 0 & 0 & h_3(x) & 0 \\ 0 & 0 & 0 & h_4(x) \end{pmatrix}.$$

Moreover,

$$\sigma(A) =]-\infty, -\mu c^2] \cup [\mu c^2, \infty[\qquad \sigma_p(A) = \emptyset.$$

10. Higher Order Equations

In this section we indicate, without proof, some results concerning abstract Cauchy problems for equations of order at least three. We also present some related results.

10.1 Consider the Cauchy problem

$$d^n u(t)/dt^n = Au(t) \qquad (t \in \mathbb{R}^+), \tag{10.1}$$

$$u^{(k)}(0) = f_k, \; k = 0, 1, \ldots, n-1. \tag{10.2}$$

The above Cauchy problem is termed *well-posed* iff (i) a dense subspace $\mathscr{D}$ of $\mathscr{X}$ exists such that (10.1), (10.2) have a unique solution for $f_0, f_1, \ldots, f_{n-1} \in \mathscr{D}$, and (ii) whenever $\{\{f_k^{(m)}: k = 0, \ldots, n = 1\} : m \in \mathbb{N}\}$ is a sequence of initial data (in $\mathscr{D}^n$) tending to zero, then the corresponding solutions $u^{(m)}(t)$ tend to zero for each $t \in \mathbb{R}^+$, uniformly for t is bounded intervals.

10.2. THEOREM. *The Cauchy problem* (10.1), (10.2) *with* $n \geq 3$ *is well-posed iff* $A \in \mathscr{B}(\mathscr{X})$.

10.3. Despite the above (negative) result, there are some higher-order Cauchy problems that are well-posed. Let P be the polynomial

$$P(s,x) = \sum_{k=0}^{n} \sum_{j=0}^{m} c_{kj} s^k x^j,$$

and consider the corresponding differential equation

$$P\left(\frac{d}{dt}, A\right) u(t) = \sum_{k=0}^{n} \sum_{j=0}^{m} c_{kj} \left(\frac{d}{dt}\right)^k A^j u(t) = 0 \tag{10.3}$$

with initial conditions as in (10.2). The *concrete Cauchy problem* corresponding to the abstract Cauchy problem (10.3), (10.2) is the one obtained by taking $A = d/dx$, $\mathscr{X} = L^2(\mathbb{R})$. Assume that A generates a (C_0) group T on $\mathscr{X}$. Suppose that the concrete Cauchy problem is well-posed; then (roughly speaking) the abstract Cauchy problem is well-posed and the solution of the abstract Cauchy problem can be obtained by a formula involving the "fundamental solution" of the concrete Cauchy problem together with the group T. Specifically, let $g_k(\cdot,\cdot)$, $k = 0, 1, \ldots, n-1$ be functions such that the solution of the concrete Cauchy problem $[(10.3), (10.2)$ with $A = d/dx$, $\mathscr{X} = L^2(\mathbb{R})]$ is given by

$$u(t,x) = \int_{-\infty}^{\infty} \sum_{k=0}^{n-1} g_k(t,s) f_k(s + x) \, dx;$$

then the solution of the abstract Cauchy problem (10.3), (10.2) is given by

$$u(t) = \int_{-\infty}^{\infty} \sum_{k=0}^{n-1} g_k(t,s) T(s) f_k \, ds.$$

10.4. To illustrate this result we consider a simple example in which $n = 1$. Let $P(s,x) = s - x^2$. Then the solution of the concrete Cauchy problem (i.e. the Cauchy problem for the one-dimensional heat equation) is given by

$$u(t,x) = \int_{-\infty}^{\infty} (4\pi t)^{-1/2} \exp(-y^2/4t) f_0(y + x) \, dy;$$

hence the solution of the abstract Cauchy problem

$$u'(t) = A^2 u(t) \qquad u(0) = f_0$$

is given by

$$u(t) = (4\pi t)^{-1/2} \int_{-\infty}^{\infty} \exp(-y^2/4t) T(y) f_0 \, dy. \tag{10.4}$$

In fact, we have the following theorem.

THEOREM. *Let A generate a (C_0) group T on a complex $\mathscr{X}$. Then $A^2 \in \mathscr{GA}(\pi/2)$ and A^2 generates a semigroup U analytic in the right half plane. U is given by*

$$U(t)f = (4\pi t)^{-1/2} \int_{-\infty}^{\infty} \exp(-y^2/4t) T(y) f \, dy \qquad f \in \mathscr{X}, \ \mathrm{Re}\, t > 0. \tag{10.5}$$

Of course, this theorem can be verified directly.

A^2 generates a cosine function C given by

$$C(t) = \frac{1}{2} [T(t) + T(-t)].$$

Inserting this in (10.5) yields

$$U(t)f = (\pi t)^{-1/2} \int_0^{\infty} \exp(-y^2/4t) C(y) f \, dy.$$

Recall that this formula is valid whenever $B = A^2$ generates a cosine function, whether or not B is a square or $\mathscr{X}$ is complex. (Compare with the Theorem in Section 8.7.)

10.5. DEFINITIONS. Let $A_1, \ldots, A_n$ be linear operators on a Banach space $\mathscr{X}$. Let

$$D = \bigcap \{\mathscr{D}(A_{i_1} \cdots A_{i_n}) : i_j \in \{1, \ldots, n\}, \ n \in \mathbb{N}\}.$$

The *iterated Cauchy problem* is

$$\prod_{i=1}^{n} \left(\frac{d}{dt} - A_i\right) w \equiv \left(\frac{d}{dt} - A_n\right) \cdots \left(\frac{d}{dt} - A_1\right) w = 0, \tag{10.6}$$

$$w^{(j-1)}(0) = f_j, \ j = 1, \ldots, n. \tag{10.7}$$

Throughout this section we interpret (noncommutative) products in the sense of (10.6). w will be called a *solution* of (10.6) if

$$\prod_{j=1}^{k} \left(\frac{d}{dt} - A_j\right) w \in C[\mathbb{R}^+, \mathscr{D}(A_{k+1})] \cap C^1(\mathbb{R}^+, \mathscr{X}) \text{ for } k = 0, 1, \ldots, n-1$$

and w satisfies (10.6) on $\mathbb{R}^+$. $[\prod_{j=1}^{0}(d/dt - A_j)w = w$ by definition.] The problem (10.6), (10.7) is said to be *well-posed* on $\mathbb{R}^+$ (i) if there is a dense set D in $\mathscr{H}$ such that whenever $f_1,\ldots,f_n \in D$, there is a unique solution of (10.6), (10.7) with $\prod_{j=1}^{k}(d/dt - A_j)w \in C^{n-k}(\mathbb{R}^+,\mathscr{X})$ for $k = 0,\ldots,n-1$, and (ii) if $\{w_m\}$ is a sequence of solutions of (10.6) with $\lim_{m\to\infty}\prod_{j=1}^{k}(d/dt - A_j)w_m|_{t=0}$ for $k = 0,\ldots,n-1$, then $\prod_{j=1}^{k}(d/dt - A_j)w_m(t)$ approaches zero as $m \to \infty$, uniformly for t in bounded subsets of $\mathbb{R}^+$, for $k = 0,\ldots,n-1$. Similarly for well-posedness on $\mathbb{R}$.

10.6. THEOREM. *Let $A_1,\ldots,A_n$ be linear operators on $\mathscr{X}$ with $\bigcap_{i=1}^{n}\rho(A_i) \neq 0$, and $D = \bigcap\{\mathscr{D}(A_{i_1}\cdots A_{i_m}): i_j \in \{1,\ldots,n\}, m \in \mathbb{N}\}$ be dense in $\mathscr{X}$. Then the iterated Cauchy problem (10.6), (10.7) is well-posed in $\mathbb{R}^+$ [resp. in $\mathbb{R}$] iff A_i generates a (C_0) semigroup for $i = 1,\ldots n$. In this case, if $f_1,\ldots,f_n \in D$, then the solution w actually belongs to $C^\infty(\mathbb{R}^+,\mathscr{X})$ [resp. $C^\infty(\mathbb{R},\mathscr{X})$].*

10.7. DISCUSSION. The problem (10.6), (10.7) is equivalent to the problem

$$u'(t) = (A + P)u(t) \qquad u(0) = g \qquad (10.8)$$

in $\mathscr{Y} = \mathscr{X}^n$, where

$$u(t) = \begin{pmatrix} u_1(t) \\ \vdots \\ u_n(t) \end{pmatrix}, A = \begin{pmatrix} A_1 & & 0 \\ & \ddots & \\ 0 & & A_n \end{pmatrix}, P = \begin{pmatrix} 0I & . & 0 \\ 0 & . & \ddot{} I \\ 0 & & \ddot{} 0 \end{pmatrix}.$$

The solution w of (10.6), (10.7) then turns out to be $w = u_1$ if $g = \begin{pmatrix} g_1 \\ \vdots \\ g_n \end{pmatrix}$

where $g_1 = f_1$ and $g_m = f_m + \sum_{k=1}^{m-1}(-1)^k \Sigma' A_{i_k} \cdots A_{i_1} f_{m-k}$ for $2 \leq m \leq n$, where Σ' denotes the sum over $i_1,\ldots,i_k$ with $1 \leq i_1 < \cdots < i_k \leq m-1$. One can use the Phillips perturbation theorem (Theorem I.6.6) to show that

$$w(t) = T_1(t)f$$
$$+ \sum_{m=1}^{n-1}\int_0^t\int_0^{t_m}\cdots\int_0^{t_2} T_1(t-t_m)\cdots T_m(t_2-t_1)T_{m+1}(t_1)g_{m+1}\, dt_1\cdots dt_m$$

where the g_i are as above and T_i is the (C_0) semigroup generated by A_i. Note that when $n = 2$, this formula shows that the solution of

$$\left(\frac{d}{dt} - A_2\right)\left(\frac{d}{dt} - A_1\right)w = 0 \text{ (i.e. } w'' - (A_1 + A_2)w' + A_2 A_1 w = 0),$$

$$w(0) = f_1, \; w'(0) = f_2$$

is given by

$$w(t) = T_1(t)f_1 + \int_0^t T_1(t-t_1)T_2(t_1)(f_2 - A_1 f_1)\, dt_1.$$

When B generates a (C_0) group U on $\mathscr{X}$ and $A_1 = B = -A_2$, the above formula reduces to

$$w(t) = U(t)f_1 + \int_0^t U(t - 2s)(f_2 - Bf_1)\, ds.$$

11. Singular Perturbations

Let A, B be positive self-adjoint (or possibly more general) operators on a complex Hilbert space. Let $\alpha \in \{1, i\}$. We study the equations

$$u''(t) + \epsilon \alpha Au'(t) + Bu(t) = 0,$$

$$\epsilon u''(t) + \alpha Au'(t) + Bu(t) = 0$$

as $\epsilon \to 0^+$. In the first equation the limiting equation is also of second order, whereas in the second equation the limiting equation is of lower order. Thus the first equation describes a *regular perturbation problem* while the second equation describes a *singular perturbation problem*. We shall study both of these and then apply the singular perturbation result to a problem in quantum mechanics. The limiting equation is of the form $\alpha Au' + Bu = 0$. We also briefly consider equations of this form.

11.1 Let $\mathscr{H}$ be a complex Hilbert space. Let A, B be commuting nonnegative self-adjoint operators on $\mathscr{H}$ (cf. Exercise 6.14.7). Suppose also that $0 \notin \sigma_p(B)$ (i.e. B is injective).

11.2. THEOREM (Regular perturbation theorem). *Let A, B, be as in 11.1 and suppose that $\sigma_p(A^2 B^{-1})$ is bounded above. For $\epsilon > 0$ (small) let $u_\epsilon \in C^2(\mathbb{R}^+, \mathscr{H})$ be the unique solution of the well-posed Cauchy problem*

$$u''_\epsilon(t) + \epsilon Au'_\epsilon(t) + Bu(t) = 0 \qquad (t \in \mathbb{R}^+), \tag{11.1}$$

$$u_\epsilon(0) = f_1,\ u'_\epsilon(0) = f_2 \in \mathscr{D}(A^2) \cap \mathscr{D}(B^2). \tag{11.2}$$

Then

$$v_\epsilon(t) = v_0(t) + O(\epsilon)$$

where $v_0 \in C^1(\mathbb{R}^+, \mathscr{H})$ satisfies

$$Av''_0(t) + Bu_0(t) = 0 \qquad (t \in \mathbb{R}^+),$$

$$u_0(0) = f_1,\ u'_0(0) = f_2.$$

Here the $O(\epsilon)$ term is uniform for t in compact intervals.

11.3. THEOREM (singular perturbation theorem). *Let A, B, be as in 11.1 and suppose further that $0 \notin \sigma_p(A)$ and $\sigma_p(A^2 B^{-1})$ is bounded away from zero. For $\epsilon > 0$ (small) let $v_\epsilon \in C^2(\mathbb{R}^+, \mathscr{H})$ be the unique solution of the well-posed Cauchy problem*

$$\epsilon v''_\epsilon(t) + Av'_\epsilon(t) + Bv_\epsilon(t) = 0 \qquad (t = \mathbb{R}^+), \tag{11.3}$$

$$v_\epsilon(0) = g_1,\ v'_\epsilon(0) = g_2 \in \mathscr{D}(A^2) \cap \mathscr{D}(B^2). \tag{11.4}$$

Then

$$v_\epsilon(t) = v_0(t) + O(\epsilon)$$

where $v_0 \in C^1(\mathbb{R}^+, \mathcal{H})$ *satisfies*

$$Av_0'(t) + Bv_0(t) = 0 \qquad (t \in \mathbb{R}^+), \ v_0(0) = g_1.$$

Here the $O(\epsilon)$ *term is uniform for t in compact intervals.*

11.4. We begin the proofs of the above two theorems with the following elementary observations. Let A, B be nonnegative real numbers with $A^2 \neq 4B$. Then the unique solution of

$$u'' + Au' + Bu = 0 \qquad (t \in \mathbb{R}^+), \tag{11.5}$$

$$u(0) = f_1, \ u'(0) = f_2 \tag{11.6}$$

is given by

$$u(t) = \exp(tR_+)\alpha + \exp(tR_-)\beta \tag{11.7}$$

where

$$R_\pm = \frac{1}{2}\left(-A \pm (A^2 - 4B)^{1/2}\right) \tag{11.8}$$

and α, β are the unique solutions of the linear system

$$\left.\begin{array}{l} \alpha + \beta = f_1 \\ R_+\alpha + R_-\beta = f_2 \end{array}\right\}. \tag{11.9}$$

Note that R_+, R_- are distinct complex numbers with nonpositive real parts. Our convention for square roots is that for $\gamma > 0$, $\sqrt{\gamma^2} = \gamma$, $\sqrt{-\gamma^2} = i\gamma$.

Now suppose A, B are commuting self-adjoint operators with B and $A^2 - 4B$ injective. Then, by a version of the spectral theorem (see Exercise 6.14.7), A and B are unitarily equivalent (via the same unitary operator) to an orthogonal direct sum of multiplication operators $\mathcal{A}$, $\mathcal{B}$ on $L^2(\mathbb{R}, \mu)$ of the form $\mathcal{A}f(x) = a(x)f(x)$, $\mathcal{B}f(x) = b(x)f(x)$ where $a: \mathbb{R} \to [0, \infty[$, $b: \mathbb{R} \to]0, \infty[$ are measurable and where μ is a finite measure on $\mathbb{R}$. There is no loss in generality in working with one component of the orthogonal direct sum; thus we may suppose $A = M_a$, $B = M_b$ on $\mathcal{H} = L^2(\mathbb{R}, \mu)$. The unique solution of (11.5), (11.6) is thus u given by (11.7) where R_+, R_-, α, β are given by (11.8) and (11.9). (Our convention for square roots is $(\int_{-\infty}^{\infty} \lambda \, dE_\lambda)^{1/2} = \int_{-\infty}^{0} i\sqrt{-\lambda} \, dE + \int_{0}^{\infty} \sqrt{\lambda} \, dE_\lambda$.) Note that (11.9) yields

$$\alpha = (A^2 - 4B)^{-1/2}(f_2 - R_- f_1) \qquad \beta = f_1 - \alpha; \tag{11.10}$$

this is where the injectivity of $A^2 - 4B$ comes in.

11.5. PROOF OF THEOREM 11.2. Let u_ϵ satisfy (11.1) and (11.2). Then by (11.7)–(11.10), we have

$$u_\epsilon(t) = \exp(tR_{+,\epsilon})\alpha_\epsilon + \exp(tR_{-,\epsilon})\beta_\epsilon$$

where

$$R_{\pm,\epsilon} = \frac{1}{2}\left(-\epsilon A \pm (\epsilon^2 A^2 - 4B)^{1/2}\right),$$

$$\alpha_\epsilon = (\epsilon^2 A^2 - 4B)^{1/2}(f_2 - R_{-,\epsilon}f_1) \qquad \beta_\epsilon = f_1 - \alpha_\epsilon.$$

For this to be valid we need $\epsilon A^2 - 4B$ to be injective. This is true iff $1/\epsilon \notin \sigma_p(A^2 B^{-1})$, which follows from our hypothesis (that $\sigma_p(A^2 B^{-1})$ is bounded above) when ϵ is sufficiently small.

By Taylor's theorem,

$$(\epsilon^2 A^2 - 4B)^{1/2}f = 2iB^{1/2}f + O(\epsilon^2),$$

$$(\epsilon^2 A^2 - 4B)^{-1/2}f = \frac{1}{2i} B^{-1/2}f + O(\epsilon^2),$$

where the $O(\epsilon^2)$ term depends on f. Consequently

$$R_{\pm,\epsilon}f = \frac{1}{2}(-\epsilon Af \pm [2iB^{1/2}f + O(\epsilon)])$$

$$= \pm iB^{1/2}f + O(\epsilon),$$

$$\alpha_\epsilon = \frac{1}{2i} B^{-1/2}f_2 + \frac{1}{2}f_1 + O(\epsilon) \qquad \beta_\epsilon = \frac{1}{2}f_1 - \frac{1}{2i} B^{1/2}f_2 + O(\epsilon).$$

Moreover,

$$\|\exp(tR_{\pm,\epsilon})f - \exp(\pm itB^{1/2})f\|$$

$$= \left\|\int_0^t \frac{d}{ds}\exp(sR_{\pm,\epsilon})\exp[\pm(t-s)iB^{1/2}]f)\,ds\right\|$$

$$\leq \int_0^t \|\exp(sR_{\pm,\epsilon})\exp[\pm(1-s)iB^{1/2}](R_{\pm,\epsilon}f \mp iB^{1/2}f)\|\,ds = O(\epsilon),$$

the $O(\epsilon)$ being uniform for t in bounded subsets of $\mathbb{R}^+$. Hence

$$u_\epsilon(t) = \exp(tR_{+,\epsilon})\alpha_\epsilon + \exp(tR_{-,\epsilon})\beta_\epsilon$$

$$= \exp(itB^{1/2})\left(\frac{1}{2}f_1 + \frac{1}{2i} B^{-1/2}f_2\right)$$

$$+ \exp(itB^{1/2})\left(\frac{1}{2}f_1 - \frac{1}{2i} B^{-1/2}f_2\right) + O(\epsilon)$$

$$= \cos(tB^{1/2})f_1 + B^{-1/2}\sin(tB^{1/2})f_2 + O(\epsilon).$$

If $u_0(t) = \cos(tB^{1/2})f_1 + B^{-1/2}\sin(tB^{1/2})f_2$, then $u_0'' + Bu_0 = 0$, $u_0(0) = f_1$, $u_0'(0) = f_2$, and $u_\epsilon(t) = u_0(t) + O(\epsilon)$.   ■

11.6.   Proof of Theorem 11.3.   The unique solution of (11.3), (11.4) is given by

$$v_\epsilon(t) = \exp(tS_{+,\epsilon})\gamma_\epsilon + \exp(tS_{-,\epsilon})\delta_\epsilon$$

where

$$S_{\pm,\epsilon} = \frac{1}{2\epsilon}(-A \pm (A^2 - 4\epsilon B)^{1/2}),$$

$$\gamma_\epsilon = \epsilon(A^2 - 4\epsilon B)^{-1/2}(g_2 - S_{-,\epsilon}g_1), \quad \delta_\epsilon = g_1 - \gamma_\epsilon.$$

Note that for $\epsilon > 0$ sufficiently small, $(A^2 - 4\epsilon B)^{-1/2}$ exists since $4\epsilon \notin \sigma_p(A^2 B^{-1})$. Next, an elementary calculation using Taylor's formula yields

$$\gamma_\epsilon = g_1 + O(\epsilon), \ \delta_\epsilon = O(\epsilon), \text{ and}$$

$$S_{\pm,\epsilon}g = \frac{1}{2a}(-Ag \pm [Ag - 2\epsilon A^{-1}Bg + O(\epsilon^2)])$$

for $g \in \mathscr{D}(A^2) \cap \mathscr{D}(B^2)$. Consequently

$$\exp(tS_{-,\epsilon})g = \exp\left(-\frac{t}{\epsilon}A\right)\exp(\epsilon A^{-1}B)g + O(\epsilon) = O(\epsilon),$$

$$\exp(tS_{+,\epsilon})g = \exp(-tA^{-1}B)g + O(\epsilon)$$

where the $O(\epsilon)$ terms (which depend on g) are uniform for t in bounded intervals in $\mathbb{R}^+$. Hence if we set $v_0(t) = \exp(-tA^{-1}B)g_1$, we have $Av_0' + Bv_0 = 0$ ($t \in \mathbb{R}^+$), $v_0(0) = g_1$, and $v_\epsilon(t) = v_0(t) + O(\epsilon)$ where the $O(\epsilon)$ term is uniform in compact subsets of $\mathbb{R}^+$. ∎

11.7. THEOREM. *Let A, B be commuting nonnegative self-adjoint operators on $\mathcal{H}$.*

(i) *If B is injective and u_ϵ is the solution of*

$$u_\epsilon''(t) + i\epsilon Au_\epsilon'(t) + Bu_\epsilon(t) = 0 \quad (t \in \mathbb{R}^+),$$
$$u_\epsilon(0) = f_1, \ u_\epsilon'(0) = f_2 \in \mathscr{D}(A^2) \cap \mathscr{D}(B^2),$$

then

$$u_\epsilon(t) = u_0(t) + O(\epsilon)$$

where

$$u_0''(t) + Bu_0(t) = 0 \quad (t \in \mathbb{R}^+),$$
$$u_0(0) = f_1, \ u_0'(0) = f_2.$$

(ii) *If A is injective and v_ϵ is the solution of*

$$\epsilon v_\epsilon''(t) + iAv_\epsilon'(t) + Bv_\epsilon(t) = 0 \quad (t \in \mathbb{R}^+), \tag{11.11}$$
$$v_\epsilon(0) = g_1, \ v_\epsilon'(0) = g_2 \in \mathscr{D}(A^2) \cap \mathscr{D}(B^2), \tag{11.12}$$

then

$$v_\epsilon(t) = v_0(t) + O(\epsilon)$$

where

$$Av_0'(t) + Bv_0(t) = 0 \qquad (t \in \mathbb{R}^+),$$

$$v_0(0) = g_1.$$

In both cases the $O(\epsilon)$ term is uniform for t in compact intervals.

This is proven just like Theorem 11.2 and 11.3 were. No assumption is needed concerning $\sigma_p(A^2 B^{-1})$ for the following reason. We wanted $\epsilon^2 A^2 - 4B$ [resp. $A^2 - 4\epsilon B$] to be injective for $\epsilon > 0$ small in Theorem 11.2 [resp. 11.3]. The analogous operators now are $-\epsilon^2 A^2 - 4B$ and $-A^2 - 4\epsilon B$, which are automatically injective. The details of the proof are left as an exercise.

11.8. APPLICATION TO QUANTUM THEORY
Now we take a case of interest in quantum mechanics. Let $\mathscr{H} = L^2(\mathbb{R}^n)$ (e.g. $n = 3$). For a single free particle of mass m travelling in $\mathbb{R}^n$, the relativistic equation governing the motion of the particle is the Klein-Gordon equation

$$\hbar^2 \frac{\partial^2 v}{\partial t^2} = \hbar^2 c^2 \, \Delta v - m^2 c^4 v.$$

Here $v(t,\cdot) = v(t) \in L^2(\mathbb{R}^n)$ is the wave function of the particle, c is the speed of light, and $\hbar$ is Planck's constant divided by 2π. Let

$$v_\epsilon(t,x) = \exp\{im\, c^2 t\hbar^{-1}\}v(t,x).$$

Then v_ϵ is a solution of

$$\frac{\hbar^2}{2mc^2} v_\epsilon''(t) + i\hbar v_\epsilon'(t) - \frac{\hbar^2}{2m} \Delta v_\epsilon(t) = 0 \qquad (t \in \mathbb{R}).$$

We identify this with (11.11) by taking $\epsilon = 1/c^2$, $\mathscr{H} = L^2(\mathbb{R}^n)$, etc. Note that A becomes a positive multiple of the identity and B an injective nonnegative self-adjoint operator. $v_\epsilon(0) = v(0)$ and $v_\epsilon'(0)$ is closely related to $v'(0)$; the exact relationship is easy to compute. It follows from Theorem 11.7 that for $T > 0$, there is a constant $C(T)$ such that

$$\int_{\mathbb{R}^n} |v_\epsilon(t,x) - v_0(t,x)|^2 \, dx \leq C(T)\epsilon \qquad |t| \leq T,$$

where v_0 is the solution of the Schrödinger equation

$$v_0'(t) = -\frac{i\hbar}{2m} \Delta v_0(t) \qquad (t \in \mathbb{R}),$$

$$v_0(0) = v(0) \in \mathscr{D}(B^2) = H^4(\mathbb{R}^n).$$

Recall that $\epsilon = 1/c^2$. This establishes that as the speed of light becomes infinite, the Klein-Gordon equation has the correct nonrelativistic limit.

11.9. REMARKS. There is substantial literature on singular perturbations involving (C_0) semigroups rather than self-adjoint operators. The key tool in such an approach is a suitable version of the approximation theorem. One such approach is sketched in the exercises.

11.10. SOBOLEV EQUATIONS. Equations of the type

$$Au'(t) = Bu(t) \qquad (t \in \mathbb{R}^+) \tag{11.13}$$

are called *Sobolev equations*. We indicate one simple approach.

11.11. THEOREM. *Let B be m-dissipative and A positive and self-adjoint on a Hilbert space $\mathcal{H}$. Suppose $0 \in \rho(A)$ and $\mathcal{D}(B) \supset \mathcal{D}(A)$ or $\mathcal{D}(A) \supset \mathcal{D}(B)$. Then the Cauchy problem for (11.13) is well-posed and is governed by a (C_0) contraction semigroup on a Hilbert space $\mathcal{K}$.*

PROOF. Let $A^{1/2}$ be the positive square root of A. Then for all $f \in \mathcal{D}(B) \cap \mathcal{D}(A^{1/2})$, $\operatorname{Re}\langle A^{1/2}(A^{-1}B)f, A^{1/2}f \rangle = \operatorname{Re}\langle Bf, f \rangle \le 0$. Thus $A^{-1}B$ is dissipative on $\mathcal{K}$, the completion of $\mathcal{D}(A^{1/2})$ in the inner product

$$\langle f, g \rangle_{\mathcal{K}} = \langle A^{1/2}f, A^{1/2}g \rangle.$$

Next let $g \in \mathcal{D}(A^{3/2})$ and $\alpha > 0$. To show that g is in the range of $I - \alpha A^{-1}B$, we solve

$$Af - \alpha Bf = Ag \in \mathcal{D}(A^{1/2}).$$

Since $A = \epsilon I + A_1$ where $A_1 = A_1^* \ge 0$ and $\epsilon > 0$, the above equation becomes

$$[I + (A_1 - \alpha B)]f = h$$

when $h = Ag \in \mathcal{D}(A^{1/2})$. $-A_1$ is *m*-dissipative on $\mathcal{H}$ and $\mathcal{D}(A) = \mathcal{D}(A_1)$. By Chapter I, Theorem 6.1, $-A_1 + \alpha B$ is *m*-dissipative for $\alpha > 0$ sufficiently small or large, depending on whether $\mathcal{D}(A) \subset \mathcal{D}(B)$ or $\mathcal{D}(A) \supset \mathcal{D}(B)$. Choose and fix such an α. It follows that the desired solution f can be found, and so Range $(I - \alpha A^{-1}B) \supset \mathcal{D}(A^{3/2})$, which is dense in $\mathcal{K}$. Thus $A^{-1}B$ is *m*-dissipative on $\mathcal{K}$, and the Cauchy problem for $u' = A^{-1}Bu$ is well-posed. The theorem follows. ∎

11.12. EXERCISES

1. Prove Theorem 11.7.
2. Write out the details of 11.8.
*3. Let A_1, A_2 generate (C_0) semigroups T_1, T_2 on $\mathcal{X}$ with $T_1(t)T_2(s) = T_2(s)T_1(t)$ for all $t, s \in \mathbb{R}^+$. Let P be a 2×2 complex matrix. Then $A = \begin{pmatrix} A_1 & 0 \\ 0 & A_2 \end{pmatrix} + P \begin{pmatrix} I & 0 \\ 0 & I \end{pmatrix}$ generates a (C_0) semigroup T on $\mathcal{Y} = \mathcal{X} \times \mathcal{X}$. When $P = 0$, $T(t) = \begin{pmatrix} T_1(t) & 0 \\ 0 & T_2(t) \end{pmatrix}$. When $P \ne 0$, one can compute T with the aid of the Phillips perturbation theorem (Chapter I, Theorem 6.6). When A_0 generates a (C_0) group T_0, $A_1 = A_0 = -A_2$, and $P = \begin{pmatrix} -a & a \\ a & a \end{pmatrix}$ where $a > 0$, then $T(t) = [T_{ij}(t)]_{i,j=1,2}$ becomes

$$T_{11}(t)f = e^{-at}\left\{ T_0(t) + \frac{a}{2}\int_{-t}^{t} \left(\frac{t+s}{t-s}\right)^{1/2} I_1\{a[(t^2 - s^2)]^{1/2}\} T_0(s)f \, ds, \right.$$

etc., where I_1 is the modified Bessel function of order one.

Let $G_\epsilon = \dfrac{1}{\epsilon^{1/2}} \begin{pmatrix} A_0 & 0 \\ 0 & -A_0 \end{pmatrix} + \dfrac{1}{\epsilon} \begin{pmatrix} -a & a \\ a & -a \end{pmatrix}$, and let $D = \dfrac{1}{4a} \begin{pmatrix} A^2 & A^2 \\ A^2 & A^2 \end{pmatrix}$.

On the diagonal of $\mathcal{Y} = \mathcal{X} \times \mathcal{X}$, which can be identified with $\mathcal{X}$, we have

$$\left\| (\lambda - G_\epsilon)^{-1} \begin{pmatrix} f \\ f \end{pmatrix} - D \begin{pmatrix} f \\ f \end{pmatrix} \right\| \to 0$$

as $\epsilon \to 0^+$ for each λ sufficiently large and each $f \in \mathcal{X}$. It follows that "e^{tG_ϵ}" converges strongly to "e^{tD}" on elements of the form $\begin{pmatrix} f \\ f \end{pmatrix}$. Each component of the limit is "$\exp[(t/2a)A^2]$" f, and the rate of convergence is $O(\epsilon)$, uniformly for t in compactar, if $f \in \mathcal{D}(A^2)$.

Fill in the missing details and obtain a theorem which contains as an application the result 11.7.

12. Mixed Problems

12.1. In previous sections we have discussed initial value problems for parabolic and hyperbolic equations where the space variable x ranged over all of $\mathbb{R}^n$. This section deals with problems where x ranges over a proper subset Ω of $\mathbb{R}^n$. In this case we must impose conditions at the boundary of Ω as well as initial conditions; thus we consider *mixed initial-boundary value problems*.

Only results for bounded domains will be discussed, although many of the results extend to certain unbounded domains.

Some of the results discussed are based on deep theorems about elliptic boundary value problems. Most of the proofs are omitted.

12.2. DEFINITIONS. Recall the multi-index and differential operator notation introduced in Section 3. Let Ω be a domain (i.e. an open, connected set) in $\mathbb{R}^n$. Let $f : \Omega \to \mathbb{C}$ be measurable. To say that the *distributional derivative* $D^\alpha f$ equals $g : \Omega \to \mathbb{C}$ means that for all $\phi \in C_c^\infty(\Omega)$,

$$\int_\Omega g(x)\phi(x)\,dx = (-1)^{|\alpha|} \int_\Omega f(x) D^\alpha \phi(x)\,dx.$$

Let $1 \le p < \infty$ and $m \in \mathbb{N}_0$. The *Sobolev space* $W^{m,p}(\Omega)$ consists of all $f \in L^p(\Omega)$ such that the distributional derivative $D^\alpha f$ belongs to $L^p(\Omega)$ for all α such that $|\alpha| \le m$. $W^{m,p}(\Omega)$ becomes a Banach space under the norm

$$\|f\|_{m,p} = \left(\sum_{|\alpha| \le m} \|D^\alpha f\|_{L^p(\Omega)}^p \right)^{1/p} \qquad 1 \le p < \infty.$$

Equivalently, $W^{m,p}(\Omega)$ is the completion of $C^m(\Omega)$ in the $W^{m,p}(\Omega)$ norm. The *Sobolev space* $W_0^{m,p}(\Omega)$ is the completion of $C_c^m(\Omega)$ [or $C_c^\infty(\Omega)$] in the $W^{m,p}(\Omega)$ norm. $W^{m,2}(\Omega)$ is usually denoted by $H^m(\Omega)$. When $\Omega = \mathbb{R}^n$ this definition reduces to the one given in Section 5. Note that $W^{m,p}(\mathbb{R}^n) = W_0^{m,p}(\mathbb{R}^n)$, but in general $W_0^{m,p}(\Omega) \subsetneqq W^{m,p}(\Omega)$. If $f \in W_0^{m,p}(\Omega)$, then, in a generalized sense, $D^\alpha f = 0$ on $\partial\Omega$ for $|\alpha| \le m - 1$. More precisely, for example, if $\partial\Omega$ is smooth and $f \in W_0^{1,p}(\Omega) \cap C(\bar{\Omega})$, then $f = 0$ on $\partial\Omega$.

Some of the basic properties of these spaces are given in the following theorem. Part (i) is usually called the *Sobolev inequality*, while part (ii) is the *Sobolev embedding theorem*.

12.3.   THEOREM. *Let Ω be a bounded domain in $\mathbb{R}^n$ with $\partial\Omega$ of class C^{1}.[†]*

(i) *Let $m \in \mathbb{N}$, $1 \leq r < \infty$. There is a constant $C = C(\Omega,m,r)$ such that for $j \in \{0, 1, \ldots, m-1\}$ and $u \in W^{m,r}(\Omega)$, we have $u \in W^{j,p}(\Omega)$ and*

$$\|u\|_{j,p} \leq C\|u\|_{m,r}$$

where $1/p = j/n + 1/r - m/n$.

(ii) $$W^{j,p}(\Omega) \subset C^m(\Omega) \text{ for } j > m + n/p.$$

That is, if u is an equivalence class in $W^{j,p}(\Omega)$, then u has a representative belonging to $C^m(\Omega)$ if $j > m + n/p$.

12.4   DEFINITIONS AND DISCUSSION.   Let Ω be a bounded domain in $\mathbb{R}^n$ with a smooth boundary $\partial\Omega$. Consider the differential operator

$$a(x,D) = \sum_{|\alpha| \leq 2m} a_\alpha(x)\, D^\alpha$$

of order $2m$. We assume for simplicity that the coefficients a_α are smooth real valued functions on $\bar{\Omega}$. (With minor changes we can allow for complex values.) $a(x,D)$ is called *strongly elliptic* if there is a $c_0 > 0$ such that

$$(-1)^{m+1} \sum_{|\alpha| = 2m} a_\alpha(x,\xi) \geq c_0|\xi|^{2m}$$

for all $\xi \in \mathbb{R}^n$ and all $x \in \bar{\Omega}$. A basic property of a strongly elliptic operator $a(x,D)$ is given by the *Gårding inequality*: there are constants $c_1 > 0, c_2 \geq 0$ such that for all $u \in H^{2m}(\Omega) \cap H_0^m(\Omega)$,

$$-\langle a(x,D)u, u\rangle_{L^2(\Omega)} \geq c_1\|u\|_{m,2}^2 - c_2\|u\|_{0,2}^2.$$

This is the main tool for solving the *elliptic boundary value problem*

$$-a(x,D)u + \lambda u = f \text{ in } \Omega,$$

$$D^\alpha u = 0 \text{ on } \partial\Omega, 0 \leq |\alpha| \leq m - 1. \tag{12.1}$$

For $\lambda \geq c_2$ and $f \in L^2(\Omega)$, one can find a unique $u \in H^{2m}(\Omega) \cap H_0^m(\Omega)$ such that $-a(x,D)u + \lambda u = f$. The *Dirichlet boundary condition* (12.1) is incorporated in the solution via $u \in H_0^m(\Omega)$. One can also consider more general boundary conditions than (12.1).

The semigroup strategy for solving parabolic equations of the form

$$\frac{\partial u}{\partial t} = a(x,D)u$$

[†] For each point $x_0 \in \partial\Omega$ there is a neighborhood N of x_0 in $\mathbb{R}^n$ such that $N \cap \partial\Omega = \{x \in N : \phi(x) = 0\}$ for some $\phi \in C^1(N,\mathbb{R})$.

is as follows. Associate with $a(x,D)$ an operator A on $L^p(\Omega)$ by $Au = a(x,D)u$ for $u \in \mathcal{D}(A)$. The domain of A will satisfy $W_0^{2m,p}(\Omega) \subset \mathcal{D}(A) \subset W^{2m,p}(\Omega)$ and will be determined by the boundary conditions associated with $a(x,D)$. For example, when $a(x,D)$ has Dirichlet conditions, $\mathcal{D}(A) = W^{2m,p}(\Omega) \cap W_0^{m,p}(\Omega)$. The theory of elliptic boundary problems provides us with theorems which can be interpreted to say that A generates an analytic semigroup on $L^p(\Omega)$. We shall state a very general version of this as Theorem 12.7.

12.5. MORE DISCUSSION. Let Ω and $a(x,D)$ be as in 12.4. $a(x,D)$ is called *formally self-adjoint* if, for all $u,v \in C_c^\infty(\Omega)$,

$$\langle a(x,D)u,v \rangle_{L^2(\Omega)} = \langle u,a(x,D)v \rangle_{L^2(\Omega)}.$$

For example, integration by parts shows this to be the case when

$$a(x,D)u = \sum_{i,j=1}^n \frac{\partial}{\partial x_i}\left(a_{ij}(x)\frac{\partial u}{\partial x_j}\right) + a_0(x)u, \tag{12.2}$$

where $(a_{ij}(x))$ is a real symmetric matrix and $a_0(x) \in \mathbb{R}$, for each $x \in \bar{\Omega}$.

Now suppose $a(x,D)$ is both strongly elliptic and formally self-adjoint. (For $a(x,D)$ as in (12.2), the strong ellipticity is equivalent to the assumption that the matrix $[a_{ij}(x)]$ is (strictly) positive definite for each $x \in \bar{\Omega}$.) The operator $A_0u = a(x,D)u$ for $u \in \mathcal{D}(A_0) = C_c^\infty(\Omega)$ is symmetric and has many self-adjoint extensions which are restrictions of $a(x,D)$ on $H^{2m}(\Omega)$. Typically these extensions are determined by imposing suitable boundary conditions. Also, typically, these extensions are semibounded, i.e. if A is such an extension, there is a $c_1 \in \mathbb{R}$ such that $A + c_1 I \leq 0$ in the sense that $\langle(A + c_1)u,u\rangle_{L^2(\Omega)} \leq 0$ for all $u \in \mathcal{D}(A)$. Let A be such an extension. Then A generates a semigroup analytic in the right half plane by the spectral theorem.

Now let

$$b(x,D) = \sum_{|\alpha| \leq 2m} b_\alpha(x) D^\alpha$$

be a strongly elliptic operator with smooth coefficients (in $\bar{\Omega}$). Assign boundary conditions to $b(x,D)$ which would make it self-adjoint if it were formally so (e.g. Dirichlet conditions). One can perturb $b(x,D)$ by a lower-order operator $c(x,D)$ so that the sum $a = b + c$, i.e.

$$a(x,D) = b(x,D) + \sum_{|\alpha| \leq 2m-1} c_\alpha(x) D^\alpha,$$

is both strongly elliptic and formally self-adjoint. [For example, if

$$b(x,D) = \sum_{j,k=1}^n b_{jk}(x)\frac{\partial^2}{\partial x_j \partial x_k} + \sum_{j=1}^n b_j(x)\frac{\partial}{\partial x_j} + b_0(x),$$

take $a_{jk}(x) = \frac{1}{2}[b_{jk}(x) + b_{kj}(x)]$, $a_0(x) = b_0(x)$. Then $a = b + c$ where a is as in (12.2) and

$$c(x,D) = \sum_{j=1}^n c_j(x)\frac{\partial}{\partial x_j}, \ c_j(x) = \sum_{i=1}^n \left[\frac{\partial}{\partial x_i}a_{ij}(x)\right] - b_j(x).]$$

Since $c(x,D)$ is a lower order perturbation of the slef-adjoint operator A [corresponding to $a(x,D)$ together with the boundary conditions for $b(x,D)$], it follows that

$$A - c(x,D) = B[\, = b(x,D) + \text{boundary conditions}]$$

is in $\mathscr{GA}(\pi)$ and so generates a semigroup analytic in the right half plane. This is by Chapter I Corollary 6.9, since $-c(x,D)$ is a Kato perturbation of A. (See Exercise 12.10.4.)

12.6. DISCUSSION.   Let Ω be a bounded domain in $\mathbb{R}^n$ with boundary $\partial\Omega$ of class C^{2m}. Let $a(x,D) = \sum_{|\alpha| \le 2m} a_\alpha(x) D^\alpha$ be a strongly elliptic partial differential operator of order $2m$. Assume $a_\alpha \in L^\infty(\Omega)$ for all α and $a_\alpha \in C(\bar\Omega)$ for $|\alpha| = 2m$.
Next let

$$b_j(x,D) = \sum_{|\beta| \le m_j} b_{j\beta}(x) D^\beta$$

be a partial differential operator on the boundary $\partial\Omega$. Specifically, suppose $j = 1,\ldots,m$, $m_j \in \mathbb{N}_0$, $m_j < 2m$, and $b_{j\beta} \in C^{2m-m_j}(\partial\Omega)$ for all β, j. We assume that $b_j(x,D)$ is of order m_j (i.e. $\sum_{|\beta| = m_j} |b_{j\beta}(x)| > 0$ for some $x \in \partial\Omega$) and that $\partial\Omega$ is nowhere characteristic with respect to $b_j(x,D)$. (Locally, the boundary $\partial\Omega$ is described by an equation of the form $\phi(x) = 0$. We are supposing that $\phi \in C^{2m}$ and, locally, $\sum_{|\alpha| = 2m} a_\alpha(x) [\partial\phi/\partial x_1(x)]^{\alpha_1} \cdots [\partial\phi/\partial x_2(x)]^{\alpha_n} \ne 0$.)
Let $x \in \partial\Omega$, let v be the unit outward normal to $\partial\Omega$ at x, and let $\xi \in \mathbb{R}^n$ be a vector parallel to $\partial\Omega$ at x. If $\lambda \in \mathbb{R}\setminus\{0\}$, the polynomial

$$p(\tau) = (-1)^m \sum_{|\alpha| = 2m} a_\alpha(x)(\xi + \tau v)^\alpha - \lambda$$

has exactly m roots $\tau_k(\xi,\lambda)$ $(k = 1,\ldots,m)$ having positive imaginary part. Suppose that the polynomials

$$q_j(\tau) = \sum_{|\beta| = m_j} b_{j\beta}(x)(\xi + \tau v)^\beta \ (j = 1,\ldots,m)$$

are linearly independent, modulo the polynomial $\prod_{k=1}^m [\tau - \tau_k(\xi,\lambda)]$. Then, writing $\mathscr{A} = a(x,D)$, $\mathscr{B} = \{b_j(x,D); j = 1,\ldots,m\}$, we shall call $\{\mathscr{A},\mathscr{B};\Omega\}$ an *elliptic boundary value system of Agmon type*. Associated with such a system are the elliptic boundary value problem

$$a(x,D)u(x) - \lambda u(x) = f(x) \qquad (x \in \Omega),$$

$$b_j(x,D)u(x) = 0 \qquad (x \in \partial\Omega, j = 1,\ldots,m),$$

and the parabolic mixed (initial value-boundary value) problem

$$\frac{\partial u}{\partial t} = a(x,D)u \qquad (x \in \Omega, t \ge 0),$$

$$b_j(x,D)u(t,x) = 0 \qquad (x \in \partial\Omega, t \ge 0, j = 1,\ldots,m),$$

$$u(0,x) = f(x) \qquad (x \in \Omega).$$

Let $\{\mathscr{A},\mathscr{B};\Omega\}$ be as in the preceding paragraph. Let $1 < p < \infty$ and let $\mathscr{X} = L^P(\Omega)$. Define $W^{2m,p}(\Omega;\mathscr{B})$ to be the completion of $\{u \in C^{2m}(\bar{\Omega}) : b_j(x,D)u(x) = 0$ for $x \in \partial\Omega$ and $j = 1,\dots,m\}$ in the $W^{2m,p}(\Omega)$ norm. Define

$$(A_p u)(x) = a(x,D)u(x)$$

for $u \in \mathscr{D}(A_p) = W^{2m,p}(\Omega;\mathscr{B})$. Note that $\mathscr{D}(A_p) = W^{2m,p}(\Omega) \cap W_0^{m,p}(\Omega)$ when $\mathscr{B}$ describes Dirichlet boundary conditions.

12.7.   THEOREM.   *Let $\{\mathscr{A},\mathscr{B};\Omega\}$ be an elliptic boundary system of Agmon type. Let $1 < p < \infty$. Then there are constants $k_p \in \mathbb{R}$ and $\theta_p \in]\pi/2,\pi]$ such that $A_p - k_p I \in \mathscr{G}\mathscr{A}_b(\theta_p)$, whence A_p generates an analytic semigroup of type $(\theta_p - \pi/2)$.*

12.8.   SECOND-ORDER EQUATIONS.   Let, as in Section 12.4 and 12.5, Ω be a bounded domain in $\mathbb{R}^n$ with smooth boundary and let $ax(,D)$ be a formally self-adjoint strongly elliptic operator with coefficients defined on $\bar{\Omega}$. Let A be a self-adjoint operator such that $H_0^{2m}(\Omega) \subset \mathscr{D}(A) \subset H^{2m}(\Omega)$, $Af = a(x,D)f$ for all f in $\mathscr{D}(A)$, and $A + c_1 I \leq 0$ for some $c_1 \in \mathbb{R}$. Let $c = c_1 - 1$. Then $-A - cI \geq I$. Consequently, by Theorem 7.4, the Cauchy problem

$$u''(t) - (c + A)u(t) = 0 \qquad (t \in \mathbb{R}),$$

$$u(0) = f_1, \, u'(0) = f_2$$

is well-posed and is governed by a (C_0) unitary group on $\mathscr{Y} = \mathscr{D}(B) \times \mathscr{H}$ where $B = (-A - cI)^{1/2}$, and the initial data should satisfy $f_1 \in \mathscr{D}(A)$, $f_2 \in \mathscr{D}(B)$. It is usually the case that $\mathscr{D}(B) = H^m(\Omega)$; for example, this is true when A has Dirichlet boundary conditions, so that $\mathscr{D}(A) = H^{2m}(\Omega) \cap H_0^m(\Omega)$.

Now let $b(x,D) = \sum_{|\alpha| \leq m} b_\alpha(x) D^\alpha$ be a differential operator of order $\leq m$ with coefficients b_α in $L^\infty(\Omega)$. Let $\gamma \in L^\infty(\Omega)$. Define $B_1 = b(x,D)$ on $\mathscr{D}(B_1) = H^m(\Omega)$, which we assume equals $\mathscr{D}(B)$. Then $\begin{pmatrix} 0 & 0 \\ B_1 & \gamma \end{pmatrix}$ is a bounded operator on $\mathscr{Y}$.

It follows that the equation

$$\frac{d}{dt}\begin{pmatrix} u(t) \\ u'(t) \end{pmatrix} = \begin{pmatrix} 0 & I \\ c + A & 0 \end{pmatrix}\begin{pmatrix} u(t) \\ u'(t) \end{pmatrix} + \begin{pmatrix} 0 & 0 \\ B_1 & \gamma \end{pmatrix}\begin{pmatrix} u(t) \\ u'(t) \end{pmatrix}$$

is governed by a (C_0) group on $\mathscr{Y}$. Consequently, absorbing the constant term c into $b_0(x)$, the mixed problem

$$\frac{\partial^2 u}{\partial t^2} = \sum_{|\alpha| \leq 2m} a_\alpha(x) D^\alpha u + \sum_{|\alpha| \leq m} b_\alpha(x) D^\alpha u + \gamma(x)\frac{\partial u}{\partial t} \quad (t \in \mathbb{R}, x \in \Omega),$$

$$D^\alpha u(t,x) = 0 \text{ for } t \in \mathbb{R}, \, x \in \partial\Omega, \, |\alpha| \leq m - 1,$$

$$u(0,x) = f_1(x), \, \frac{\partial u}{\partial t}(0,x) = f_2(x)$$

with $f_1 \in H^{2m}(\Omega) \cap H_0^m(\Omega)$, $f_2 \in H^m(\Omega)$ is well-posed. Moreover, the solution satisfies

$$u(t,\cdot) \in H^{2m}(\Omega) \cap H_0^m(\Omega) \qquad u_t(t,\cdot) \in H^m(\Omega)$$

for all $t \in \mathbb{R}$. Again, boundary conditions other than the Dirichlet one can be considered. As a special case of this we have the following result ($m = 1$).

12.9. THEOREM.   *Let Ω be a bounded domain in $\mathbb{R}^n$ with a C^2 boundary $\partial\Omega$. Let $[a_{ij}(x)]$ be, for each $x \in \bar{\Omega}$, a (strictly) positive definite $n \times n$ symmetric matrix whose coefficients a_{ij} are in $C(\bar{\Omega})$. Let $a_i \in L^\infty(\Omega)$, $i = 0, 1, \ldots, n+1$. Then the mixed problem*

$$\frac{\partial^2 u}{\partial t^2} = \sum_{i,j=1}^n a_{ij}(x) \frac{\partial^2 u}{\partial x_i \partial x_j} + \sum_{i=1}^n a_i(x) \frac{\partial u}{\partial x_i} + a_0(x)u$$

$$+ a_{n+1}(x)\frac{\partial u}{\partial t} \qquad (t \in \mathbb{R}, x \in \Omega),$$

$$u(0,x) = f_1(x), \, u_t(0,x) = f_2(x) \qquad (x \in \Omega)$$

$$u(t,x) = 0 \qquad (t \in \mathbb{R}, x \in \partial\Omega)$$

is well-posed and is governed by a (C_0) group on $H^1(\Omega) \oplus L^2(\Omega)$. The initial data should be in the domain of the generator, i.e. $f_1 \in H^2(\Omega) \cap H_0^1(\Omega)$ and $f_2 \in H^1(\Omega)$.

12.10   EXERCISES.

*1. Let $\mathcal{H} = L^2([0,1])$. Let $\alpha_1, \ldots, \alpha_4$, $\beta_1, \ldots, \beta_4$ be fixed real numbers such that the vectors $(\alpha_1, \ldots, \alpha_4)$ and $(\beta_1, \ldots, \beta_4)$ are linearly independent (in $\mathbb{R}^4$). Let $\mathcal{D}(A_0) = \{u \in C^2([0,1]) : \alpha_1 u(0) + \alpha_2 u'(0) + \alpha_3 u(1) + \alpha_4 u'(1) = 0, \, \beta_1 u(0) + \beta_2 u'(0) + \beta_3 u(1) + \beta_4 u'(1) = 0\}$. Let $A_0 = d^2/dx^2$ on $\mathcal{D}(A_0)$.
   (i) Show that A_0 is symmetric iff

$$\alpha_3 \beta_4 - \alpha_4 \beta_3 = \alpha_1 \beta_2 - \alpha_2 \beta_1.$$

   (ii) When A_0 is symmetric, A_0 is automatically semibounded and its closure is self-adjoint.
2. The results of the above exercise hold for $A_1 u = (d/dx)[p(x)(d/dx)u] + q(x)$ on $\mathcal{D}(A_1) = \mathcal{D}(A_0)$ where p, p', q are continuous and real-valued on $[0,1]$ with p positive and $p(0) = p(1)$.
3. Let $a(x,D)$ be a strongly elliptic partial differential operator of order $2m$ with coefficients defined on a smooth bounded domain Ω in $\mathbb{R}^n$. Let A be the operator on $L^2(\Omega)$ associated with $a(x,D)$ and with Dirichlet boundary conditions. Use the Gårding inequality to conclude that $A - \lambda I$ is m-dissipative for some $\lambda \in \mathbb{R}$.
*4. Let Ω be a bounded domain in $\mathbb{R}^n$ with a C^2 boundary. Let $j \in \mathbb{N}$ and $1 \leq p < \infty$. Then for every $\epsilon > 0$ there is a constant $C_\epsilon = C_\epsilon(\epsilon, j, p, \Omega)$ such that

$$\|u\|_{j-1,p} \leq \epsilon\|u\|_{j,p} + C_\epsilon\|u\|_{0,p}$$

holds for all $u \in C^j(\bar{\Omega})$, hence for all $u \in W^{j,p}(\Omega)$. (Hints. First let $n = 1$ so that Ω is an interval. Next let Ω be a cube. Etc.)
5. Let Ω be as above and let A acting in $L^P(\Omega)$, $1 \leq p < \infty$, be a strongly elliptic operator of order $2m$. Use the preceding problem to show that any lower-order operator with bounded measurable coefficients is a Kato perturbation of A.
6. Let $\mathcal{H} = L^2([0,1])$ and let $A_0 = id/dx$ on $\mathcal{D}(A_0) = \{u \in C^2([0,1]) : \alpha u(0) + \beta u(1) = 0\}$ where α, β are fixed complex numbers which are not both zero. For which values of α, β is the closure of A_0 self-adjoint?

13. Time Dependent Equations

13.1 Consider the abstract Cauchy problem

$$du(t)/dt = A(t)u(t) \qquad (t \geq s \geq 0),\ u(s) = f. \tag{13.1}$$

Suppose that the problem is well-posed in the sense that for data f in a dense set, there exists a unique solution of (13.1) which depends continuously on the data. Let $U(t,s)$ map the solution $u(s) = f$ at time s to the solution $u(t)$ at time t. The uniqueness implies that the family of operators $\{U(t,s) : t \geq s \geq 0\} \subset \mathscr{B}(\mathscr{X})$ satisfies

$$U(t,s)U(s,r) = U(t,r) \qquad (t \geq s \geq r \geq 0), \tag{13.2}$$

$$U(t,t) = I \qquad (t \in \mathbb{R}^+), \tag{13.3}$$

$$U(\cdot,\cdot)f \text{ is (strongly) continuous for each } f \in \mathscr{X}. \tag{13.4}$$

(In case $A(t) \equiv A$ independent of t, then $U(t,s) = T(t-s)$ depends only on $t-s$, and $\{T(t) : t \in \mathbb{R}^+\}$ is a (C_0) semigroup.) Formally,

$$U(t,s) = \exp\left[\int_s^t A(\tau)\, d\tau\right];$$

at least this is valid if, say, $\mathscr{X} = \mathbb{K}$ and $A : \mathbb{R}^+ \to \mathbb{K}$ is continuous. Introduce Riemann sums for $\int_s^t A(\tau)\, d\tau$; using obvious notation,

$$\exp\left[\int_s^t A(\tau)\, d\tau\right] = \lim_{\text{mesh } \pi \to 0} \exp\left\{\sum_{j=1}^n A(\tau_j')(\tau_j - \tau_{j-1})\right\}$$

$$= \lim_{\text{mesh } \pi \to 0} \prod_{j=1}^n \{\exp(\tau_j - \tau_{j-1})A(\tau_j')\}$$

[Here $\pi : s = \tau_0 < \tau_1 < \cdots < \tau_n = t$, $\tau_j' \in [\tau_{j-1}, \tau_j]$, $j = 1, \ldots, n$, mesh $\pi = \max_j(\tau_j - \tau_{j-1})$.] Now assume that for each $\tau \in \mathbb{R}^+$, $A(\tau)$ generates a (C_0) contraction semigroup on $\mathscr{X}$ which we denote by $\{\exp\{sA(\tau)\} : s \in \mathbb{R}^+\}$. If π is a partition of $[s,t] \subset \mathbb{R}^+$ as above, define

$$U_\pi(t,s) = \exp\{(\tau_n - \tau_{n-1})A(\tau_n')\} \ldots \exp\{(\tau_1 - \tau_0)A(\tau_1')\}. \tag{13.5}$$

Then, hopefully, $U(t,s)f$ exists and equals $\lim_{\text{mesh } \pi \to 0} U_\pi(t,s)f$ for each $f \in \mathscr{X}$. The order of the factors in (13.5) is important since the operators do not commute in general. To see why the ordering is chosen as in (13.5), we approximate (13.1) by

$$du(r)/dr = A_\pi(r)u(r) \qquad (r \geq s),\ u(s) = f \tag{13.6}$$

where A_π is the "step function": $A_\pi(r) = A(\tau_j')$ for $\tau_{j-1} \leq r < \tau_j$, $1 \leq j \leq n$. Solving (13.6) amounts first to solving $u'(r) = A(\tau_1')u(r)$, $u(s) = f$ for $s = \tau_0 \leq r \leq \tau_1$; let $f_2 = u(\tau_1)$. Next solve $u'(r) = A(\tau_2')u(r)$, $u(\tau_1) = f_2$ for $\tau_1 \leq r \leq \tau_2$. Continuing in this way we finally get

$$u(t) = u(\tau_n) = U_\pi(t,s)f$$

where U_π is given by (13.5). Thus, using classical jargon, $U_\pi(t,s)f$ is a *Cauchy-Peano polygonal approximation* for $U(t,s)f$.

We therefore expect that if for each $t \in \mathbb{R}^+$, $A(t)$ generates a (C_0) contraction semigroup, if $\bigcap_{t \in \mathbb{R}^+} \mathscr{D}[A(t)]$ is dense in $\mathscr{X}$, and if $A(t)$ depends on t in a suitably "smooth" manner, then (13.1) can be solved uniquely for appropriate choices of initial data f, and the solution is given by $u(t) = U(t,s)f$ where the *evolution operators* $\{U(t,s): t \geq s \geq 0\}$ satisfy (13.2)–(13.4).

Note that it is enough to construct the evolution operators $U(t,s)$ for $0 \leq s \leq t \leq T$ where $T > 0$ is fixed but otherwise arbitrary.

13.2. The following theorem gives a simple sufficient condition for the Cauchy problem (13.1) to be well-posed.

THEOREM. *Suppose that for each $t \in \mathbb{R}^+$, $A(t)$ generates a (C_0) contraction semigroup on $\mathscr{X}$ such that $\exp\{sA(t)\}$ and $\exp\{\sigma A(\tau)\}$ commute for all $s, t, \sigma, \tau \in \mathbb{R}^+$. Let $\mathscr{D} \subset \bigcap_{t \geq 0} \mathscr{D}[A(t)]$ be a dense subspace of $\mathscr{X}$ such that $A(\cdot)f \in C(\mathbb{R}^+, \mathscr{X})$ for each $f \in \mathscr{D}$. Then there exists a family $\{U(t,s): t \geq s \geq 0\}$ of contraction evolution operators satisfying (13.2)–(13.4), and for all $f \in \mathscr{D}$, $t \geq s \geq 0$,*

$$\frac{\partial}{\partial t} U(t,s)f = A(t)U(t,s)f = U(t,s)A(t)f,$$

$$\frac{\partial}{\partial s} U(t,s)f = -U(t,s)A(s)f = -A(s)U(t,s)f.$$

$u(\cdot) = U(\cdot,s)f$ *is the unique solution of* (13.1) *for* $f \in \mathscr{D}$.

Proof. We shall sketch the proof. Since $\|U_\pi(t,s)\| \leq 1$ and $\bar{\mathscr{D}} = \mathscr{X}$, to prove the existence of $U(t,s)$ it suffices to show: for each $\epsilon > 0$ and $f \in \mathscr{D}$ there is a $\delta > 0$ such that

$$\|U_{\pi_1}(t,s)f - U_{\pi_2}(t,s)f\| < \epsilon$$

if mesh π_1, mesh $\pi_2 < \delta$. Let π_3 be the common refinement of π_1, π_2. Then, using the semigroup property, we can write

$$U_{\pi_j}(t,s) = \prod_{k=1}^{m} \exp\{(\tau_k^{(3)} - \tau_{k-1}^{(3)})A(s_k^{(j)})\}, j = 1, 2,$$

where each $s_k^{(j)}$ is a $\tau_l^{(j)'}$, i.e. a "choice point" associated with the partition π_j.
    For $f \in \mathscr{D}$,

$$\|U_{\pi_2}(t,s)f - U_{\pi_1}(t,s)f\|$$

$$= \left\| \prod_{k=1}^{m} \exp\{(\tau_k^{(3)} - \tau_{k-1}^{(3)})A(s_k^{(2)})\}f - \prod_{k=1}^{m} \exp\{(\tau_k^{(3)} - \tau_{k-1}^{(3)})A(s_k^{(1)})\}f \right\|$$

$$\leq \sum_{k=1}^{m} \|\exp\{(\tau_k^{(3)} - \tau_{k-1}^{(3)})A(s_k^{(2)})\}f - \exp\{(\tau_k^{(3)} - \tau_{k-1}^{(3)})A(s_k^{(1)})\}f\|$$

by Exercise 13.15.1 (below)

$$\leq \sum_{k=1}^{m} (\tau_k^{(3)} - \tau_{k-1}^{(3)}) \| A(s_k^{(2)})f - A(s_k^{(1)})f \| \tag{13.7}$$

by Lemma I.2.9. Since $A(\cdot)f$ is uniformly continuous on $[s,t]$, given $\epsilon > 0$ there is a $\delta > 0$ such that $\| A(v)f - A(w)f \| < \epsilon(t - s)^{-1}$ if $v,w \in [s,t]$ and $|v - w| < \delta$. Inserting this in (13.7) yields

$$\| U_{\pi_2}(t,s)f - U_{\pi_1}(t,s)f \| < \epsilon$$

if mesh π_1, mesh $\pi_2 < \delta$. Hence $U(t,s)$ exists. The rest of the proof is straightforward but not entirely trivial; the details are omitted.   ∎

13.3 EXAMPLE. Let $q(\cdot,\cdot):\mathbb{R}^+ \times \mathbb{R}^n \to \mathbb{C}$ be continuous and satisfy $\operatorname{Re} q(t,x) \leq 0$ for all $(t,x) \in \mathbb{R}^+ \times \mathbb{R}^n$. Let $A(t) = \mathscr{F}^{-1} M_{q(t,\cdot)} \mathscr{F}$ on $\mathscr{X} = L^2(\mathbb{R}^n)$. Then all the hypotheses of Theorem 13.2 are satisfied with $\mathscr{D} = C_c^\infty(\mathbb{R}^n)$.

This enables us to solve some Cauchy problems for parabolic equations whose coefficients depend on time but not on spatial variables. To allow dependence on both t and x, we need an existence theorem in which $\exp\{sA(t)\}$ and $\exp\{\sigma A(\tau)\}$ are not required to commute.

13.4. Here we briefly sketch another method of attack on (13.1). Let $U(\cdot,s)$ denote the (operator-valued) solution of

$$dV(t)/dt = A(t)V(t) \qquad (t \geq s), \ V(s) = I.$$

$Z(\cdot) = \exp\{(\cdot - s)A(s)\}$ satisfies

$$dZ(t)/dt = A(s)Z(t) \qquad (t \geq s), \ Z(s) = I.$$

It follows that

$$W(\cdot) = U(\cdot,s) - \exp\{(\cdot - s)A(s)\}$$

satisfies the inhomogeneous Cauchy problem

$$dW(t)/dt = A(t)W(t) + [A(t) - A(s)]\exp\{(t - s)A(s)\}, \ W(s) = 0.$$

The variation of parameters formula of Section 1 suggests that the solution W is given by

$$U(t,s) - \exp\{(t - s)A(s)\} = \int_s^t U(t,\tau)[A(\tau) - A(s)]\exp\{(\tau - s)A(s)\} \, ds. \tag{13.8}$$

Together with the (terminal) condition $U(t,t) = I$, (13.8) can be viewed as a Volterra integral equation for $U(t,s)$ (with t fixed) to be solved for $s \in [0,t]$. One solves this integral equation by successive approximations. This method is particularly successful when each $A(t)$ generates an analytic semigroup, as the following theorem indicates.

13.5. THEOREM. *Let $T > 0$ and let R be an open convex neighborhood in $\mathbb{C}$ of the interval $[0,T]$. Let $M \geq 1$, $\theta \in \,]\pi/2,\pi]$, and let $\{A(t):t \in R\} \subset \mathscr{GA}_b(\theta,M)$. Suppose that for all $f \in \mathscr{X}$, $t \to [I - A(t)]^{-1}f$ is an analytic function from R to $\mathscr{X}$.*

Finally let $g: R \to \mathscr{X}$ *be analytic. Then the evolution operator family* $\{U(t,s): 0 \le s \le t \le T\}$ *governing* (13.1) (*on* $t \le T$) *exists. Moreover,* $U(t,s)$ *has an analytic continuation to complex values of* t,s *such that it is analytic for* $s,t \in R$ *with* $s \ne t$ *and* $|\arg(t - s)| < \theta - \pi/2$ *and strongly continuous up to* $s = t$. *Equations* (13.2)–(13.4) *hold* (*for* $0 \le s \le t \le T$) *and furthermore*

$$\frac{\partial}{\partial t} U(t,s)f = A(t)U(t,s)f \text{ for all } f \in \mathscr{X} \text{ and } t > s,$$

$$\frac{\partial}{\partial t} U(t,s)f = A(t)U(t,s)f \text{ for all } f \in \mathscr{D}[A(s)] \text{ and } t \ge s,$$

$$\frac{\partial}{\partial s} U(t,s)f = -U(t,s)A(s)f \text{ for all } f \in \mathscr{D}[A(s)] \text{ and } t \ge s.$$

Finally, if

$$u(t) = U(t,s)f + \int_s^t U(t,\tau)g(\tau)\, d\tau,$$

then $u(s) = f$, $u'(t) = A(t)u(t) + g(t)$ *for* $s < t \le T$ *and for* $t = s$ *if also* $f \in \mathscr{D}[A(s)]$, *and* u *has an analytic continuation into* R.

13.6. EXAMPLE. Let Ω be a bounded domain in $\mathbb{R}^n$. Let R be as in Theorem 13.5. For each $t \in [0,T]$ let $\{\mathscr{A}(t) = a(t,x,D),\ \mathscr{B}(t) = \{b_j(t,x,D): j = 1,\ldots,m\}: \Omega\}$ be an elliptic boundary system of Agmon type. We suppose that the order $2m$ of $\mathscr{A}(t)$ and the order m_j of $b_j(t,x,D)$ do not depend on t. Finally assume that the coefficients $a_\alpha(t,x)$, $b_{j\beta}(t,x)$ all extend to $R \times \bar{\Omega}$ so that they are all analytic in $t \in R$ for each fixed $x \in \bar{\Omega}$, and every t derivative of each of these coefficients is jointly continuous on $R \times \bar{\Omega}$. Then, using Theorem 12.7, one can check that the hypotheses of Theorem 13.5 hold with $\mathscr{X} = L^p(\Omega)$ for any p, $1 < p < \infty$. Thus we can solve the mixed problem

$$\frac{\partial u}{\partial t} = a(t,x,D)u + g(t,x) \qquad (x \in \Omega,\ t \in \mathbb{R}^+),$$

$$b_j(t,x,D)u = 0 \qquad (x \in \partial\Omega,\ t \in \mathbb{R}^+,\ j = 1,\ldots,m),$$

$$u(0,x) = f(x) \qquad (x \in \Omega).$$

The solution will be an analytic function of t. Note that the domain $\mathscr{D}[A_p(t)] = W^{2m,p}[\Omega; \mathscr{B}(t)]$ will vary quite substantially with t. If all the coefficients $a_\alpha(x)$, $b_{j\alpha}(t,x)$, and $g(t,x)$ are in $C^\infty(\mathbb{R}^+ \times \bar{\Omega})$, then the solution $u = u(t,x)$ will satisfy $u \in C^\infty(]0,\infty[\times \bar{\Omega})$ if $f \in L^p(\Omega)$.

13.7. We now turn to some general theorems in which each $A(t)$ generates a (C_0) semigroup which need not be analytic.

EXISTENCE THEOREM. *Let* $A(t)$ *be a densely defined linear operator on* $\mathscr{X}$ *for each* $t \in \mathbb{R}^+$. *Assume the following two conditions.*

(i) *For each* $t \in \mathbb{R}^+$ *there is a* $Q(t) \in \mathscr{B}(\mathscr{X})$ *with* $Q(t)^{-1} \in \mathscr{B}(\mathscr{X})$ *such that* $Q(t)A(t)Q(t)^{-1}$ *is m-dissipative, and* $Q(\cdot)f \in C^1(\mathbb{R}^+, \mathscr{X})$ *for each* $f \in \mathscr{X}$.

(ii)   *There is an $R: \mathbb{R}^+ \to \mathscr{B}(\mathscr{X})$ with $R(\cdot)f \in C^2(\mathbb{R}^+, \mathscr{X})$ for each $f \in \mathscr{X}$ such that for each $t \in \mathbb{R}^+$, $R(t)^{-1} \in \mathscr{B}(\mathscr{X})$ and $\mathscr{D}(R(t)A(t)R(t)^{-1})$ is independent of t.*

Then for every $f \in \mathscr{D}[A(s)]$ there is a unique (strongly continuously differentiable) solution of

$$du(t)/dt = A(t)u(t) \qquad (t \geq s), u(s) = f.$$

When $R(t) \equiv I$, $\mathscr{D}[A(t)]$ becomes independent of t. R is the device that allows $\mathscr{D}[A(t)]$ to depend on t. We shall briefly comment on the significance of Q when we discuss second-order equations. (See the discussion following Theorem 13.10.) For now, the reader may take $Q(t) \equiv I$.

13.8.   EXAMPLE.   Consider the mixed problem

$$\gamma \frac{\partial u}{\partial t} = \frac{\partial}{\partial x}\left[p(t,x)\frac{\partial u}{\partial x}\right] - q(t,x)u \qquad (0 \leq x \leq 1, s \leq t < \infty),$$

$$\frac{\partial u}{\partial x}(t,0) = \alpha(t)u(t,0), \frac{\partial u}{\partial x}(t,1) = \beta(t)u(t,1),$$

$$u(s,x) = f(x).$$

Here p, q, α, β are smooth real-valued functions with p positive. γ is 1 (the parabolic case) or $-i$ (the Schrödinger case). We can apply Theorem 13.7 in $\mathscr{H} = L^2([0,1])$ taking $Q(t) \equiv I$ and $R(t)$ to be multiplication by $\rho(t,s)$, a function which can be determined from α and β.

13.9.   PERTURBATION THEOREM.   *Let A, Q, R be as in Theorem 13.7. Let $P_1(t)$, $P_2(t)$ be linear operators satisfying the following two conditions.*

(iii) $P_2(\cdot)f \in C^1(\mathbb{R}^+, \mathscr{X})$ *for each $f \in \mathscr{X}$.*
(iv)   *For each $t \in \mathbb{R}^+$, $\mathscr{D}[P_1(t)] \supset \mathscr{D}[A(t)]$, $Q(t)P_1(t)Q(t)^{-1}$ is dissipative, and there are constants $a(t) < 1$ and $b(t) \geq 0$ such that*

$$\|Q(t)P_1(t)f\| \leq a(t)\|Q(t)A(t)f\| + b(t)\|f\|$$

for all $f \in \mathscr{D}[A(t)]$ and all $t \in \mathbb{R}^+$.
      Then for each $f \in \mathscr{D}[A(s)]$ there is a unique solution of

$$du(t)/dt = [A(t) + P_1(t) + P_2(t)]u(t) \qquad (t \geq s), u(s) = f.$$

13.10.   This can be applied to second-order time-dependent equations by writing them as first-order systems. In this context there is the following result.

THEOREM.   *Let $\mathscr{H}$ be a complex Hilbert space. For each $t \in \mathbb{R}$ let $A(t)$ be a self-adjoint operator with domain $\mathscr{D}$ independent of t and suppose*

$$\langle A(t)f, f \rangle \geq c_1(t)\langle f, f \rangle$$

holds for each $f \in \mathscr{D}$ and each $t \in \mathbb{R}$, for some $c_1 \in C(\mathbb{R},]0, \infty[)$. Let $\mathscr{X}$ be the Hilbert space $\mathscr{D}(A(0)^{1/2})$ with norm $\|f\|_{\mathscr{X}} = \|A(0)^{1/2}f\|$. Let $P_j(t)$, $j = 0, 1, 2$, $t \in \mathbb{R}$ be linear operators satisfying the following conditions. $P_2(\cdot)f \in C^1(\mathbb{R}, \mathscr{X})$

for each $f \in \mathscr{X}$ *and* $P_0(\cdot)f, P_1(\cdot)f \in C^1(\mathbb{R}, \mathscr{X})$ *for each* $f \in \mathscr{Z}$; $P_1(t)$ *is a dissipative Kato perturbation of* $A(0)^{1/2}$ *for all* $t \in \mathbb{R}$. *Then for each* $f_1 \in \mathscr{D}$ *and* $f_2 \in \mathscr{Z}$ *there is a unique solution of*

$$u''(t) - [P_1(t) + P_2(t)]u'(t) + [A(t) + P_0(t)]u(t) = 0$$

$$(t \in \mathbb{R}), \quad (13.9)$$

$$u(0) = f_1, u'(0) = f_2.$$

Recall (cf. Lemma 7.7) that

$$\begin{pmatrix} 0 & I \\ -A(t) & 0 \end{pmatrix}$$

is skew-adjoint with respect to a norm that depends on t. When writing $u'' + A(t)u = 0$ [or more generally (13.9)] as a first-order system in $\mathscr{Z} \times \mathscr{X}$ we must, in order to use Theorem 13.9 (or 13.7), choose a fixed norm (which does not depend on t). This can be done by constructing $Q(\cdot)$ properly. For the above theorem, we have $R(t) \equiv I$. A nontrivial choice of R allows $\mathscr{D}[A(t)]$ to vary with t. This is illustrated in the following application.

13.11. EXAMPLE. Consider the mixed problem

$$\frac{\partial^2 u}{\partial t^2} = a\frac{\partial^2 u}{\partial x^2} + b\frac{\partial u}{\partial x} + c\frac{\partial u}{\partial t} + \delta u + g \qquad (13.10)$$

$$(t \in \mathbb{R}, x \in [0,1]),$$

$$\frac{\partial u}{\partial x}(t,0) = \alpha(t)u(t,0), \frac{\partial u}{\partial x}(t,1) = \beta(t)u(t,1), \qquad (13.11)$$

$$u(s,x) = f_1(x), u_t(s,x) = f_2(x).$$

Here $a,b,c,\delta,g,\alpha,\beta$ are smooth real-valued functions with a positive. The boundary conditions (13.11) are incorporated into $\mathscr{D}[A(t)]$. In order to have $f_1 \in \mathscr{D}[A(s)]$, $f_2 \in \mathscr{D}(A(s)^{1/2})$ we need obvious consistency conditions, namely, $f_1'(0) = \alpha(0)f_1(0)$, $f_1'(1) = \beta(1)f_1(1)$, $f_2'(0) = \alpha(0)f_2(0) + \alpha'(0)f_1(0)$, $f_2'(1) = \beta(0)f_2(1) + \beta'(0)f_1(1)$. By restricting t to belong to $[a_0,b_0]$, where $-\infty < a_0 \leq s \leq b_0 < \infty$, we may add a term of the form $\epsilon h \, \partial^2 u / \partial t \, \partial x$ to the right-hand side of (13.10), where $h: \mathbb{R} \times [0,1[\to \mathbb{R}$ is smooth, ϵ is real, and $|\epsilon| < \epsilon_0 = \epsilon_0(a, h, a_0, b_0)$.

13.12. The next theorem is Kato's recent definitive result. Prior to its statement some preliminaries are needed. For $\{B(t) : t \in [0,T]\} \subset \mathscr{B}(\mathscr{X})$, the time-ordered product $\Pi_{j=1}^k B(t_j)$ is $B(t_1)B(t_2)\ldots B(t_k)$ if $0 \leq t_1 \leq t_2 \leq \cdots \leq t_k \leq T$. From now on all products are time ordered. Consider a family $A = \{A(t) : t \in [0,T]\}$ of (C_0) semigroup generators. A is called *quasi-stable* if

$$\left\| \prod_{j=1}^k [\lambda_j - A(t_j)]^{-1} \right\| \leq M \prod_{j=1}^k [\lambda_j - \beta(t_j)]^{-1}$$

whenever $k \in \mathbb{N}, 0 \leq t_1 < t_2 < \cdots < t_k \leq T$, and $\lambda_j > \beta(t_j)$ for $j = 1,\ldots,k$; here M is a constant and the upper Riemann integral of $\beta: [0,T] \to \mathbb{R}$ is finite. This is equivalent to

$$\left\| \prod_{j=1}^{k} \exp\{-s_j A(t_j)\} \right\| \leq M \exp\{\beta(s_1) + \cdots + \beta(s_k)\}$$

for all t_j as above and $s_j \geq 0, j = 1,\ldots,k$. [Here $\{\exp\{sA(t)\} : s \in \mathbb{R}^+\}$ denotes the semigroup generated by $A(t)$.]

13.13. KATO'S EXISTENCE THEOREM. *Let* $A = \{A(t): t \in [0,T]\}$ *be generators of* (C_0) *semigroups on* $\mathscr{X}$ *satisfying the following three conditions.*

(i)   *A is quasi-stable.*

(ii)   *There is a Banach space* $\mathscr{Y}$, *continuously and densely imbedded in* $\mathscr{X}$, *and a family*

$S = \{S(t): t \in [0,T]\} \subset \mathscr{B}(\mathscr{Y},\mathscr{X})$ *with* $S(t)^{-1} \in \mathscr{B}(\mathscr{X},\mathscr{Y})$ *for all* $t \in [0,T]$ *such that*

$$S(t)A(t)S(t)^{-1} = A(t) + B(t)$$

where $B(t) \in \mathscr{B}(\mathscr{X})$ *for a.e.* t, $B(\cdot)f : \mathbb{R} \to \mathscr{X}$ *is strongly measurable for each* $f \in \mathscr{X}$, *and the upper Riemann integral* $\int_0^T \|B(t)\|_{\mathscr{B}(\mathscr{X})} dt$ *is finite. Furthermore, there is a function* $\dot{S}: [0,T] \to \mathscr{B}(\mathscr{Y},\mathscr{X})$ *such that* $\dot{S}(\cdot)f$ *is strongly measurable for each* $y \in \mathscr{Y}$, *the upper Riemann integral* $\int_0^T \|\dot{S}(t)\|_{\mathscr{B}(\mathscr{Y},\mathscr{X})} dt$ *is finite, and* $S(t)f = S(0)f + \int_0^t \dot{S}(s)f\, ds$ *for all* $t \in [0,T]$ *and all* $f \in \mathscr{Y}$.

(iii) $\mathscr{Y} \subset \bigcap\{\mathscr{D}[A(t)]: t \in [0,T]\}$ *and* $A \in C[[0,T], \mathscr{B}(\mathscr{Y},\mathscr{X})]$. *Then there is a unique family of evolution operators* $\{U(t,s):(t,s) \in \Delta = \{(r,p): 0 \leq p \leq r \leq T\} \subset \mathscr{B}(\mathscr{X})$ *such that*

(a) $U(t,s)U(s,r) = U(t,r), U(t,t) = I$ *for* $0 \leq r \leq s \leq t \leq T$,

(b) $U(\cdot,\cdot)f \in C(\Delta,\mathscr{X})$ *for each* $f \in \mathscr{X}$,

(c) $U(\cdot,\cdot)f \in C(\Delta,\mathscr{Y})$ *for each* $f \in \mathscr{Y}$,

(d)   *for each* $f \in \mathscr{Y}$,

$$\frac{d}{dt} U(t,s)f = A(t)U(t,s)f, \frac{d}{ds} U(t,s)f = -U(t,s)A(s)f,$$

and both of these functions are in $C(\Delta,\mathscr{X})$.

We omit the proof, which is based on the product integral construction discussed in 13.1 and 13.2. We also omit statements of related perturbation and approximation theorems.

13.14 EXAMPLE.   The above theorem can be applied to symmetric hyperbolic systems. Let Ω be a bounded domain in $\mathbb{R}^n$ with a C^3 boundary $\partial\Omega$. The equation (or rather system of equations) under consideration is

$$\frac{\partial u}{\partial t} + \sum_{j=1}^{n} a_j(t,x) \frac{\partial u}{\partial x_j} + b(t,x)u = 0$$

where $u = (u_1,\ldots,u_m):[0,T] \times \bar{\Omega} \to \mathbb{R}^m$, and for each $t \in [0,T]$ and $x \in \bar{\Omega}$, $a_j(t,x)$ and $b(t,x)$ are real $m \times m$ matrices with $a_j(t,x)$ symmetric, $j = 1,\ldots,n$. Moreover, the entries of each $a_j(\cdot,\cdot)$ [resp. $b_j(\cdot,\cdot)$] are in $C^2([0,T] \times \bar{\Omega})$ [resp.

$C^1([0,T] \times \bar{\Omega})]$. The initial condition is

$$u(0,x) = f(x)$$

and the boundary conditions are

$$u(t,x) \in \mathscr{P}(t,x) \text{ for all } (t,x) \in [0,T] \times \partial\Omega.$$

The boundary subspace $\mathscr{P}$ is defined as follows. Let $v = (v_1, \ldots, v_m)$ be the unit outward normal to $\partial\Omega$ at x. The boundary matrix

$$a_v(t,x) = \sum_{j=1}^{n} v_j a_j(t,x)$$

is assumed to be nonsingular on $[0,T] \times \partial\Omega$. $\mathscr{P}(t,x)$ is a subspace (i.e. subvector space) of $\mathbb{R}^m$ which varies in a C^3 manner as (t,x) varies over $[0,T] \times \partial\Omega$, such that for each $\xi \in \mathscr{P}(t,x)$,

$$a_v(t,x)\xi \cdot \xi \geq 0,$$

and $\mathscr{P}(t,x)$ is not contained in any larger subspace of $\mathbb{R}^m$ having this property.

Theorem 13.13 applies to this situation. One takes $\mathscr{X} = \{v = (v_1, \ldots, v_m): v_j \in L^2(\Omega) \text{ for } 1 \leq j \leq m\} = [L^2(\Omega)]^m$.

Next, a suitable change of variables reduces the problem to the case when $\mathscr{P}(t,x)$ is independent of t; then one takes $\mathscr{Y}$ to be the completion of

$$\{v \in \mathscr{X}: v_j \in C^1(\bar{\Omega}), v_j(x) \in \mathscr{P}(0,x) \text{ for all } x \in \partial\Omega, 1 \leq j \leq m\}$$

in the $[H^1(\Omega)]^m$ norm. The (complicated) details are omitted.

We remark that given $S \in \mathbb{N}$, by assuming that $\partial\Omega$, a_j, b, and f are sufficiently smooth, the solution u can be shown to belong to $[H^s(\Omega)]^m$ (if certain consistency conditions hold).

13.15 EXERCISES

1. Let $T_1, \ldots, T_{2m} \in \mathscr{B}(\mathscr{X})$ be commuting contractions. Then for all $f \in \mathscr{X}$,

$$\|(T_1 \cdots T_m)f - (T_{m+1} \cdots T_{2m})f\| \leq \sum_{j=1}^{m} \|T_j f - T_{j+m} f\|.$$

2. Complete the proof of Theorem 13.2.
3. Verify that Theorem 13.7 applies to Example 13.8. In particular, find $\rho(t,x)$.

14. Scattering Theory

14.1. Abstract scattering theory is concerned with the asymptotic equivalence of two (C_0) unitary groups on a Hilbert space. The main result presented here is the Kato-Kuroda theorem, which says for two-body quantum mechanical problems, the wave operators exist and are complete and the scattering operator is unitary if the potential dies down as fast as $(1 + |x|)^{-1-\epsilon}$ for some $\epsilon > 0$. We prove this as a consequence of the theory of local smoothness of operators and a result of Agmon in partial differential equations; we develop the former theory but merely quote Agmon's result. We also briefly introduce the Lax-Phillips theory, and we discuss a few miscellaneous topics.

14.2. DISCUSSION. Let $\mathcal{H}$ be the complex space $L^2(\mathbb{R}^n)$ and H_0 the self-adjoint operator $-\Delta$ with domain $H^2(\mathbb{R}^n)$, the Sobolev space. Let $V: \mathbb{R}^n \to \mathbb{R}$ satisfy $V = V_1 + V_2$ where $V_1 \in L^p(\mathbb{R}^n)$, $V_2 \in L^\infty(\mathbb{R}^n)$ where $p > n/2, p \geq 2$. Then $H_1 = H_0 + M_V$ is self-adjoint and bounded from below, and $\mathscr{D}(H_1) = H^2(\mathbb{R}^n)$ (compare Exercises 6.14.4 and 6.14.11). H_1 has the following interpretation. (See Chapter I, Section 8.13.) H_1 is the *Hamiltonian* of a spinless, nonrelativisitic quantum mechanical particle travelling in $\mathbb{R}^n$ under the influence of a *potential* V. The *Schrödinger equation* for the particle is

$$iu'(t) = H_1 u(t) \qquad (t \in \mathbb{R}),$$

and the solution $u(t) = \exp\{-itH_1\}u(0)$ is the *wave function* of the particle at time t. H_1 is also the Hamiltonian (and $u(t)$ the wave function) of the relative motion of two such particles in $\mathbb{R}^n$ when there is no external force and where V is the potential describing the force the particles exert on one another. (When the first particle is at $x (\in \mathbb{R}^n)$ and the second at y, the force is assumed to be a function of $y - x$.) H_0 corresponds to $V \equiv 0$ and describes *free motion*, where the two bodies aren't influenced by each other. Let $U_j(t) = \exp\{-itH_j\}, j = 0, 1, t \in \mathbb{R}$. Let $f \in \mathcal{H} = L^2(\mathbb{R}^n)$. f is said to be an *asymptotically free state* of H_1 as $t \to -\infty$, if there is a $g \in \mathcal{H}$ such that

$$\|U_1(t)f - U_0(t)g\| \to 0 \qquad (14.1)$$

as $t \to -\infty$. Similarly for $t \to +\infty$. Thus an asymptotically free state is the initial condition for the *perturbed equation* (i.e. $iu' = H_1 u$) such that the solution looks like a solution of the free equation as $t \to -\infty$ (or $t \to +\infty$).

Let $\mathcal{H}_{af,+}(H_1)$, $\mathcal{H}_{af,-}(H)$ denote the asymptotically free states of H_1 as $t \to +\infty, t \to -\infty$. Rewrite equation (14.1) as

$$\|f - U_1(-t)U_0(t)g\| \to 0$$

as $t \to -\infty$. We define the *wave operators* $W_\pm$ by

$$W_\pm g = \lim_{t \to \pm\infty} U_1(-t)U_0(t)g,$$

whenever these limits exist. Clearly the range of $W_\pm$ is $\mathcal{H}_{af,\pm}(H_1)$.

When $V(x) \to 0$ sufficiently rapidly as $|x| \to \infty$ we expect one of two things to happen. Either the particles move independently of one another for large $|t|$ at large relative distances, or else the particles stay close together. This leads us to define the *scattered states* for H_1, $\mathcal{H}_{scat}(H_1)$, and the *bound states* for H_1, $\mathcal{H}_{bd}(H_1)$ to be

$$\mathcal{H}_{scat}(H_1) = \left\{ f \in \mathcal{H}: \lim_{t \to \pm\infty} \int_{|x| \leq r} |U_1(t)f(x)|^2 \, dx = 0 \text{ for each } r > 0 \right\},$$

$$H_{bd}(H_1) = \Big\{ f \in \mathcal{H}: \text{For each } \epsilon > 0 \text{ there is an } r > 0 \text{ such that}$$

$$\int_{|x| > r} |U_1(t)f(x)|^2 \, dx < \epsilon \text{ for each } t \in \mathbb{R} \Big\}.$$

We want to relate these notions to the intrinsic structure of H_1 as an operator on Hilbert space. This will be done in Theorem 14.4 after some preliminary definitions.

14.3. DEFINITIONS. Let H be a self-adjoint operator on Hilbert space $\mathcal{H}$. Write $H = \int_{-\infty}^{\infty} \lambda \, dE_\lambda$. For $f \in \mathcal{H}$, the function $m_f : \lambda \to \|E_\lambda f\|^2$ is bounded and nondecreasing. Let $\mathcal{H}_{ac}(H) = \{f \in \mathcal{H} : m_f \text{ is absolutely continuous}\}$. Similarly define $\mathcal{H}_{sc}(H)$, $\mathcal{H}_s(H)$, $\mathcal{H}_c(H)$, $\mathcal{H}_d(H)$. Here sc stands for singular continuous, s for singular, c for continuous, and d for discrete (i.e. $f \in \mathcal{H}_d(H)$ iff m_f is constant except for jumps). We have

$$\mathcal{H}_c(H) = \mathcal{H}_{sc}(H) \oplus \mathcal{H}_{ac}(H), \quad \mathcal{H}_s(H) = \mathcal{H}_{sc}(H) \oplus \mathcal{H}_d(H),$$

$$\mathcal{H} = \mathcal{H}_{ac}(H) \oplus \mathcal{H}_{sc}(H) \oplus \mathcal{H}_d(H). \tag{14.2}$$

The last equation is the self-adjoint operator analogue of the *Lebesgue decomposition theorem* (see Kato [12, p. 516]).

14.4 THEOREM. *Let $H_0 = -\Delta$, $H_1 = -\Delta + M_V$ on $\mathcal{H} = L^2(\mathbb{R}^n)$ where V satisfies the conditions of the first paragraph of 14.2. Then*

$$\mathcal{H}_{bd}(H_1) = \mathcal{H}_d(H_1),$$

and if $\mathcal{H}_{sc}(H) = \{0\}$,

$$\mathcal{H}_{scat}(H_1) = \mathcal{H}_{ac}(H_1).$$

The proof is omitted; see Wilcox [3], Amrein-Georgescu [1], and Ruelle [1].

14.5. DEFINITION. Let $V = V_1 + V_2 : \mathbb{R}^n \to \mathbb{R}$ where $V_1 \in L^P(\mathbb{R}^n)$, $V_2 \in L^\infty(\mathbb{R}^n)$, with $p > n/2$, $p \geq 2$. V will be called a *short-range potential* if, in addition, $V(x) = O(1/|x|^{1+\epsilon})$ as $|x| \to \infty$ for some $\epsilon > 0$.

14.6. THEOREM (Kato-Kuroda). *Let V be a short-range potential. Then $W_\pm$ both exist as isometric operators defined on $\mathcal{H} = L^2(\mathbb{R}^n)$ and have range equal to*

$$\mathcal{H}_{af,+}(H_1) = \mathcal{H}_{af,-}(H_1) = \mathcal{H}_{scat}(H_1) = \mathcal{H}_{ac}(H_1).^\dagger$$

Moreover $\mathcal{H}_{sc}(H_1) = \{0\}$.

An outline of the proof will be given shortly.

14.7. DISCUSSION. Motivated by the preceding discussion, we introduce the abstract formalism of scattering theory. Let H_1, H_0 be self-adjoint operators on a complex Hilbert space $\mathcal{H}$. Let $U_j = \{U_j(t) = \exp(-itH_j) : t \in \mathbb{R}\}$ be the (C_0) unitary group generated by $-iH_j$, and let P_j be the orthogonal projection onto $\mathcal{H}_{ac}(H_j)$ for $j = 0, 1$. The *generalized wave operators* are defined by

$$W_\pm f = W_\pm(H_1, H_0)f = \lim_{t \to \pm\infty} U_1(-t)U_0(t)P_0 f \qquad f \in \mathcal{H}.$$

When these exist, they are said to be *complete* if the range of $W_\pm$ is $\mathcal{H}_{ac}(H_1)$. [When $W_\pm$ exists, it is isometric from $\mathcal{H}_{ac}(H_0)$ to its range $R_\pm$ which is automatically contained in $\mathcal{H}_{ac}(H_1)$.]

† $\mathcal{H}_{af,\pm}(H_1)$ denotes the asymptotically free states of H_1 as $t \to \pm\infty$.

The intertwining relationship

$$U_1(s)W_\pm = W_\pm U_0(s) \qquad s \in \mathbb{R},$$

which is trivial to verify, yields

$$H_1 W_\pm \supset W_\pm H_0, \quad H_0 W_\pm^* \supset W_\pm^* H_1.$$

If $W_\pm$ exist and are complete, and if H_j^{ac} denotes the restriction of H_j to $\mathcal{H}_{ac}(H_j)$, then the self-adjoint operators H_1^{ac} and H_0^{ac} are unitarily equivalent via the unitary operator $W_\pm$ from $\mathcal{H}_{ac}(H_0)$ to $\mathcal{H}_{ac}(H_1)$.

When $W_\pm$ exist and are complete, the *scattering operator* is defined to be $S = W_+^{-1}W_-$.[†] It has the following interpretation. Let $f \in \mathcal{H}_{ac}(H_1)$. If $u(t) = U_1(t)f$ then $\|u(t) - U_0(t)W_-^{-1}f\| \to 0$ as $t \to \infty$. Thus the (abstract) experimenter, who can detect "free" motion, "measures" $f_- = W_-^{-1}f \in \mathcal{H}_{ac}(H_0)$ at the "beginning" of the experiment (i.e. when $t \to -\infty$). Since we also have $\|u(t) - U_0(t)W_+^{-1}f\| \to 0$ as $t \to +\infty$, the experimenter measures the state $f_+ = W_+^{-1}f$ at the "end" of the experiment (i.e. as $t \to +\infty$). Since

$$f_+ = W_+^{-1}W_- f_- = Sf_-,$$

we see that the scattering operator sends the *initial (free) state* f_- to the *final (free) state* f_+ and so corresponds to what the experimenter measures.

When $W_\pm$ exist and are complete, S is unitary on $\mathcal{H}_{ac}(H_0)$ and S commutes with the "free" group U_0.

When $H_0 = -\Delta$ on $L^2(\mathbb{R}^n)$ it is easy to check that $\mathcal{H}_{ac}(H_0) = \mathcal{H} = L^2(\mathbb{R}^n)$. Thus, according to the Kato-Kuroda theorem, when V is short range, the wave operators are complete and the scattering operator S is unitary on $L^2(\mathbb{R}^n)$.

14.8. We begin our discussion of the abstract theory of scattering with the following simple result.

PROPOSITION. *Let H_0, H_1 be self-adjoint operators on $\mathcal{H}$. Let $\mathscr{D} \subset \mathcal{H}$ have a linear span that is dense in $\mathcal{H}_{ac}(H_1)$ and suppose that for each $f \in \mathscr{D}$, there is an $s \in \mathbb{R}^+$ such that $v(t) = e^{-itH_0}f \in \mathscr{D}(H_1) \cap \mathscr{D}(H_0)$ for $t \geq s$ and $(H_1 - H_0)v(\cdot) \in L^1([s,\infty);\mathcal{H})$. Then $W_+(H_1,H_0)$ exists.*

Proof. Let $W(t)f = e^{itH_1}e^{-itH_0}f$. It is enough to show that $W(t)f$ is Cauchy as $t \to \infty$ for each $f \in \mathscr{D}$. For $f \in \mathscr{D}$ and $t \geq \tau \geq s$,

$$\|W(t)f - W(\tau)f\| \leq \int_\tau^t \left\| \frac{d}{dr} W(r)f \right\| dr$$

$$= \int_\tau^t \|(H_1 - H_0)e^{-irH_0}f\| \, dr \to 0 \text{ as } t,\tau \to \infty,$$

and the result follows.   ∎

[†] We are viewing $W_\pm$ as unitary operators from $\mathcal{H}_{ac}(H_0)$ to $\mathcal{H}_{ac}(H_1)$, so their inverses make sense.

14.9. EXAMPLE. Let $\mathscr{H} = L^2(\mathbb{R}^3)$, $H_0 = -\Delta$, $H_1 = -\Delta + M_V$ where $V \in L^2(\mathbb{R}^3)$. Let

$$f(x) = \exp\{-|x - a|^2/2b\} \qquad x \in \mathbb{R}^3.$$

As a varies over $\mathbb{R}^3$ and b over $]0,\infty[$, the set of such fs has dense linear span, since for all $g \in L^2(\mathbb{R}^3)$,

$$g(x) = L^2(\mathbb{R}^3) - \lim_{b \to 0^+} (2\pi b)^{-3/2} \int_{\mathbb{R}^3} \exp -\left\{\frac{|x - y|^2}{2b}\right\} g(y)\, dy.$$

Next, taking Fourier transforms, for $\xi \in \mathbb{R}^3$,

$$\hat{f}(\xi) = b^{3/2} \exp\{-b|\xi|^2/2\}\, e^{i\xi \cdot a},$$

$$(e^{-itH_0}f)^{\wedge}(\xi) = b^{3/2} e^{i\xi \cdot a} \exp\left\{-|\xi|^2 \left(\frac{b}{2} + it\right)\right\}.$$

Consequently, by the Fourier inversion formula [3.10 and 3.9(6)],

$$e^{-itH_0}f(x) = \left(\frac{b}{2it + b}\right)^{3/2} \exp\{-|x - a|^2/2(b + 2it)\}.$$

Therefore, for $|t| \geq 1/2$,

$$\|(H_1 - H_0)e^{-itH_0}f\| \leq \|V\|_{L^2} \|e^{-itH_0}f\|_{L^\infty} \leq \frac{\text{constant}}{|t|^{3/2}}.$$

If follows that $W_\pm(H_1, H_0)$ both exist.

If we assume $|V(x)| \leq C(1 + |x|^\alpha)^{-1}$, then $(1 + |\cdot|^\alpha)^{-1} \in L^2(\mathbb{R}^3)$ is sufficient for $V \in L^2(\mathbb{R}^3)$. But, using polar coordinates, we see that

$$\int_{\mathbb{R}^3} (1 + |x|^\alpha)^{-2}\, dx = \int_{S^2} \int_0^\infty (1 + r^\alpha)^{-2} r^2\, dr\, d\omega < \infty$$

holds iff $\alpha > 3/2$. Therefore $W_\pm(H_0 + M_V, H_0)$ exists if $V = V_1 + V_2$ where $V_1 \in L^2(\mathbb{R}^3)$ and $|V_2(x)| \leq C(1 + |x|)^{-(3/2)-\epsilon}$ for some $\epsilon > 0$, i.e. $V \in L^2_{\text{loc}}(\mathbb{R}^3)$ and $V(x) = O(|x|^{-(3/2)-\epsilon})$ as $|x| \to \infty$.

14.10. It is not difficult to strengthen the conclusion of the last sentence to $W_\pm(H_0 + M_V, H_0)$ exists whenever $V \in L^2_{\text{loc}}(\mathbb{R}^3)$ and $V(x) = O(|x|^{-1-\epsilon})$ as $|x| \to \infty$. What is hard is to prove the completeness. For this the following lemma is useful.

CHAIN RULE. *Let H_j be self adjoint on $\mathscr{H}$, for $j = 0,1,2$. If $W_\pm(H_2, H_1)$ and $W_\pm(H_1, H_0)$ exist, then $W_\pm(H_2, H_0)$ exists and equals $W_\pm(H_2, H_1)\, W_\pm(H_1, H_0)$.*

The easy proof is omitted. By taking $H_2 = H_0$, since $W_\pm(H_0, H_0) = P_0$, we see that if the wave operators $W_\pm(H_1, H_0)$ exist, then they are complete iff $W_\pm(H_0, H_1)$ exist. Proposition 14.8 is a useful tool to apply when $U_0(t)$ is explicitly known. In spite of the Feynman path integral formula of Chapter I, Section 8.13, we do not have an expression for $\exp\{-it(-\Delta + M_V)\}$ which allows us to apply Proposition 14.8 with H_0 and H_1 interchanged.

14.11 We shall sketch a conceptual, abstract approach to the proof of the Kato-Kuroda theorem (Theorem 14.6).

14.12. DEFINITIONS. Let H be self-adjoint on $\mathcal{H}$ and let $U(t) = e^{-itH}, t \in \mathbb{R}$. A linear operator A on $\mathcal{H}$ is called H-*smooth* if there is a real number C such that

$$\int_{-\infty}^{\infty} \|AU(t)f\|^2 \, dt \leq C\|f\|^2$$

holds for a dense set of vectors f in $\mathcal{H}$ (and hence for all f in $\mathcal{H}$).

Write $H = \int_{-\infty}^{\infty} \lambda \, dE_\lambda$ by the spectral theorem and let $E(\Gamma) = \int_\Gamma dE_\lambda$ for Γ a Borel set in $\mathbb{R}$. A is called H-*smooth on* Γ if $AE(\Gamma)$ is H-smooth.

We shall not attempt to relate this notion of smoothness to the usual notion of smoothness. We simply declare that "H-smooth" is now standard terminology.

14.13. PROPOSITION. *Let A be H-smooth on Γ. Then for every $f \in \mathscr{D}(A^*)$. $E(\Gamma)A^*f \in \mathscr{H}_{ac}(H)$.*

Proof. For $f \in \mathscr{D}(A^*)$ let F be the Fourier transform of the measure $d\sigma$ determined by the monotone function

$$\sigma(\lambda) = \langle E_\lambda E(\Gamma)A^*f, E(\Gamma)A^*f \rangle = \|E_\lambda g\|^2$$

where $g = E(\Gamma)A^*f$. Then, for $\xi \in \mathbb{R}$,

$$\sqrt{2\pi}\, F(\xi) = \int_{-\infty}^{\infty} e^{i\lambda\xi} \, d\langle E_\lambda g, g \rangle$$

$$= \langle e^{i\xi H} E(\Gamma)A^*f, g \rangle$$

$$= \langle f, AE(\Gamma)e^{-i\xi H} g \rangle,$$

whence

$$|F(\xi)| \leq \frac{1}{\sqrt{2\pi}} \|f\| \|AE(\Gamma)U(\xi)g\|,$$

$$\int_{-\infty}^{\infty} |F(\xi)|^2 \, d\xi \leq \frac{1}{2\pi} \|f\|^2 \int_{-\infty}^{\infty} \|AE(\Gamma)U(\xi)g\|^2 \, d\xi < \infty$$

since A is H-smooth on Γ. Thus $F \in L^2(\mathbb{R})$, and by the unitarity of the Fourier transform, $F = \hat{G}$ for some unique $G \in L^2(\mathbb{R})$. But $F = (d\sigma)^\wedge$, and the Fourier transform of a measure uniquely determines it, hence σ is absolutely continuous and $\sigma' = G \in L^1(\mathbb{R}) \cap L^2(\mathbb{R})$. Consequently $g \in \mathscr{H}_{ac}(H)$.   ∎

14.14 Let H_j be self-adjoint, $j = 0,1$. Write $H_j = \int_{-\infty}^{\infty} \lambda \, dE_\lambda^j$, $U_j(t) = e^{-itH_j}$.

PROPOSITION. *Suppose $H_1 = H_0 + A_1^* A_0$ in the sense that $\mathscr{D}(H_1) \subset \mathscr{D}(A_1)$, $\mathscr{D}(H_0) \subset \mathscr{D}(A_0)$, and*

$$\langle H_1 f, g \rangle = \langle f, H_0 g \rangle + \langle A_1 f, A_0 g \rangle$$

holds for all $f \in \mathcal{D}(H_1), g \in \mathcal{D}(H_0)$. Let Γ be an open interval, and suppose A_j is H_j-smooth on Γ for $j = 0,1$. Then the local wave operators

$$W_\pm(\Gamma)h = \lim_{t \to \pm\infty} U_1(-t)U_0(t)E^0(\Gamma)h,$$

$$\Omega_\pm(\Gamma)h = \lim_{t \to \pm\infty} U_0(-t)U_1(t)E^1(\Gamma)h$$

exist for all $h \in \mathcal{H}$, and

$$W_\pm(\Gamma)^* = \Omega_\pm(\Gamma), \ W_\pm(\Gamma)^*W_\pm(\Gamma) = E^0(\Gamma), \ W_\pm(\Gamma)W_\pm(\Gamma)^* = E^1(\Gamma).$$

Proof. Let $w(t) = E^1(\Gamma)U_1(-t)U_0(t)E^0(\Gamma)h$ for $h \in \mathcal{H}$. Because of the chain rule and other elementary considerations it is enough to show

(i) $\lim_{t \to \pm\infty} w(t)$ exists (and equals $W_\pm(\Gamma)h$, say),
(ii) $\lim_{t \to \pm\infty} E^1(\mathbb{R} \setminus \Gamma)U_1(-t)U_0(t)E^0(\Lambda)h = 0$ for every compact interval $\Lambda \subset \Gamma$.

For (i), suppose $h \in \mathcal{D}(H_0)$ and $k \in \mathcal{D}(H_1)$. Then

$$\frac{d}{dt}\langle k,w(t)\rangle = -i[\langle H_1 U_1(t)E^1(\Gamma)k, U_0(t)E^0(\Gamma)h\rangle - \langle U_1(t)E^1(\Gamma)k, H_0 U_0(t)h\rangle]$$

$$= -i\langle A_1 E^1(\Gamma)U_1(t)k, A_0 E^0(\Gamma)U_0(t)h\rangle$$

by hypothesis. Thus for $t > s$,

$$|\langle k,w(t) - w(s)\rangle| \le \int_s^t |\langle A_1 E^1(\Gamma)U_1(\tau)k, A_0 E^0(\Gamma)U_0(\tau)h\rangle|\, d\tau$$

$$\le \left(\int_s^t \|A_1 E^1(\Gamma)U_1(\tau)k\|^2\, dt\right)^{1/2}\left(\int_s^t \|A_0 E^0(\Gamma)U_0(t)h\|^2\, d\tau\right)^{1/2}$$

$$\le C\|k\|\left(\int_s^t \|A_0 E^0(\Gamma)U_0(\tau)h\|^2\, d\tau\right)^{1/2}$$

for some constant C since A_1 is H_1-smooth on Γ. Taking the supremum over such k with $\|k\| \le 1$ and letting $t,s \to \pm\infty$ yields (i).

Now let Λ be as in (ii). Let L be a smooth contour in $\mathbb{C}$ which encloses Λ and is disjoint from $\overline{\mathbb{R} \setminus \Gamma}$. By the Cauchy integral formula, for $k \in \mathcal{H}$,

$$\oint_L E^1(\mathbb{R}\setminus\Gamma)(z - H_1)^{-1}k\, dz = 0, \qquad \oint_L E^0(\Lambda)(z - H_0)^{-1}k\, dz = 2\pi ik.$$

Thus, taking $k = U_1(-t)U_0(t)E^0(\Lambda)h$ for $h \in \mathcal{H}$, we have

$$2\pi i E^1(\mathbb{R}\setminus\Gamma)U_1(-t)U_0(t)E^0(\Lambda)h$$

$$= -\oint_L E^1(\mathbb{R}\setminus\Gamma)[(z - H_1)^{-1} - (z - H_0)^{-1}]U_1(-t)U_0(t)E^0(\Lambda)h\, dz.$$

Since the integrand is uniformly bounded it suffices to establish

$$\lim_{t \to \pm\infty} \|[(z - H_1)^{-1} - (z - H_0)^{-1}]U_1(-t)U_0(t)E^0(\Lambda)h\| = 0 \qquad (14.3)$$

for all $z \in \mathbb{C}\setminus\mathbb{R}$. But for $k = U_1(-t)U_0(t)E^0(\Lambda)h$ and for all $k_1 \in \mathcal{H}$,

$$|\langle k_1,[(z - H_1)^{-1} - (z - H_0)^{-1}]k\rangle| = |\langle A_1(z - H_1)^{-1}k_1, A_0(z - H_0)^{-1}k\rangle|$$

$$\leq \|A_1(z - H_1)^{-1}\|\,\|k_1\|\,\|A_0(z - H_0)^{-1}k\|.$$

Thus it suffices to show that for

$$F(t) = \|A_0(z - H_0)^{-1}k\| = \|A_0(z - H_0)^{-1}U_1(-t)U_0(t)E^0(\Lambda)h\|,$$

$\lim_{t \to \pm\infty} F(t) = 0$. But since A_0 is H_0-smooth, $F \in L^2(\mathbb{R})$. Since also the derivative $(F^2)'$ is bounded, F^2 is uniformly continuous and so $\lim_{t \to \pm\infty} F^2(t) = 0$. Equation (14.3) now follows. ∎

14.15. PROPOSITION. *Let $H_1 = H_0 + A_1^* A_0$ as in Proposition 14.14. Let $\{\Gamma_n : n \in \mathbb{N}\}$ be a sequence of open intervals in $\mathbb{R}$ such that for each $n \in \mathbb{N}$ and $j = 0,1$, A_j is H_j-smooth on Γ_n, and $\sigma(H_j)\setminus\bigcup_{n=1}^{\infty}\Gamma_n$ is at most countable. Then the generalized wave operators $W_\pm(H_1, H_0)$ exist and are complete.*

Proof. Let $\Gamma = \bigcup_{n=1}^{\infty}\Gamma_n$. By Proposition 14.14,

$$\lim_{t \to \pm\infty} U_1(-t)U_0(t)h$$

exists for all $h \in E^0(\Gamma)(\mathcal{H})$. But $E^0(\Gamma)(\mathcal{H}) \supset \mathcal{H}_{ac}(H_0)$ by the hypothesis that $\sigma(H_0)\setminus\Gamma$ is at most countable. Thus the generalized wave operators $W_\pm(H_1, H_0)$ exist. Interchanging the subscripts 0, 1 and appealing to the chain rule completes the proof. ∎

14.16 DISCUSSION. Let $V = V_1 + V_2 : \mathbb{R}^n \to \mathbb{R}$ where $V_1 \in L^p(\mathbb{R}^n)$ ($p > n/2$, $p \geq 2$), $V_2 \in L^\infty(\mathbb{R}^n)$. Let $V_\rho(x) = V(x)$ or 0, according as $|x| \leq \rho$ or $|x| > \rho$. If $H_0 = -\Delta$, then $M_{V_\rho}(\lambda - H)^{-1}$ is compact by Rellich's theorem, (i.e. by the Hilbert space version of the Ascoli-Arzela theorem). It follows by a classical (1909) theorem of Weyl (cf. Kato's book [12, p. 244]) that $H_{1_\rho} = H_0 + M_{V_\rho}$ and H_0 have the same essential spectrum, i.e. they have the same spectrum except for isolated eigenvalues of finite multiplicity. The same is true for $H_1 = H_0 + M_V$ if also $V(x) \to 0$ as $|x| \to \infty$, which we assume. Thus the spectrum $\sigma(H_1) \cap\,]-\infty,0[$ consists of at most countably many eigenvalues of finite multiplicity which are bounded below and which converge to zero if they are infinite in number.

Let V be a short-range potential. That is, in addition to the above properties suppose $V(x) = O(|x|^{-1-\epsilon})$ as $|x| \to \infty$ for some $\epsilon > 0$. Then one can show that the positive eigenvalues of $H_1 = -\Delta + M_V$ must be of finite multiplicity and can cluster only at 0 and ∞. Moreover, there are none if V is bounded. Let Γ be a compact interval in $]0,\infty[$ which does not contain any eigenvalue of H_1. Let $W = |V_1|^{1/2}$, $U = \text{sign}(V_1)|V_1|^{1/2}$ where $\text{sign}(V_1)(x) = V_1(x)/|V_1(x)|$ or 0, according as $V_1(x) \neq 0$ or $= 0$. Let $A_1 = M_W$, $A_0 = M_U$. Agmon [2] showed that A_j is H_j-smooth on Γ, $j = 1, 2$. This deep result, together with Proposition 14.15 and the absolute continuity of H_0, implies the Kato-Kuroda theorem 14.6. For Agmon's arguments see Agmon [2] or Reed-Simon [4].

14.17. THEOREM (Kato). *Let H_0 and A be self-adjoint. Suppose that $\|A(\lambda - H_0)^{-1}A\|$ is uniformly bounded for $\lambda \in \mathbb{C}\backslash\mathbb{R}$ in the sense that*

$$K \equiv \sup_{\lambda \in \mathbb{C}\backslash\mathbb{R}} \sup_{\substack{f,g \in \mathscr{D}(A) \\ \|f\|, \|g\| \leq 1}} |\langle(\lambda - H_0)^{-1}Af, Ag\rangle| < \infty.$$

Let $B \in \mathscr{B}(\mathscr{H})$ be a contraction such that BA^2 is self-adjoint. Then for $-K < \alpha < K$, $H_\alpha = H_0 + \alpha BA^2$ is self-adjoint and A, BA are both H_α-smooth. The wave operators

$$W_\pm(\alpha)f = \lim_{t \to \pm\infty} e^{-itH_\alpha}e^{itH_0}f$$

exist for all $f \in \mathscr{H}$, $W_\pm(\alpha)$ are unitary on $\mathscr{H}$, and H_α is unitarily equivalent to H_0.

An example of this is when $\mathscr{H} = L^2(\mathbb{R}^n)$, $n \geq 3$, $V \in L^p(\mathbb{R}^n) \cap L^q(\mathbb{R}^n)$ where $1 \leq p < n/2 < q \leq \infty$, and $A = M_{|V|^{1/2}}$, $B = M_{\text{sign}(V)}$, so that $BA^2 = M_V$. Then all the above hypotheses are satisfied, and $K \geq K_1$, where

$$K_1^{-1} = \frac{\delta}{4\pi} \|V\|_p^\beta \|V\|_q^\gamma,$$

where $\beta = (2pq - np)/4n(q - p)$, $\gamma = (nq - 2pq)/4n(q - p)$, and $\delta = 4npq(q - p)/(np + nq - 2pq)$. In particular $H_\alpha = -\Delta + M_{\alpha V}$ is absolutely continuous (i.e. $\mathscr{H}_{ac}(H_\alpha) = \mathscr{H}$) for $|\alpha|$ sufficiently small.

The proof is omitted. See Kato [12] or Reed-Simon [4].

14.18. REMARKS. Let H_0, A be self-adjoint nonnegative operators on $\mathscr{H}$. Suppose also that $H_1 = H_0 - A$ is self-adjoint and that $A(\lambda - H_0)^{-1}$ is compact for $\lambda \in \mathbb{C}\backslash\mathbb{R}$. Then, by Weyl's theorem (cf. 14.16), $\sigma(H_1) \cap \,]-\infty,0[$ consists of isolated eigenvalues of finite multiplicity. Suppose now that $\mathscr{H} = L^2(\mathbb{R}^n)$, $H_0 = -\Delta$, and $A = M_V$ where $V \in L^1(\mathbb{R}^n) \cap L^\infty(\mathbb{R}^n)$, say, V is nonnegative, $\|V\|_1 > 0$, and $V(x) \to 0$ as $|x| \to \infty$. Let $\Lambda_\alpha = \sigma(H_0 - \alpha A) \cap \,]-\infty,0[$. Can Λ_α be nonempty for small $\alpha > 0$? By Theorem 14.17, the answer is no when $n \geq 3$. However, the answer is yes when $n = 1$ or 2. We show this now.

$H_0 + \epsilon I$ and $A + \epsilon_1 I$ are invertible for $\epsilon, \epsilon_1 > 0$. $\Lambda_\alpha = \emptyset$ iff $H_0 + \epsilon - \alpha(A + \epsilon_1) \geq 0$ for every $\epsilon > 0$ and $0 < \epsilon_1 < \epsilon_2(\alpha, \epsilon)$, i.e. ϵ_1 is sufficiently small. Let $K = H_0 + \epsilon$, $L = \alpha(A + \epsilon_1)$. Thus $\Lambda_\alpha = \emptyset$ iff

$$\langle Kf - Lf, f\rangle \geq 0 \text{ for every } f \in \mathscr{D}(H_0) \tag{14.4}$$

and appropriate ϵ and ϵ_1. But (14.4) (with $g = L^{1/2}f$) is equivalent to

$$\langle L^{-1/2}KL^{-1/2}g - g, g\rangle \geq 0$$

for each $g \in L^{1/2}[\mathscr{D}(H_0)]$. This is equivalent to $L^{-1/2}KL^{-1/2} \geq I$, or $L^{-1/2}K^{-1}L^{1/2} \leq I$. Using $H_0 = -\Delta$, $A = M_V$, etc., as above, we have

$$\|L^{1/2}K^{-1}L^{1/2}\| = \sup_{\|f\|=1} \langle L^{1/2}K^{-1}L^{1/2}f,f \rangle$$

$$= \sup_{g \in \mathscr{M}} \langle K^{-1}g,g \rangle \text{ where } \mathscr{M} = \{L^{1/2}f : \|f\| \leq 1\}$$

$$= \sup_{g \in \mathscr{M}} \langle (K^{-1}g)^\wedge,\hat{g} \rangle \text{ by 3.12}$$

$$= \sup_{g \in \mathscr{M}} \int_{\mathbb{R}^n} \frac{|\hat{g}(\xi)|^2}{|\xi|^2 + \epsilon} d\xi.$$

Choose f such that $\|f\| \leq 1$ and $g = L^{1/2}f$ satisfies $\hat{g}(\xi) \geq \eta > 0$ for $|\xi| < \delta$ where δ, η are positive constants which do not depend on ϵ or ϵ_1. Then, using polar coordinates,

$$\|L^{1/2}K^{-1}L^{1/2}\| \geq \int_{\mathbb{R}^n} \frac{|\hat{g}(\xi)|^2}{|\xi|^2 + \epsilon} d\xi$$

$$\geq \int_{|\omega|=1} \int_0^\delta \frac{\eta}{r^2 + \epsilon} r^{n-1} dr\, d\omega.$$

As $\epsilon \to 0^+$, the right-hand side tends to infinity when $n = 1$ or 2. Thus negative potentials always produce negative eigenvalues in one and two dimensions. See also Exercise 14.23.1.

14.19. THE TWO HILBERT SPACE THEORY
For $j = 0, 1$, let U_j be the (C_0) unitary group which governs the abstract wave equation

$$u''(t) + A_j^2 u(t) = 0 \qquad (t \in \mathbb{R}) \tag{14.5}$$

where A_j is self-adjoint on $\mathscr{H}$ (compare Section 7). To construct the wave operators associated with U_1, U_0 we want to use the formula

$$W_\pm f = \lim_{t \to \pm\infty} U_1(-t)U_0(t_1)P_0 f. \tag{14.6}$$

However, U_j is unitary on the space $\mathscr{H}_j$, which is the completion of $[\mathscr{D}(A_j)/\mathscr{N}(A_j)] \oplus \mathscr{H}$ in the norm $\left\| \binom{f_1}{f_2} \right\|_j = (\|A_j f_1\|^2 + \|f_2\|^2)^{1/2}$. In general, $\mathscr{H}_0 \neq \mathscr{H}_1$, so that the expression (14.6) does not make sense. Formula (14.6) is modified as follows. If U_j is a (C_0) unitary group on $\mathscr{H}_j$, $j = 0,1$, we define generalized wave operators with respect to an "identification map" $J \in \mathscr{B}(\mathscr{H}_0,\mathscr{H}_1)$ by

$$W_\pm f = \lim_{t \to \pm\infty} U_1(-t)JU_0(t)P_0 f \qquad f \in \mathscr{H}_0.$$

Refer to (14.5). When $A_j^2 \geq \epsilon I$ for some $\epsilon > 0$ and $\mathscr{D}(A_0^2) = \mathscr{D}(A_1^2)$ then $\mathscr{H}_0$ and $\mathscr{H}_1$ consist of the same vectors but have different (though equivalent) norms. J is then chosen to be a unitary operator from $\mathscr{H}_0$ to $\mathscr{H}_1$ constructed with the aid of the Riesz representation theorem.

When comparing the wave equation in the exterior of an obstacle[†] with Dirichlet boundary condition with the wave equation in the whole space $\mathbb{R}^n$, we may take J to be the map which restricts functions on $\mathbb{R}^n$ to Ω.

The two-space theory, with a suitable identification map, can be developed following the one-space theory of the preceding sections. It allows for scattering theory to be applied to a number of additional important vases. See for instance Kato [15] or Lax-Phillips [2].

14.20. THE S-MATRIX
Let $S^{n-1} = \{x \in \mathbb{R}^n : |x| = 1\}$ be the unit sphere in $\mathbb{R}^n$, equipped with its usual (Lebesgue) surface measure $d\omega$. The Fourier transform diagonalizes $H_0 = -\Delta$ in the sense that for $\phi \in \mathscr{D}(H_0) = H^2(\mathbb{R}^n)$, $(H_0\phi)^\wedge(\xi) = |\xi|^2 \hat{\phi}(\xi)$. Write $\lambda = |\xi|^2$, $\omega = \xi/|\xi|$, and let

$$\tilde{\phi}(\lambda,\omega) = c\lambda^{1/2}\hat{\phi}(\xi).$$

Choosing c appropriately shows that the map $\phi(x) \to \tilde{\phi}(\lambda,\omega)$ is unitary from $L^2(\mathbb{R}^n)$ to $L^2[\mathbb{R}^+, L^2(S^{n-1})]$, the norm in the latter space being

$$\|\tilde{\phi}\| = \left[\int_0^\infty \left(\int_{S^{n-1}} |\tilde{\phi}(\lambda,\omega)|^2 \, d\omega \right) d\lambda \right]^{1/2}.$$

The correspondence between $\mathscr{H} = L^2(\mathbb{R}^n) \ni \phi(x)$ and $\tilde{\phi}(\lambda,\omega) \in L^2[\mathbb{R}^+, L^2(S^{n-1})]$ will be denoted by

$$\phi \leftrightarrow \{\tilde{\phi}(\lambda)\}.$$

Then, in this representation, H_0 becomes

$$H_0\phi \leftrightarrow \{\lambda\tilde{\phi}(\lambda)\}.$$

Since the scattering operator S commutes with H_0 (cf. 14.7), it takes the form

$$S\phi \leftrightarrow \{S(\lambda)\tilde{\phi}(\lambda)\},$$

where, for each $\lambda \in \mathbb{R}^+$, $S(\lambda) \in \mathscr{B}[L^2(S^{n-1})]$ is unitary. $S(\lambda)$ is called the *S-matrix*. When $n = 1$, $S(\lambda)$ is a 2×2 matrix for each $\lambda \in \mathbb{R}^+$ (since $S^0 = \{-1,1\}$). $S(\lambda)$ is determined, in this case, by the transmission and reflection coefficients. When $n > 1$, $S(\lambda) - I$ can be shown to be an integral operator with a Hilbert-Schmidt kernel.

14.21. THE LAX-PHILLIPS THEORY. This theory is especially well-suited to scattering problems of classical physics as opposed to quantum physics, i.e. to scattering problems involving the wave equation, Maxwell's equations, or other symmetric hyperbolic systems.

[†] That is, in $\mathbb{R}^n \backslash \Omega$ where Ω is a bounded domain.

Let U_1 be the (C_0) group on $\mathscr{H} = [L^2(\mathbb{R}^n)]^m$ which governs the symmetric hyperbolic system

$$\frac{\partial u}{\partial t}(= iH_1 u) = \sum_{j=1}^{n} A_j(x)\frac{\partial u}{\partial x_j} + B(x)u,$$

where A_j is a real symmetric $m \times m$ matrix-valued function on $\mathbb{R}^n$ such that $A_j(x)$ is a constant matrix, A_j^0, for $|x| \geq \rho$. For H_1 to be self-adjoint we require

$$B(x) + B(x)^* = \sum_{j=1}^{n} \frac{\partial}{\partial x_j} A_j(x) \qquad |x| \leq \rho$$

and $B(x) = 0$ for $|x| > \rho$. Let U_0 be the corresponding "free" (C_0) group, i.e. U_0 governs

$$\frac{\partial u}{\partial t} = \sum_{j=1}^{n} A_j^0 \frac{\partial u}{\partial x_j} (= iH_0 u).$$

We make several technical assumptions, namely n is odd and $n \geq 3$, all propagation speeds are positive, i.e. $\sum_{j=1}^{n} A_j(x)\omega_j$ has only nonzero eigenvalues for all $x \in \mathbb{R}^n$ and $\omega \in \mathbb{R}^n \backslash \{0\}$ (i.e. iH_1 is elliptic); and iH_1 has the unique continuation property, i.e. if $H_1 f = 0$ in an open connected set Ω and if $f = 0$ in a nonempty open subset of Ω then $f = 0$ in Ω.

Let $c > 0$ be the smallest propagation speed of iH_0, and let

$$\mathscr{D}_{\pm} = \{ f \in \mathscr{H} : U_1(t)f(x) = 0 \text{ for } |x| \leq \rho \pm ct, \ \pm t \in \mathbb{R}^+ \}.$$

The subspace $\mathscr{D}_+$ [resp. $\mathscr{D}_-$] is *outgoing* [resp. *incoming*] for U_1, i.e.

$$U_1(t)(\mathscr{D}_{\pm}) \subset \mathscr{D}_{\pm} \text{ for } \pm t \in \mathbb{R}^+,$$

$$\bigcap \{ U_1(t)(\mathscr{D}_{\pm}): t \in \mathbb{R} \} = \{0\},$$

$$\overline{\bigcup \{ U_1(t)(\mathscr{D}_{\pm}): t \in \mathbb{R} \}} = \mathscr{H} \ominus \mathscr{N}(H_1)$$

where $\mathscr{N}(H_1)$ is the null space of H_1. [Necessarily $\mathscr{N}(H_1) = \mathscr{H}_s(H_1)$.]

Let $\mathscr{H}_1 = \mathscr{H} \ominus \mathscr{N}(H_1)$. Using the Radon transform, Lax and Phillips showed that there is a unitary mapping V_+ [resp. V_-] from $\mathscr{H}_1$ to $L^2[\mathbb{R}, L^2(S^{n-1})]$, such that in the representation defined by V_+ [resp. V_-], $V_1(t)$ becomes translation to the right by t units and $\mathscr{D}_+$ [resp. $\mathscr{D}_-$] becomes $L^2[[0,\infty[, L^2(S^{n-1})]$ [resp. $L^2(]-\infty, 0], L^2(S^{n-1}))$]. Moreover, $\mathscr{H}_1 = \mathscr{H}_{ac}(H_1)$.

For $f \in \mathscr{H}_1$, let f_+, f_- be the outgoing and incoming representers of f, i.e. $f_{\pm} = V_{\pm}f$. Then the map

$$S: f_- \to f_+$$

turns out to be (unitarily equivalent to) the scattering operator for the pair H_1, H_0.

The Lax-Phillips approach gives an explicit representation for the S-matrix, and the basic properties of the S-matrix can be derived from a detailed study of a semigroup Z constructed as follows. For $\mathscr{D}_{\pm}$ as above, it can be shown (with n odd) that $\mathscr{D}_+ \perp \mathscr{D}_-$; let $\mathscr{K}_1 = \mathscr{H} \ominus (\mathscr{D}_+ \oplus \mathscr{D}_-)$. Let $P_{\pm}$ be the orthogonal projection onto $(\mathcal{D}_\pm)^\perp$. Then

$$Z(t) = P_+ U(t) P_- \qquad t \in \mathbb{R}^+$$

defines a (C_0) contraction semigroup Z on $\mathcal{K}_1$ which, roughly speaking, contains the behavior of U_1 near the perturbed region. In contrast to $-iH_1$, the generator B of Z has pure point spectrum. When the S-matrix is extended to a meromorphic function $S(z)$ on the complex plane, the poles of $S(z)$ correspond to the eigenvalues of B.

The Lax-Phillips approach also works in a number of other contexts, e.g. the wave equation in the exterior of an obstacle. The main limitation of this approach is that the perturbation must lie in a bounded region.

14.22. CONNECTIONS WITH EQUIPARTION OF ENERGY
(i) Let A be an injective self-adjoint operator on $\mathcal{H}$. Then

$$u''(t) + A^2 u(t) = 0 \qquad (t \in \mathbb{R}) \tag{14.7}$$

is governed by a (C_0) group U_0 on $\mathcal{K}$, the completion of $\mathcal{D}(A) \oplus \mathcal{H}$ in the norm $\left\| \begin{pmatrix} f_1 \\ f_2 \end{pmatrix} \right\|_{\mathcal{K}} = (\|Af_1\|^2 + \|f_2\|^2)^{1/2}$. Define the *Riemann-Lebesgue set* of A to be

$$\mathcal{H}_{RL}(A) = \{ f \in \mathcal{H} : \langle e^{itA} f, f \rangle \to 0 \text{ as } t \to \pm\infty \}.$$

According to Theorem 7.12, Equation (14.7) admits equipartition of energy iff $\mathcal{H}_{RL}(A) = \mathcal{H}$. Let $iH_0 = \begin{pmatrix} 0 & I \\ -A^2 & 0 \end{pmatrix}$ on $\mathcal{K}$, so that $U_0(t) = e^{itH_0}$, $t \in \mathbb{R}$. The following assertions hold:

$$\mathcal{H}_{ac}(A) \subset \mathcal{H}_{RL}(A) \subset \mathcal{H}_c(A), \tag{14.8}$$

$$\mathcal{H}_{ac}(A) = \mathcal{H} \text{ iff } \mathcal{K}_{ac}(H_0) = \mathcal{K}.$$

Moreover, each containment in (14.8) can be strict.

One can easily check that for $H = -\Delta$ on $\mathcal{H} = L^2(\mathbb{R}^n)$, $\mathcal{H}_{ac}(H) = \mathcal{H}$. Exercise 7.14.7(i) easily follows this observation together with Theorem 7.12.

(ii) Let H_0 be a self-adjoint, absolutely continuous operator on a Hilbert space $\mathcal{K}$ (i.e. $\mathcal{K}_{ac}(H_0) = \mathcal{K}$). The algebra of operators defined by

$$\mathcal{L}(H_0) = \{ V \in \mathcal{B}(\mathcal{K}): \lim_{t \to \pm\infty} U_0(-t) V U_0(t) f \text{ exists for all } f \in \mathcal{K} \},$$

where $U_0(t) = e^{-itH_0}$, has connections with scattering theory. To see this, let V be bounded and self-adjoint, and suppose that the wave operators $W_\pm = W_\pm(H_0 + V, H_0)$ exist and are unitary on $\mathcal{K}$. Then, for $H_1 = H_0 + V$ and $U_1(t) = e^{-itH_1}$,

$$U_0(-t) V U_0(t) f = U_0(-t) U_1(t) H_1 U_1(-t) U_0(t) f - H_0 f$$

$$\to W_\pm^{-1} H_1 W_\pm f - H_0 f$$

as $t \to \pm\infty$ for each $f \in \mathcal{D}(H_0)$. It follows that $V \in \mathcal{L}(H_0)$. Conversely, a study of $\mathcal{L}(H_0)$ can give insight into which perturbations H_1 of H_0 have the property that

$W_\pm(H_1,H_0)$ exist. We do not pursue this here; rather we shall relate the operators $U_0(-t)VU_0(t)$ to equipartition of energy.

Let A, H_0, $\mathscr{K}$ be as in (i) above. Let $P_1\begin{pmatrix} f_1 \\ f_2 \end{pmatrix} = \begin{pmatrix} f_1 \\ 0 \end{pmatrix}$, $P_2 = I - P_1$. If A is absolutely continuous, then so is H_0, and the conclusion of Theorem 7.12 can be restated as

$$\lim_{t \to \pm\infty} \langle U_0(-t)P_jU_0(t)f,f \rangle = \frac{1}{2}\|f\|^2 \tag{14.9}$$

for all $f \in \mathscr{K}$ and $j = 1,2$. However, $P_j \notin \mathscr{L}(H_0)$. To see this, the assumption that $P_j \in \mathscr{L}(H_0)$ implies that the operator W_j defined by

$$W_jf = \lim_{t \to \infty} U_0(-t)P_jU_0(t)f, f \in \mathscr{K}$$

would satisfy $W_j^2 = W_j$, whereas (14.9) implies $W_j = \frac{1}{2}I$, a contradiction. Thus scattering theory is connected with the strong convergence of $U_0(-t)VU_0(t)f$, whereas equipartition of energy is connected with its weak convergence.

14.23. EXERCISES

1. Let $-\infty < a < b < \infty$ and $\alpha > 0$. Let $J_1 = \,]-\infty,a[$, $J_2 = [a,b]$, $J_3 = \,]b,\infty[$, and let $V(x) = -\alpha$ or 0 according as x belongs to J_2 or not. Set

   $$f(x) = \beta_je^{k_jx} + \gamma_je^{-k_jx}, \qquad x \in J_j, j = 1,2,3.$$

   Choose the constants β_j, γ_j, k_j so that f is an eigenvector corresponding to a negative eigenvalue of $-d^2/dx^2 - M_V$. (Hint: This calculation is done in several quantum mechanics texts.) As a consequence show that if $W:\mathbb{R} \to \,]-\infty,0]$ satisfies $W = W_1 + W_2$ where $W_1 \in L^2(\mathbb{R})$, $W_2 \in L^\infty(\mathbb{R})$, and $W(x) < 0$ at some continuity point x of W, then $-d^2/dx^2 - M_W$ has a negative eigenvalue.
2. Let $\mathscr{H} = L^2(\mathbb{R})$, $H_0 = M_{id}$, $H_1 = id/dx$. Let $g \in \mathscr{H}\setminus\{0\}$ and let $Af = [\int_{-\infty}^\infty f(x)m(x)\,dx]g$, $f \in \mathscr{H}$.
   (i) $A \in \mathscr{B}(\mathscr{H})$ iff A is H_0-smooth iff $m \in \mathscr{H}$.
   (ii) M_V is H_1-smooth if $V \in \mathscr{H}$. (Note that in this case M_V can be unbounded.)
   (iii) If A is H-smooth, then $A[\mathscr{H}_d(H)] = \{0\}$.
*3. Let H_0 be self-adjoint on $\mathscr{H}_1$ and define $\mathscr{L}(H_0)$ as in 14.22. Then $\mathscr{L}(\mathscr{H}_0) = \mathscr{B}(\mathscr{H})$ iff $H_0 = \lambda I$ for some real number λ.
4. Show that the Dirac operator A of Exercise 9.5.4 is absolutely continuous, i.e. for $\mathscr{H} = L^2[\mathbb{R}^3,\mathscr{B}(\mathbb{C}^4)]$, $\mathscr{H}_{ac}(A) = \mathscr{H}$.
5. Let A and H be bounded self-adjoint operators on $\mathscr{H}$. Let $[H,A]$ be the commutator $HA - AH$, which is skew-adjoint. If $i[H,A] \geq 0$ and if $0 \leq B = B^* \leq i[H,A]$, then B is H-smooth. (Hint. For $f \in \mathscr{H}$ let $\phi(t) = \langle e^{itH}Ae^{-itH}f,f \rangle$, which is bounded. Calculate $\phi'(t)$ and relate it to $\|Be^{-itH}f\|^2$.)

15.   Further Applications

15.1.   In this section we treat the Cayley transform of a dissipative operator on Hilbert space and, in connection with this, discuss boundary conditions. Other main topics are random evolutions and the Navier-Stokes equations of fluid dynamics. Miscellaneous topics include functional differential equations, controllability, and a discussion of why the differential operators which are generators of positivity-preserving semigroups are second order at most.

15.2 Let $\mathbb{T} = \{z \in \mathbb{C} : |z| = 1\}$ be the unit circle in the complex plane. The mapping

$$\phi : z \to \frac{z+1}{z-1}$$

defines a bijective correspondence between the imaginary axis $i\mathbb{R}$ and the punctured circle $\mathbb{T} \setminus \{1\}$. Moreover, ϕ maps the open left half plane $\{z \in \mathbb{C} : \mathrm{Re}\, z < 0\}$ onto the open unit disc $\{z \in \mathbb{C} : |z| < 1\}$. The spectral mapping theorem suggests that if A is a skew-symmetric [resp. dissipative] operator on a complex Hilbert space $\mathcal{H}$, then $\phi(A)$ is an isometry (resp. a contraction]. Here is a precise formulation of this.

15.3. PROPOSITION. *Let A be a dissipative operator on a complex Hilbert space $\mathcal{H}$. Let $C \equiv (I + A)(I - A)^{-1}$ be the Cayley transform of A.*

(i)   *C is closed iff A is closed.*
(ii) *$\mathcal{D}(C) = \mathcal{H}$ iff $I - A$ is surjective (i.e. A is m-dissipative).*
(iii) *Range $(I + C) = \mathcal{D}(A)$, $I + C$ is injective, and*

$$A = -(I - C)(I + C)^{-1};$$

   thus the map $A \to C$ is injective.
(iv) *A is skew-symmetric iff C is an isometry.*
(v)   *A is skew-adjoint iff C is unitary.*
(vi) *Conversely, let C be a contraction with $I + C$ injective. Then C is the Cayley transform of a dissipative operator A on $\mathcal{H}$. Moreover, A is skew-symmetric [resp. skew-adjoint] iff C is isometric [resp. unitary].*

The proof, which is elementary, is left as an exercise. (See Exercises 15.26.1.) The proof of the following companion theorem is also left to the reader.

15.4. THEOREM. *Let A be a densely defined dissipative operator on a complex Hilbert space $\mathcal{H}$, and let $C = (I + A)(I - A)^{-1}$ be the Cayley transform of A. An operator $\tilde{A}$ is a dissipative extension of A iff the Cayley transform $\tilde{C}$ of $\tilde{A}$ is a contraction extension of C. All m-dissipative extension $\tilde{A}$ of A correspond to all contraction extensions $\tilde{C}$ of C with $\mathcal{D}(\tilde{C}) = \mathcal{H}$.*

In particular, any dissipative operator A on $\mathcal{H}$ has an m-dissipative extension. To see this, note first that $\bar{A}$, the closure of A, is dissipative. Let C be its Cayley transform which is a closed contraction. For $f \in \mathcal{D}(C)$ and $g \in \mathcal{D}(C)^{\perp}$ define $\tilde{C}(f + g) = Cf$ [i.e. $\tilde{C} = C$ on $\mathcal{D}(C)$, $\tilde{C} = 0$ on $\mathcal{D}(C)^{\perp}$]. Then $\tilde{C}$ is a contraction extension of C to all of $\mathcal{H}$, whence its inverse Cayley transform $\tilde{A}$ is an m-dissipative extension of A.

15.5 BOUNDARY CONDITIONS. Let A be a dissipative operator on $\mathcal{H}$. We may and do assume that A is closed (if not look at $\bar{A}$) and densely defined (if not extend

A to be 0 on $\mathscr{D}(A)^{\perp}$). A is m-dissipative iff A^* is. If A is not m-dissipative, then A^* will not be dissipative. Since $A \subset B$ iff $B^* \subset A^*$, finding m-dissipative extensions of $\mathscr{A}$ is equivalent to finding m-dissipative restrictions of $\mathscr{A}^*$. Thus the domain of A^* must be shrunk (in one of many ways) to find an m-dissipative restriction; equivalently, the domain of A must be enlarged to get an m-dissipative extension of A. The prescription for doing this is usually referred to as specifying the *boundary conditions*. Theorem 15.4 tells us in principle how to find all such boundary conditions. For a specific example, let $A = d/dx$ on $\mathscr{D}(A) = C_c^{\infty}]0,1[\subset \mathscr{H} = L^2[0,1]$. Then $A^* = -d/dx$ on $\mathscr{D}(A^*) = \{u \in \mathscr{H} : u$ is absolutely continuous on $[0,1]$, $u' \in \mathscr{H}\} = H^1[0,1]$. For each $\theta \in \mathbb{R}$, the restriction of A^* to $\{u \in H^1[0,1] : u(0) = e^{i\theta} u(1)\}$ is a skew-adjoint extension of A.

15.6 A CONNECTION BETWEEN SECOND-ORDER DIFFERENTIAL OPERATORS AND POSITIVITY. Let A generate a (C_0) positivity preserving contraction semigroup T on a Banach lattice $\mathscr{X}$ (see Chapter I, Section 9.13). If A is an elliptic partial differential operator and if $\mathscr{X}$ is a space of functions on $\Omega \subset \mathbb{R}^n$, then Chapter I, Exercise 9.21.21 suggests that if $n = 1$, the order of A is at most two. (Recall that the first derivative generates the translation semigroup, which is positivity preserving.) We verify that this is indeed the case in dimension $n \geq 1$ under some additional assumptions.

Let $\mathscr{X} = L^2(\Omega)$ be a Hilbert space of complex-valued functions and let $A = -B$ where $B = B^* \geq 0$. Let $f \in \mathscr{X}$. By looking at the real and imaginary parts of f separately, we may assume f is real-valued. Since

$$|T(t)f|^2 = \max(|T(t)f_+|^2, |T(t)f_-|^2) \leq |T(t)f_+|^2 + |T(t)f_-|^2,$$

we have

$$\|T(t)f\|^2 \leq \|T(t)f_+\|^2 + \|T(t)f_-\|^2.$$

Let $g(t) = \|T(t)f\|^2$, $h(t) = \|T(t)f_+\|^2 + \|T(t)f_-\|^2$. Since $g,h: \mathbb{R}^+ \to \mathbb{R}^+$ are continuous and nonincreasing, $g'(0)$, $h'(0)$ exist in $[0,\infty]$, and since $g(0) = h(0)$ and $g(t) \leq h(t)$ for all t it follows that $g'(0) \leq h'(0)$. But this says that

$$-\langle Bf, f \rangle \leq -\langle Bf_+, f_+ \rangle - \langle Bf_-, f_- \rangle.$$

We interpret this inequality as follows. If $f \in \mathscr{D}(B)$ [or even $f \in \mathscr{D}(B^{1/2})$], so that $(\langle Bf, f \rangle =) \|B^{1/2}f\|^2 < \infty$, then $\|B^{1/2}f_{\pm}\|^2$ are also finite. Consequently, $f \in \mathscr{D}(B)$ implies $f_{\pm} \in \mathscr{D}(B^{1/2})$ and hence $|f| \in \mathscr{D}(B^{1/2})$.

If A is of order $2m$ (where $m \in \mathbb{N}$), then $H_0^{2m}(\Omega) \subset \mathscr{D}(A) = \mathscr{D}(B) \subset H^{2m}(\Omega)$. It can be shown that necessarily $\mathscr{D}(B^{1/2}) = H^m(\Omega)$. Thus for each real f in $H_0^{2m}(\Omega)$, we have $|f| \in H^m(\Omega)$. This implies $m = 1$. (Think about $f(x) = x$ near $x = 0$ in one dimension.)

15.7. RANDOM EVOLUTIONS. Let $\xi = \{\xi(t) : t \in \mathbb{R}^+\}$ be an n state stationary Markov chain;[†] the state space will be denoted by S and for convenience we take $S = \{1, 2, \ldots, n\}$. The underlying probability space is (Ω, Σ, P). Let $\tau_j(\omega)$ be the time

---

[†] Compare Chapter I, Section 9.11. A Markov chain is a Markov process whose state space S is discrete.

of the jth jump of the sample path $t \to \xi(t,\omega)$ and let $N(t,\omega)$ be the number of jumps taken up to time t. Finally let $T_i = \{T_i(t): T \in \mathbb{R}^+\}$ be a (C_0) semigroup on a Banach space $\mathscr{X}$ for $i \in S$. The (ordered) product

$$M(t) = T_{\xi(0)}(\tau_1) T_{\xi(\tau_1)}(\tau_2 - \tau_1) \cdots T_{\xi(\tau_{N(t)})}(t - \tau_{N(t)})$$

is called the *random evolution* determined by $\{\xi: T_i, i \in S\}$. (As usual, the ω is suppressed.) On the space $\mathscr{X}^n$ consisting of n copies of $\mathscr{X}$ one can define the *expectation semigroup* T by the formula

$$T(t)f = \{(T(t)f)_i: i = 1, \ldots, n\},$$

$$[T(t)f]_i = E_i[M(t)f_{\xi(t)}]$$

for $t \in \mathbb{R}^+$, $f \in \mathscr{X}^n$, $i \in S$; here E_i means expectation under the condition that $\xi(0) = i$. These notions were introduced by Griego and Hersh [2], who proved the following result.

15.8.   THEOREM.   *For ξ, S, T_i, T as above, T is a (C_0) semigroup on $\mathscr{X}^n$. It is a contraction semigroup if each T_i is. Let A_i be the generator of T_i and let $Q = (q_{ij})$ be the generator of the Markov chain ξ (thus Q is an $n \times n$ matrix). Then the generator of T is*

$$A = \begin{pmatrix} A_1 & & \mathbf{0} \\ & \ddots & \\ \mathbf{0} & & A_n \end{pmatrix} + QI$$

with $\mathscr{D}(A) = \mathscr{D}(A_1) \times \cdots \times \mathscr{D}(A_n)$. In other words, A is an $n \times n$ matrix whose ijth entry is $\delta_{ij} A_i + q_{ij}I$, δ_{ij} being the Kronecker symbol, i.e. $\delta_{ij} = 1$ if $i = j$, $\delta_{ij} = 0$ if $i \neq j$.

15.9.   DISCUSSION.   Let A_0 generate a (C_0) group, $n = 2$, $A_1 = -A_2 = A_0$, and $Q = \begin{pmatrix} -a & a \\ a & -a \end{pmatrix}$ where $a > 0$. Suppose $\begin{pmatrix} u_1 \\ u_2 \end{pmatrix} = T(t) \begin{pmatrix} f_1 \\ f_2 \end{pmatrix}$ where T is the expectation semigroup of the random evolution and $f_1, f_2 \in \mathscr{D}(A_0^2)$. Then $u = u_1 + u_2$ is the unique solution of the abstract (telegraph equation) Cauchy problem

$$u''(t) + 2a\,u'(t) = A_0^2 u \qquad (t \geq 0),$$

$$u(0) = f_1 + f_2, \, u'(0) = A_0(f_1 - f_2).$$

For our choice of Q, $\{N(t)\}$ is a Poisson process with intensity a. If w is the unique solution of

$$w'' = A_0^2 w \qquad (t \geq 0),$$

$$w(0) = f, \, w'(0) = A_0 g$$

where $f, g \in \mathscr{D}(A_0^2)$, then the (expectation semigroup) formula for the solution u of

$$u''(t) + 2au'(t) = A_0^2 u \qquad (t \geq 0),$$

$$u(0) = f, \, u'(0) = A_0 g$$

can be shown to reduce to

$$u(t) = E_1\left[w\left(\int_0^1 (-1)^{N(s)} \, ds \right) \right], \ t \in \mathbb{R}^+.$$

This provides an interesting application of probability theory to hyperbolic problems.

Replacing a by λa and A by $\sqrt{\lambda} \, A$ where $\lambda > 0$, the equation

$$u'' + 2au' = A_0^2 u$$

becomes

$$\frac{1}{\lambda} u'' + 2au' = A_0^2 u.$$

This suggests that as $\lambda \to \infty$, the solution u_λ of

$$\frac{1}{\lambda} u_\lambda'' + 2au_\lambda' = A_0^2 u_\lambda \qquad (t \in \mathbb{R}^+),$$

$$u_\lambda(0) = f, \ u_\lambda'(0) = A_0 g$$

converges to the unique solution u_0 of

$$2au_0' = A_0^2 u_0 \qquad (t \in \mathbb{R}^+),$$

$$u_0(0) = f.$$

Moreover, since the solution of this problem is known to be

$$u_0(t) = \left(\frac{a}{2\pi t} \right)^{1/2} \int_{-\infty}^{\infty} T_0(s) f \, e^{-as^2/2t} \, ds$$

(cf. Theorem 10.4), we can interpret the limit as

$$u_0(t) = E_{N(o,t/a)}[T_0(s)f],$$

the expectation of $T_0(s)f$ as s varies over $\mathbb{R}$ in accordance with the normal distribution with mean 0 and variance t/a. This suggests that the convengence of u_λ to u_0 can be obtained as a consequence of the central limit theorem. Griego and Hersh [2] proved this to be the case.

15.10. More Discussion. There are many limit theorems and applications associated with the notion of a random evolution. We shall state one of them. For this purpose we shall allow the stochastic process ξ to be more general than a Markov chain.

Let $\xi = \{\xi(t) : t \in \mathbb{R}^+\}$ be a pure jump process i.e. for a.e. $\omega \in \Omega$ and all $t \in \mathbb{R}^+$, $\xi(t + h, \omega) = \xi(t, \omega)$ for sufficiently small positive h. Suppose that the state space S of ξ is a locally compact separable metric space, and assume there is a probability measure μ on the Borel sets of S such that

$$P\left\{ \lim_{t \to \infty} \frac{1}{t} \int_0^t g[\xi(s)] \, ds = \int_S g(x)\mu(dx) \right\} = 1$$

for every $g \in BUC(S)$. (When ξ is a finite state stationary Markov chain the existence of μ comes from the ergodic theorem for Markov chains; see Chung [1].) For each $x \in S$ let $T_x = \{T_x(t) : t \in \mathbb{R}^+\}$ be a (C_0) contraction semigroup on $\mathcal{X}$ with generator A_x. Assume both

$$D = \{f \in \mathcal{X} : x \to A_x f \in BUC(S; \mathcal{X})\}$$

and $(A - \alpha I)(D)$ are dense in $\mathcal{X}$ for some $\alpha > 0$, where

$$Af = \int_S A_x f \, \mu(dx)$$

for $f \in D = \mathscr{D}(A)$. Finally, let, as before, τ_j be the time of the jth jump of ξ and $N(t)$ be the number of jumps taken up to time t. For $\lambda > 0$ let

$$T_\lambda(t) = T_{\xi(0)}\left(\frac{1}{\lambda}\tau_1\right)T_{\xi(\tau_1)}\left[\frac{1}{\lambda}(\tau_2 - \tau_1)\right]\cdots T_{\xi(\tau_{N(t)})}\left[\frac{1}{\lambda}(t - \tau_{N(t)})\right]$$

be the random evolution determined by $\{\xi : T_x, x \in S\}$ and λ.

15.11. THEOREM. *Suppose that the assumptions of the above paragraph hold. Then the closure $\bar{A}$ of A is m-dissipative on $\mathcal{X}$, and the (C_0) contraction semigroup T generated by $\bar{A}$ satisfies*

$$P\left\{\lim_{\lambda \to \infty} T_\lambda(t, \omega)f = T(t)f\right\} = 1$$

for all $f \in \mathcal{X}$ and $t \in \mathbb{R}^+$.

15.12. EXAMPLE. Taking $S = \{1, 2\}$ and letting $\xi(t) = 1$ when $2n \le t < 2n + 1$, $\xi(t) = 2$ for $2n + 1 \le t < 2n + 2$ $(n \in \mathbb{N}_0)$, the resulting random evolution becomes

$$T_\lambda(t) = [T_1(t/2n) T_2(t/2n)]^n$$

where $\lambda = n/t$. If $\mathscr{D}(A_1) \cap \mathscr{D}(A_2)$ and $[\mu I - \frac{1}{2}(A_1 + A_2)][\mathscr{D}(A_1) \cap \mathscr{D}(A_2)]$ are both dense in $\mathcal{X}$, the random evolution $T_\lambda(t)f$ converges to $T(t)f$, where T is the (C_0) semigroup generated by the closure of $\frac{1}{2}(A_1 + A_2)$. This is the Trotter product formula (see Chapter I, Section 8.12) in a stochastic setting.

More generally, let V be an operator depending on a stationary Markov process $\{\xi(t) : t \in \mathbb{R}^+\}$. If M_x is a Banach space-valued random variable satisfying $M_x'(t) = V[\xi(t)]M_x(t)$, $M_x(0) = f(x)$, then

$$u(t, x) = E[M_x(t)]$$

defines a formal solution of the problem

$$\frac{du}{dt} = V(x)u + Au \qquad u(0, x) = f(x).$$

Regarding $u(t, \cdot)$ as $T(t)f$, $\{T(t) : t \in \mathbb{R}^+\}$ will be a (C_0) semigroup in many cases, and the representation $T(t)f(x) = E[M_x(t)]$ can be regarded as a generalized Feynman-Kac formula.

15.13. CONTROLLABILITY AND STABILIZABILITY. The inhomogeneous equation

$$du/dt = Au(t) + Bc(t) \qquad (15.1)$$

often describes a problem in control theory. Take A to generate a (C_0) semigroup T on a Hilbert space $\mathcal{H}$, suppose that the control function c takes values in a Hilbert space $\mathcal{H}_0$, and let $B \in \mathcal{B}(\mathcal{H}_0, \mathcal{H})$. The solution is

$$u(t) = T(t)u(0) + \int_0^t T(t-s)Bc(s)\,ds.$$

The initial data $u(0)$ is given, and the problem is to choose the control c in such a way as to minimize some functional of the solution, which can be interpreted as a cost. For instance, we might want to minimize $\int_0^1 \|u(t)\|^2\,dt$ or $\{t > 0 : u(t) = 0\}$ or something else.

We first state a result about controllability. Let

$$\Omega(t) = \left\{ \int_0^t T(t-s)\,Bc(s)\,ds : c \in L^1([0,t], \mathcal{H}_0) \right\}.$$

The system $\{A,B\}$ or the equation (15.1) is called *controllable* provided that

$$\bigcup\{\Omega(t) : t \in \mathbb{R}^+\} \qquad \text{is dense in } \mathcal{H}.$$

15.14. THEOREM.   *Equation* (15.1) *is controllable iff* $\bigcup\{Range[T(t)B] : t \in \mathbb{R}^+\}$ *is dense in* $\mathcal{H}$ *iff* $\int_0^t T(s)BB^*T(s)^*f\,ds = 0$ *for all* $t \in \mathbb{R}^+$ *and some* $f \in \mathcal{H}$ *implies* $f = 0$.

15.15. DEFINITIONS.   The semigroup T on $\mathcal{H}$ is *weakly stable* [resp., *strongly stable; uniformly stable; exponentially stable*] if as $t \to \infty$, $\langle T(t)u,v \rangle \to 0$ for all $u,v \in \mathcal{H}$ [resp., $\|T(t)u\| \to 0$ for all $u \in \mathcal{H}$; $\|T(t)\| \to 0$; $\|T(t)\| \le Me^{-\omega t}$ where M and ω are both positive]. The control system $\{A,B\}$ is called *stabilizable* if there is a $G \in \mathcal{B}(\mathcal{H}, \mathcal{H}_0)$ such that $A + BG$ generates a stable (C_0) semigroup U on $\mathcal{H}$, and then G is called a stabilizing feedback. Thus we have four notions of stabilizability in view of the four notions of stability.

Stabilizability and controllability are closely connected. The following result illustrates this.

15.16. THEOREM.   *Let the semigroup T be a contraction semigroup on $\mathcal{H}$.*

(i)   *The system $\{A,B\}$ [or (15.1)] is weakly stabilizable iff*

$$\bigcup\{\Omega(t) : t \in \mathbb{R}^+\}^\perp \subset \left\{ f \in \mathcal{H} : \text{weak } \lim_{t \to \infty} T(t)f = 0 \right\},$$

and $G = -B^*$ is a stabilizing feedback.

(ii)   *$\mathcal{H}$ has a maximal closed subspace $\mathcal{H}_1$ on which T is (i.e. restricts to) a unitary semigroup. Moreover, both T and T^* are strongly stable on $\mathcal{H}_1^\perp$ iff*

$P = Q$ is a projection where

$$Pf = \lim_{t\to\infty} T(t)^*T(t)f = \lim_{t\to\infty} T(t)T(t)^*f = Qf$$

for $f \in \mathcal{H}$ (these limits exist). The range of $P = Q$ is then $\mathcal{H}_1$.

(iii) Let $\mathcal{H}_1$ be as above. If $\{A,B\}$ is strongly stabilizable with the stabilizing feedback $G = -B^*$, then $\mathcal{H}_1$ is controllable for $\{A,B\}$ or for $\{A^*,B\}$, i.e. $\bigcup\{\int_0^t S(t-s)Bc(s)\,ds : c \in L^1([0,t],\mathcal{H}_0), t \in \mathbb{R}^+\}$ is dense in $\mathcal{H}_1$ for $S = T$ or $S = T^*$. On the other hand, if $B\mathcal{H}_0 = \mathcal{H}$ or if $\mathcal{H}_1$ is controllable for either $\{A,B\}$ or for $\{A^*,B\}$, or if $\{A,B\}, \{A^*,B\}$ are both controllable, then $\{A,B\}$ is strongly stabilizable with stabilizing feedback $G = -B^*$.

Note that T cannot be strongly stable on any subspace containing a nonzero vector in $\mathcal{H}_1$.

15.17. REMARK A feedback controller monitors measurements $y(t) = Cu(t)$ of the solution of the system (15.1) and uses these to produce control commands $c(t)$ to stabilize the system and/or to improve its performance. This idea is behind many of the industrial applications of control theory.

15.18. FUNCTIONAL DIFFERENTIAL EQUATIONS
Usually the evolution of a physical system in time is given by a differential equation of the form $du(t)/dt = Au(t)$. Here A is a (linear) operator. But what if $du(t)/dt$ depends not only on $u(t)$ but also on $u(s)$ for $t - r \le s < t$? This is the case in delay differential equations or, more generally, functional differential equations.

As an introductory example consider the delay ordinary differential equation

$$du(t)/dt = cu(t - r)$$

for $t \in \mathbb{R}^+$ where $c \in \mathbb{R}, r > 0$ are fixed constants. Any $u \in \mathcal{Y} = C([-r,0],\mathbb{K})$ has a unique extension to $[-r,\infty[$ satisfying the equation; clearly

$$u(t) = u(0) + c \int_{-r}^{t-r} u(s)\,ds$$

holds for $0 \le t \le r$, and this can be iterated. For $t \in \mathbb{R}^+$ let $T(t)$ map a given member of $\mathcal{Y}$, viewed as an initial datum, to the restriction of the solution to $[t-r,t]$, which we identify with a member of $\mathcal{Y}$ by translating the independent variable. This defines a (C_0) semigroup T on $\mathcal{Y}$. For $0 \le t \le r$ and for $f \in \mathcal{Y}$,

$$(T(t)f)(s) = \begin{cases} f(s+t) \text{ if } s \in [-r,-t], \\ f(0) + c\int_0^{s+t} f(y-r)\,dy \text{ if } s \in [-t,0]; \end{cases}$$

and for $t > r$, $T(t)$ is determined from $\{T(s): 0 \le s \le r\}$ by the semigroup property. The generator A of T is given by

$$(Af)(s) = \begin{cases} f'(s) \text{ if } s \in [-r,0[, \\ cf(-r) \text{ if } s = 0 \end{cases}$$

on $\mathcal{D}(A) = \{f \in C^1([-r,0],\mathbb{K}) : f'(0) = cf(-r)\}$.

For an analogous example involving a delay partial differential equation take

$$\frac{\partial}{\partial t} w(t,x) = \frac{\partial^2}{\partial x^2} w(t,x) + w(t-1,x) \qquad (0 \le x \le \pi, t \in \mathbb{R}^+),$$

$$w(t,0) = w(t,\pi) = 0 \qquad (t \in \mathbb{R}^+),$$

$$w(t,x) = f(t,x) \qquad (0 \le x \le \pi, -1 \le t \le 0).$$

Let $\mathcal{X}$ be a Banach space and $r > 0$. Let $\mathcal{Y} = C([-r,0],\mathcal{X})$; thus a point in $\mathcal{Y}$ can be regarded as a curve in $\mathcal{X}$. If $u \in C([-r,\infty[,\mathcal{X})$ and $t \in \mathbb{R}^+$ let $u_t(\theta) = u(t + \theta)$ for $-1 \le \theta \le 0$. Thus $u_t \in \mathcal{Y}$. The equation describing the evolution of u in time we take to be

$$u(t) = F(u_t) \qquad (t \in \mathbb{R}^+),$$

$$u_0 = f.$$

Here F is assumed to be a bounded linear operator from $\mathcal{Y}$ to $\mathcal{X}$. The condition $u_0 = f$ means $u(\theta) = f(\theta)$ for $-1 \le \theta \le 0$. This problem is governed by a (C_0) semigroup on $\mathcal{Y}$. More precisely, let $D\phi = \phi'$ for ϕ in $\mathcal{D}(D) = \{v \in \mathcal{Y} : v' \in \mathcal{Y}, v'(0) = F(v)\}$. Then D generates a (C_0) semigroup T on $\mathcal{Y}$, and $T(t)f = u_t$; thus $T(t)f(0) = u(t)$ describes the solution of the functional differential equation as t varies over $\mathbb{R}^+$.

This result is due to Hale [1]. We shall discuss this in a nonlinear context in the next volume.

15.19.   THE NAVIER-STOKES EQUATIONS.   Let Ω be a bounded domain in $\mathbb{R}^3$ having a sufficiently smooth boundary $\partial\Omega$. (Sufficient is that $\partial\Omega$ is of class C^3.) The flow of an incompressible fluid in Ω is governed by the Navier-Stokes equations. *For* $u = \begin{pmatrix} u_1 \\ u_2 \\ u_3 \end{pmatrix} = u(t,x)$ the velocity vector of the fluid at the point $x \in \Omega$ at time t and $p = p(t,x)$ the pressure, the Navier-Stokes equations are

$$\left. \begin{aligned} \frac{\partial u_j}{\partial t} - \frac{1}{\rho}\Delta u_j &= -\frac{\partial p}{\partial x_j} - \sum_{k=1}^{3} u_k \frac{\partial u_j}{\partial x_k} + g_j(x) \\ &\qquad (t \in \mathbb{R}^+, x \in \Omega), \\ \sum_{j=1}^{3} \frac{\partial u_j}{\partial x_j} &= 0 \qquad (t \in \mathbb{R}^+, x \in \Omega), \\ u(t,x) &= 0 \qquad (t \in \mathbb{R}^+, x \in \partial\Omega). \end{aligned} \right\} \tag{15.2}$$

Here the unknowns are u and p; $g = \begin{pmatrix} g_1 \\ g_2 \\ g_3 \end{pmatrix}$ is the external force, ρ is a positive constant called the Reynolds' number ($1/\rho$ is the viscosity of the fluid), and the density of the fluid is 1. In vector notation (15.2) becomes

$$\frac{\partial u}{\partial t} - \frac{1}{\rho} \Delta u = -\nabla p - (u \cdot \nabla)u + g,$$

$$\operatorname{div} u = 0 \text{ in } \Omega,$$

$$u = 0 \text{ on } \partial\Omega.$$

We also consider the initial condition

$$u(0,x) = f(x) \qquad (x \in \Omega). \tag{15.3}$$

We shall write (15.2) as an equation of the form $du/dt = Au + h(u)$ and solve it (locally in time) by the successive approximations technique used in Section 2.

Let $\mathscr{H} = [L^2(\Omega)]^3 = L^2(\Omega, \mathbb{C}^3)$. Let $\mathscr{H}_\sigma$ be the closure of

$$\{u \in C^1(\bar{\Omega}, \mathbb{C}^3): \operatorname{div} u = 0 \text{ in } \Omega, u = 0 \text{ on } \partial\Omega\}$$

in $\mathscr{H}$. Vectors in $\mathscr{H}_\sigma$ are called solenoidal vector fields, thus explaining the subscript σ. Let $\mathscr{H}_\nabla$ be the closure in $\mathscr{H}$ of

$$\{h: h = \nabla\phi \text{ for some } \phi \in C^1(\bar{\Omega})\}.$$

15.20.   LEMMA. $\mathscr{H} = \mathscr{H}_\sigma \oplus \mathscr{H}_\nabla$.

Proof.   Let $u \in C^1(\bar{\Omega}, \mathbb{C}^3)$ satisfy $\operatorname{div} u = 0$ in Ω and $u = 0$ on $\partial\Omega$. Let $\phi \in C^1(\bar{\Omega})$. Then

$$\langle \nabla\phi, u \rangle = \sum_{j=1}^{3} \int_\Omega \frac{\partial\phi}{\partial x_j} u_j \, dx = \int_\Omega \nabla\phi \cdot u \, dx.$$

But

$$\operatorname{div}(\phi u) = \phi \operatorname{div} u + \nabla\phi \cdot u,$$

and the divergence theorem gives

$$\int_\Omega \operatorname{div}(\phi u) \, dx = \int_{\partial\Omega} \phi u \cdot n \, ds = 0,$$

where n is the unit outer normal to $\partial\Omega$. Since also $\operatorname{div} u = 0$ it follows that $\int_\Omega \nabla\phi \cdot u \, dx = 0$, and so $\mathscr{H}_\sigma \perp \mathscr{H}_\nabla$.

Next let $u \in C_c^\infty(\Omega, \mathbb{C}^3)$ (which is dense in $\mathscr{H}$). It is enough to show that $u = v + \nabla\phi$ where $v \in \mathscr{H}_\sigma$ and $\phi \in C^1(\bar{\Omega})$. If this were true, then it would follow that

$$\operatorname{div} u = \operatorname{div} v + \operatorname{div} \nabla\phi = \Delta\phi$$

and $\partial\phi/\partial n = \nabla\phi \cdot n = 0$ on $\partial\Omega$. Let $h = \operatorname{div} u \in C_c^\infty(\Omega)$, which is known. The conditions $\Delta\phi = h$ in Ω, $\partial\phi/\partial n = 0$ on $\partial\Omega$ determine $\phi \in C^1(\bar{\Omega}, \mathbb{C})$ uniquely by a standard result in the classical theory of elliptic partial differential equations. (Recall that $\partial\Omega$ is assumed to be smooth.) Having determined ϕ let $v = u - \nabla\phi$, and the desired result follows.   ∎

15.21. Let P be the orthogonal projection of $\mathcal{H}$ onto $\mathcal{H}_\sigma$. Define A to be $(1/\rho)P\Delta$ where Δ is equipped with Dirichlet boundary conditions, i.e. $Au = (1/\rho)P\Delta u$ for $u \in \mathcal{D}(A)$, i.e. $u = \begin{pmatrix} u_1 \\ u_2 \\ u_3 \end{pmatrix}$ and $u_j \in H^2(\Omega) \cap H_0^1(\Omega)$, the Sobolev space, for $j = 1,2,3$. Standard results concerning Δ give the following lemma.

LEMMA.   *The operator A is a negative self-adjoint operator. There is a $\epsilon > 0$ such that $A \leq -\epsilon I$.*

15.22.   REDUCTION TO AN ABSTRACT CAUCHY PROBLEM.   If u is a solution of (15.2) and $v(t) = Pu(t)$, then v maps $\mathbb{R}^+$ into $\mathcal{H}_\sigma$ and satisfies

$$\frac{dv}{dt} = Av + N(v) + h. \tag{15.4}$$

Here $h = Pg$ and $Nv = -P(v \cdot \nabla)v$. The condition $\operatorname{div} u = 0$ has been absorbed into the space $\mathcal{H}_\sigma$, and the pressure p has disappeared since $P(\nabla p) = 0$ by Lemma 15.20. We shall solve (15.4) together with the initial condition

$$v(0) = f; \tag{15.5}$$

see (15.3) and note that $Pf = f$ is required by the condition $\operatorname{div} u = 0$ when $t = 0$. Once we solve (15.4), (15.5) uniquely for v, we set $u(t,\cdot) = v(t) \in \mathcal{H}_\sigma$, and then the first equation of (15.2) can be solved for ∇p, which determines p uniquely except for an additive constant.

   The following lemma, which is a variant of the results of Section 2, enables us to solve (15.4), (15.5).

15.23.   LEMMA.   *Let A_0 on a Hilbert space $\mathcal{K}$ be self-adjoint and satisfy $A_0 \leq -\epsilon I$ for some $\epsilon > 0$. Let $\mathcal{K}_\alpha$ be the Hilbert space $\mathcal{D}[(-A_0)^\alpha]$ equipped with the norm $\|f\|_\alpha = \|(-A_0)^\alpha f\|$ where $\alpha \in [0,1[$ is fixed. Let F be a locally Lipschitzian function from an open subset U of $\mathcal{K}_\alpha$ to $\mathcal{K}$, i.e. for $f \in U$ there is a neighborhood V of f and a constant K such that $\|F(f_1) - F(f_2)\| \leq K\|f_1 - f_2\|_\alpha$ for all $f_1,f_2 \in V$. Then for each $f \in U$ there is a $\tau = \tau(f) > 0$ such that*

$$\frac{dv}{dt} = A_0 v + F(v), v(0) = f$$

has a unique mild solution on $[0,\tau]$.

Proof.   (The case $\alpha = 0$ was done in Section 2.) The estimate

$$\|(-A_0)^\alpha e^{tA_0}\| \leq 1/t^\alpha \text{ for } t > 0 \tag{15.6}$$

will be used. This follows by representing A_0 as a multiplication operator (by the spectral theorem) and noting that $\sup\{s^\alpha e^{-s}: s > 0\} = (\alpha/e)^\alpha \leq 1$. Let $L > 0$, and let $\delta > 0$ be such that $\|f_1 - f\| \leq \delta$, $\|f_2 - f\| \leq \delta$ imply $\|F(f_1) - F(f_2)\| \leq K\|f_1 - f_2\|_\alpha$ for some fixed constant K. Now define

$$\mathcal{X}_{\delta,L} = \{u \in C([0,L],\mathcal{K}): u(0) = (-A_0)^\alpha f, \|u(t) - u(0)\| \leq \delta$$

$$\text{for } 0 \leq t \leq L\},$$

and define

$$S : \mathcal{X}_{\delta,L} \to C([0,L],\mathcal{K})$$

by

$$(Su)(t) = T(t)(-A_0)^{\alpha} f + \int_0^+ (-A_0)^{\alpha} T(t - s) F [(-A_0)^{-\alpha} u(s)] \, ds$$

for $0 \leq t \leq L$. For $u,v \in \mathcal{X}_{\delta,L}$,

$$\|Su - Sv\|_{\infty} = \sup\{\|Su(t) - Sv(t)\| : 0 \leq t \leq L\}$$

is the norm of $Su - Sv$ in the Banach space $C([0,L],\mathcal{K})$ and

$$\|Su(t) - Sv(t)\| \leq \int_0^t \|(-A_0)^{\alpha} T(t - s)\| \, \|F[(-A_0)^{-\alpha} u(0)]$$
$$- F[(-A_0)^{-\alpha} v(s)]\| \, ds$$
$$\leq \int_0^t (t - s)^{-\alpha} K \|(-A_0)^{-\alpha}[u(s) - v(s)]\|_{\alpha} \, ds$$

by (15.6) and the definition of K

$$= \int_0^t K(t - s)^{-\alpha} \|u(s) - v(s)\| \, ds$$

by the definition of $\|\cdot\|_{\alpha}$

$$\leq K \|u - v\|_{\infty} \frac{t^{1 - \alpha}}{1 - \alpha} \leq K \|u - v\|_{\infty} \frac{L^{1 - \alpha}}{1 - \alpha}$$

since $\alpha < 1$. Choose $L > 0$ sufficiently small so that $KL^{1-\alpha}(1 - \alpha)^{-1} < 1$. Next, if $u \in \mathcal{X}_{\delta,L}$, similar reasoning gives

$$\|Su(t) - (-A_0)^{\alpha} f\| \leq \|[T(t) - I](-A_0)^{\alpha} f\|$$
$$+ \int_0^t \|(-A_0)^{\alpha} T(t - s)\{F[(-A_0)^{-\alpha} u(s)] - F(f)\}\| \, ds$$
$$+ \int_0^t \|(-A_0)^{-\alpha} T(t - s) F(f)\| \, ds$$
$$\leq \sup_{0 \leq t \leq L} \|(T(t) - I)(-A_0)^{\alpha} f\| + \int_0^t (t - s)^{-\alpha} K \|u(s)$$
$$- (-A_0)^{\alpha} f\| \, ds + \int_0^t (t - s)^{-\alpha} \|F(f)\| \, ds$$
$$\leq \sup_{0 \leq t \leq L} \|(T(t) - I)(-A_0)^{\alpha} f\| + KL^{1-\alpha}(1 - \alpha)^{-1} \delta$$
$$+ L^{1-\alpha}(1 - \alpha)^{-1} \|F(f)\| \leq \delta$$

if $0 < L = L(A_0, f, F, \alpha, \delta)$ is chosen to be sufficiently small. Thus, for this choice of L, $F(\mathcal{X}_{\delta,L}) \subset \mathcal{X}_{\delta,L}$. The Banach-Picard fixed point theorem (Theorem 2.2) implies that F has a unique fixed point w in $\mathcal{X}_{\delta,L}$. w thus satisfies the integral equation

$$w(t) = T(t)(-A_0)^{\alpha} f + \int_0^t (-A_0)^{\alpha} T(t - s) F[(-A_0)^{-\alpha} w(s)] \, ds, \quad (15.7)$$

$0 \le t \le L$. Now let v be the unique mild solution of the linear inhomogeneous problem

$$dv(t)/dt = A_0 v(t) + F[(-A_0)^{-\alpha} w(t)] \qquad 0 \le t \le L,$$

$$v(0) = f.$$

In other words,

$$v(t) = T(t)f + \int_0^t T(t - s) F[(-A_0)^{-\alpha} w(s)] \, ds \qquad 0 \le t \le L.$$

Applying $(-A_0)^{\alpha}$ to both sides of this equation gives

$$(-A_0)^{\alpha} v(t) = T(t)(-A_0)^{\alpha} f + \int_0^t (-A_0)^{\alpha} T(t - s) F[(-A_0)^{-\alpha} w(s)] \, ds.$$

But since w is the *unique* solution of this equation [compare (15.7)], it follows that $w(t) = (-A_0)^{\alpha} v(t)$, $0 \le t \le L$. In other words, v satisfies

$$v(t) = T_0(t)f + \int_0^t T(t - s) F[v(s)] \, ds \qquad 0 \le t \le L.$$

Taking τ to be this choice of L completes the proof. ∎

15.24. REMARK. It is not difficult to show that the function w, constructed above, satisfies a Hölder condition in compact subintervals of $]0,\tau]$. Using Theorem 1.4 we deduce that the unique mild solution v of Lemma 15.23 belongs to $C([0,\tau], \mathcal{K}) \cap C^1(]0,\tau], \mathcal{K})$ and is thus a solution of $dv/dt = A_0 v + F(v)$, $v(0) = f$ in the usual sense.

15.25. THE END OF THE NAVIER-STOKES DISCUSSION. We shall apply Lemma 15.23 with $\mathcal{K} = \mathcal{H}_\sigma$, $A_0 = A$, and $F(v) = N(v) + h$ (in the notation of 15.22). The only thing to be checked is the local Lipschitz condition of F, but since $F(u) - F(v) = N(u) - N(v)$, we see that h is irrelevant.

Using the trivial inequality

$$\int_\Omega |h_1 h_2|^2 \, dx \le \|h_1\|_{L^\infty}^2 \int_\Omega |h_2|^2 \, dx$$

we obtain, for $v \in \mathcal{D}(A)$,

$$\|N(v)\| = \|P(v \cdot \nabla)v\| \le \|(v \cdot \nabla)v\| \le \|v\|_{L^\infty} \|\nabla v\|.$$

Next we *claim* that for $3/4 < \alpha < 1$ there is a constant C_α such that

$$\left.\begin{array}{c} \|v\|_{L^\infty} \le C_\alpha \|v\|_\alpha \\ \|\nabla v\| \le C_\alpha \|v\|_\alpha \end{array}\right\} \tag{15.8}$$

for $v \in \mathscr{D}(A)$. Assuming this to be true it follows that for $u, v \in \mathscr{D}(A)$,

$$\begin{aligned} \|N(v) - N(u)\| &\le \|(v \cdot \nabla)v - (u \cdot \nabla)u\| \\ &= \|(v \cdot \nabla)(v - u) + ((v - u) \cdot \nabla)u\| \\ &\le \|v\|_{L^\infty} \|\nabla(v - u)\| + \|v - u\|_{L^\infty} \|\nabla u\| \\ &\le \text{Constant } \|u - v\|_\alpha \end{aligned}$$

by (15.8) where the constant depends on α and is uniform for u, v in bounded subsets of $\mathscr{K}_\alpha$. Since $\mathscr{D}(A)$ is dense in $\mathscr{K}_\alpha$, it follows that the hypotheses of Lemma 15.23 are satisfied by F. Thus it only remains to establish the inequalities (15.8).

These inequalities follow from a suitable version of the Sobolev inequalities, which we sketch now. First, if $u \in H^2(\Omega) \cap H^1_0(\Omega)$, then (after correction on a null set) u satisfies

$$|u(x) - u(y)| \le C_\Omega \|\Delta u\| \, |x - y|^{1/2} \tag{15.9}$$

for all $x, y \in \Omega$, where the constant C_Ω depends only on Ω. To see this first suppose $u \in C_c^\infty(\Omega)$. (The general case follows from this one by an easy approximation argument.) From classical potential theory we know

$$u(x) = \frac{1}{4\pi} \int_\Omega \frac{\Delta u(z)}{|x - z|} \, dz, \; x \in \Omega.$$

Thus for $x, y \in \Omega$,

$$\begin{aligned} |u(x) - u(y)|^2 &\le (4\pi)^{-2} \left\{ \int_\Omega \Delta u(z)(|x - z|^{-1} - |y - z|^{-1}) \, dz \right\}^2 \\ &\le (4\pi)^{-2} \int_\Omega |\Delta u(z)|^2 \, dz \cdot \int_\Omega (|x - z|^{-1} - |y - z|^{-1})^2 \, dz \\ &\le C_\Omega \|\Delta u\|^2 |x - y|, \end{aligned}$$

and (15.9) follows. Next, there is a constant C_1 such that for $u \in H^2(\Omega) \cap H^1_0(\Omega)$,

$$\|u\|_{L^\infty} \le C_1 \|Lu\|^{3/4} \|u\|^{1/4}, \tag{15.10}$$

where L is the Laplacian with Dirichlet boundary conditions. Note that $-L \ge \epsilon_0 I$ for some $\epsilon > 0$. To prove (15.10) note that u is uniformly continuous on Ω by (15.9), and so $u \in C(\bar{\Omega})$ and $u = 0$ on $\partial\Omega$. If $\|u\|_{L^\infty} > 0$ select $x_0 \in \Omega$ such that $|u(x_0)| = \|u\|_{L^\infty}$; and for definiteness suppose $\|u\|_{L^\infty} = 1$. Then for x in the ball

$$B = \{x \in \mathbb{R}^3 : |x - x_0| < (C_\Omega \|Lu\|)^{-2}\},$$

we have

$$|u(x)| \geq |u(x_0)| - |u(x) - u(x_0)| \geq 1 - |x - x_0|^{1/2} C_\Omega \|Lu\| > 0$$

by (15.9). Thus $B \subset \Omega$ and

$$\|u\|^2 \geq \int_B |u(x)|^2 \, dx \geq \int_B (1 - |x - x_0|^{1/2} C_\Omega \|Lu\|)^2 \, dx$$

$$(\text{set } y = x - x_0, \ s = |y|, \ R = C_\Omega \|Lu\|)$$

$$= 4\pi \int_0^{R^{-2}} (1 - s^{1/2} R)[s^2] \, ds$$

$$\left(\text{since } dy = s^2 ds \, d\omega \text{ and } \int d\omega = 4\pi; \text{ set } t = sR^2 \right)$$

$$= 4\pi \int_0^1 (1 - t^{1/2})[t^2 R^{-4}] dt \, R^{-2} = C_* R^{-6}$$

where C_* is a fixed positive constant. Hence for some constant $C_1 > 0$ (depending on Ω),

$$\|u\|^2 \geq C_1 \|Lu\|^{-6} = C_1 \|Lu\|^{-6} \|u\|_{L^\infty}^k$$

holds for all $u \in H^2(\Omega) \cap H_0^1(\Omega)$ such that $\|u\|_{L^\infty} = 1$. Now we replace $0 \neq u \in H^2 \cap H_0^1 \cap L^\infty$ by $v = u/\|u\|_{L^\infty}$. Homogeneity necessitates that $k = 8$. Thus (15.10) follows.

Next, using the formula

$$(-L)^{-\alpha} v = \frac{\sin \pi\alpha}{\pi} \int_0^\infty \lambda^{-\alpha} (\lambda - L)^{-1} v \, d\lambda$$

(cf. Section I.9.2) together with $-L \geq \epsilon_0 I$ we obtain, for $v \in L^2(\Omega)$,

$$\|(\lambda - L)^{-1} v\| \leq (\epsilon_0 + \lambda)^{-1} \|v\|,$$

$$\|L(\lambda - L)^{-1} v\| \leq \|v\|,$$

$$\|(\lambda - L)^{-1} v\|_{L^\infty} \leq C_1 \|L(\lambda - L)^{-1} v\|^{3/4} \|(\lambda - L)^{-1} v\|$$

by (15.10), and

$$\|(-L)^{-\alpha} v\|_{L^\infty} \leq \frac{\sin \pi\alpha}{\pi} \int_0^\infty \lambda^{-\alpha} \|(\lambda - L)^{-1} v\|_{L^\infty} \, d\lambda$$

$$\leq \frac{C_1 \sin \pi\alpha}{\pi} \|v\| \int_0^\infty \lambda^{-\alpha} (\epsilon_0 + \lambda)^{-1/4} \, d\lambda$$

$$= C_2 \|v\|$$

where C_2 depends on α, C_1, ϵ_0, and the fact that $\alpha > 3/4$. By taking $v = (-L)^\alpha u$, the estimate

$$\|u\|_{L^\infty} \leq C_2 \|(-L)^\alpha u\|$$

follows. This gives the first inequality of (15.8). The second one is obtained by noting that

$$\|\nabla v\|^2 = \langle -Av, v \rangle = \|(-A_0)^{1/2}v\|^2 = \|v\|_{1/2}^2 \le C_3 \|v\|_\alpha^2$$

for $\alpha \ge 1/2$, where $C_3 = \epsilon_0^{2\alpha - 1}$; this follows by representing A_0 as a multiplication operator (by the spectral theorem).   ∎

15.26.   Exercises

*1. Prove Proposition 15.3.
*2. Prove Theorem 15.4.
 3. Verify the final assertion of the paragraph following Theorem 15.4.
 4. Find all m-dissipative extensions of $A = d^2/dx^2$ on $\mathscr{D}(A) = C_c^\infty\,]0,\infty[\in \mathscr{H} = L^2(\mathbb{R}^+)$.
*5. Construct a symmetric operator with no self-adjoint extension. (Hint: Construct a nonsurjective isometry C on $\mathscr{H}$ with $I + C$ injective.)
 6. Prove the assertions in Remark 15.24.
 7. Let $\mathbb{T}^n = \mathbb{R}^n/\mathbb{Z}^n$ be the n-dimensional torus. Define A by

$$Af = \sum_{i,j=1}^{n} a_{ij}(x) \frac{\partial^2 f}{\partial x_i \partial x_j} + \sum_{i=1}^{n} b_i(x) \frac{\partial f}{\partial x_i} + c(x)f$$

for $f \in \mathscr{D}(A) = C^2(\mathbb{T}^n) \subset \mathscr{X} \equiv C(\mathbb{T}^n)$. Show that the closure of A is m-dissipative if all the coefficients are in $C^\infty(\mathbb{T}^n)$, if $c(x) \le 0$ for all $x \in \mathbb{T}^n$, and if the matrix $[a_{ij}(x)]$ is symmetric and (strictly) positive definite for all $x \in \mathbb{T}^n$.
 8. Let $A_n = n^2[\sin(nx)]\, d/dx$ on $\mathscr{X} = C_0(\mathbb{R})$. Show that A_n generates a (C_0) semigroup T_n on $\mathscr{X}$. Show also that for each $f \in \mathscr{X} \setminus \{0\}$, $\lim_{n \to \infty} \|T_n(t)f - f\| = 0$ for each $t \in \mathbb{R}^+$ while $\lim_{n \to \infty} \|A_n f\| = \infty$.
*9. Let A generate a uniformly bounded (C_0) group T on $\mathscr{X}$. Then

$$S(t)f = \frac{t}{\pi} \int_{-\infty}^{\infty} (t^2 + s^2)^{-1} T(s)f\, ds$$

defines a uniformly bounded (C_0) semigroup which, if $\mathscr{X}$ is complex, has an analytic extension into the right half plane. Moreover, for $f \in \mathscr{D}(A^2)$, $u(t) = S(t)f$ defines the unique solution u of the (two-point boundary) problem

$$\left(\frac{d^2}{dt^2} + A^2 \right) u(t) = 0 \qquad (t \in \mathbb{R}^+),$$

$$u(0) = f,\ \lim_{t \to \infty} \sup\|u(t)\| < \infty.$$

 10. Discuss this further when $\mathscr{X} = BUC(\mathbb{R})$ or $L^p(\mathbb{R})$ $(1 \le p < \infty)$ and $A = d/dx$. (In this case u satisfies as elliptic equation, not a hyperbolic one. In general, this is true in an abstract sense.)
*11. Let A generate a (C_0) group T on $\mathscr{X}$, and let $f \in \mathscr{D}(A^2)$. Then u defined by

$$u(t) = \frac{\Gamma(\rho + 1)/2]}{\Gamma(\rho/2)\Gamma(1/2)} \int_{-1}^{1} (1 - s^2)^{\rho/2 - 1} T(st)f\, ds$$

satisfies the abstract Euler-Poisson-Darboux equation

$$u''(t) + \frac{\rho}{t} u'(t) - A^2 u(t) = 0 \qquad (t > 0)$$

for $\rho > 0$. Moreover, $u(0) = f$, $u'(0) = 0$.

16. Historical Notes and Remarks

Section 1.   The abstract Cauchy problem

$$du/dt = Au + f(t) \qquad (t \geq 0), u(0) = u_0 \tag{16.1}$$

can be treated directly, with or without semigroup methods. Books treating it include, among others, Carroll [1,2], Da Prato [9], Fattorini [19], A. Friedman [1], Henry [1], S. Kreĭn [1,2], Lions [3], Lions and Magenes [1], Pazy [5], Showalter [11], Tanabe [5], Yosida [10], and Zaidman [2]. A small sampling of authors who have written on this subject includes Agmon, Beals, Fattorini, Goldstein, S. Kreĭn, Ljubič, Nirenberg, Plamenevskiĭ, and Zaidman. The approach of Section I goes back to Hille; see his book [1] which contains references to the earlier work.

Theorem 1.2 and equivalent versions of it are due to Hille [4] and Phillips [5]. The equivalent versions involve variants of Definition 1.1. For a comprehensive discussion see Neubrander [2]. Theorem 1.3 is due to Phillips [4]. See Crandall and Pazy [1] for generalizations of Theorem 1.4. Exercises 1.5.2,3 are due, respectively, to Ljubič [1] and Webb [1]. See Fattorini [1] and Neubrander [2] for a discussion of Exercise 1.5.4 and related matters. Other authors to consult on regularity matters for (16.1) include Da Prato, Iannelli, Kato, Sinestrari, and Travis, among others. See also Ball [1].

Section 2.   This section contains standard material from the theory of ordinary differential equations. Texts such as those of Coddington and Levinson [1] and Hartman [1] may be consulted. The Picard-Banach fixed point theorem (Theorem 2.2) has many aliases, viz. the contraction mapping principle, the method of successive approximations, etc. The term *mild solution* (Definition 2.3) was coined by F. E. Browder. Our treatment follows Segal [1].

Section 3.   Many texts discuss the Fourier transform. The books of Stein and Weiss [1] and Bochner and Chandrasekharan [1] are especially nice. The Schwartz space (Definition 3.1) is named after L. Schwartz, one of the founders of the theory of distributions.

Section 4, 5.   The elementary treatment in these sections follows that of Goldstein [9]. The spaces H^v are named after S. L. Sobolev, who like Schwartz was one of the founders of distribution theory. For more general Sobolev inequalities see Adams [1]. One can also consult Stein and Weiss [1] or other books on partial differential equations such as Bers, John, and Schechter [1], Dunford and Schwartz [2], A. Friedman [1], and so on. Lax and Milgram [1] remains a useful reference. Vast generalizations of the results of these sections are stated in Section 12; the notes for that section contain further references. The example in Remark 5.10 is due to J. R. Dorroh [personal communication].

Section 6.   Bochner proved Theorem 6.2 in 1932 [1]. The spectral theorem (Theorem 6.9) was first proved by D. Hilbert for bounded self-adjoint operators; the extension to unbounded self-adjoint operators was made in the late twenties independently by J. von Neumann and M. H. Stone. Incidentally, this extension was motivated by the then new quantum theory. Our proof of the spectral theorem follows the ideas of Lax and Phillips [2]. For much more on self-adjoint operators see the books of Akhiezer and Glazman [1], Davies [8], Dunford and Schwartz [2], Faris [3], Kato [12], Nelson [7], Reed and Simon [1,2], Riesz and Sz-Nagy [1], Yosida [10], and many others as well. See Kato [12] and Reed and Simon [2] for more results like Theorem 6.13, which is due to Kato [1].

Section 7. Our treatment follows Goldstein [2]. Theorems 7.12 and 7.13 are due to Goldstein, [5] and [8] respectively. For much more on equipartition of energy see the joint work of Goldstein and Sandefur and the references contained therein. Exercise 7.14.7 is from Lax and Phillips [2] and Duffin [1]. Exercise 7.14.8 is from Rosencrans [1] and is connected with the sharp form of the Schwarzschild criterion for the linear instability of a fluid layer in hydrostatic equilibrium.

Section 8. The seminal articles on cosine functions were written in the late sixties by Sova [1], Da Prato and Giusti [1], and Fattorini [1,2]. Theorem 8.2 is due to Fattorini while Theorem 8.3 is due to Sova [1] and Da Prato and Giusti [1]. Theorem 8.5 is due to B. Nagy [4] and Travis and Webb [3,5]. Theorem 8.6 is due, independently, to Goldstein [13] and Konishi [1]. Theorem 8.7 is due to Romanoff [1]; see also Griego and Hersh [1] and Fattorini [1,2]. For the falsity of the converse assertion see Goldstein [20, p. 372] where the argument uses the result of Littman [1]. For Theorem 8.8 see Kisyński [5,6]. Remark 8.9 is due to Fattorini [2], while counterexample 8.10 is due to B. Nagy [4] and Kisyński [7]. Remarks 8.11 and 8.12 are due, respectively, to Dettman [2] and Fattorini [5]. For much more on cosine functions see Fattorini's book [14] or the survey article of Travis and Webb [3]. Other contributors to cosine function theory include Kurepa, Lutz, Okazawa, Takenaka, and others as well. Connected with this is the work done on *related differential equations* by Bragg and Dettman. Finally we mention the following formula of Bragg [8]. For $j = 1,2$ let B_j generate a (C_0) group on a Banach space X; let C_j be the cosine function generated by $A_j = B_j^2$; and suppose B_1, B_2 commute in the sense that $C_1(t)C_2(s) = C_1(s)C_1(t)$ for all real t and s. If $A_3 = A_1 + A_2$ generates a cosine function C_3, then it is given by

$$C_3(t)f = \frac{d}{dt}\left\{\frac{1}{2\pi}\int_0^{t^2}(t^2-s^2)^{-1/2}\int_0^s C_1[(s-r)^{1/2}]C_2(r^{1/2})f[r(s-r)]^{1/2}\,dr\,ds\right\}$$

$$= \frac{d}{dt}\left\{\frac{2}{\pi}\int_0^{t^2}(t^2-u^2)^{-1/2}\int_0^{\pi/2}C_1(u\cos\theta)C_2(u\sin\theta)f\,d\theta\,du\right\}.$$

Section 9. Our treatment follows Wilcox [2], which contains interesting applications. The theory goes back to Friedrichs. For more general results see any of a number of sources, including Lax and Phillips [2,3], Fattorini [14], and Tanabe [5]. Exercise 9.5.3 is due to Brenner [1].

Section 10. Theorem 10.2 is Hille's [4]; see also Fattorini [1,2]. The remarks in 10.3 are due to Hersh [2]. Theorem 10.2 has already been discussed in the context of Theorem 8.7. Theorem 10.6 is due to Sandefur [1].

Section 11. There is a large literature on singular perturbations. Authors include Bobisud, Butcher, Davies, Donaldson, Ellis, Hersh, Kisyński, Nur, Pinsky, Tanabe, M. Watanabe, and many others. Our presentation of the basic theorems (11.2, 11.3, 11.7) is new in the sense that it was developed from scratch, but all of the ideas have been well-known for a long time. The neat result of 11.8 is the work of Schoene [1]; see also Veselić [1] and the references cited there. For more on Sobolev equations see articles by Showalter. Exercise 11.12.3 is due to Kac [2]. For much more on this see Griego and Hersh [1,2] and the discussion below in Sections 15.7–15.12.

Section 12. For Sobolev spaces, see Adams [1] and the other references mentioned above in the notes for Sections 4 and 5. General results on elliptic boundary problems which yield estimates so that semigroup generation theorems can be applied were obtained by Agmon, Douglis, and Nirenberg [1]. These results are very complicated. For a readable treatment of the second-order case see Fattorini [14]. A classical reference is Dunford and Schwartz [2].

Section 13. There is an enormous literature on temporally inhomogeneous (or time dependent) problems, i.e. equation (13.1) in which A depends on t. The first important existence theorems were obtained by Kato [2,4]. Many other authors obtained related results, including J. Elliot, Goldstein, Hackman, Heyn, Kisyński, Mizohata, Poulsen, and Yosida. The simple result of Theorem 13.2 is due to Goldstein [1]. The method of 13.4 is due, independently, to Tanabe [1] and Sobolevskiĭ [1]. The sharp result in the analytic semigroup context, Theorem 13.5, is due to Kato and Tanabe [1,2]. See also Masuda [1]. Theorem 13.7, is Kato's [4]; Theorems 13.9 and 13.10 are due to Goldstein [12]. While Theorem 13.5 gives sharp results for parabolic problems, no result known by the late sixties did the same for hyperbolic problems. This was remedied in the early seventies by Kato [16,22]. Other authors who worked on aspects of Kato's theorem 13.13 include Arosio, Da Prato, Dorroh, Grisvard, Iannelli, Kobayasi, Massey, and Rauch. See also the books of Pazy [5] and Tanabe [5] for more information.

Section 14. Scattering theory is an enormous subject, and while the Reference section refers to many articles on scattering, it only includes a small percentage of the literature. Books dealing with quantum mechanics and with scattering theory include, among many others, Amrein [1], Amrein, Jauch, and Sinha [1], Eastham and Kalf [1], Kato [12], Perry [1], Prugovečki [1], and Reed and Simon [1–4]. Kato and Kuroda first published Theorem 14.6 in [1] (see also [2]). Proposition 14.8 is the Cook [1]–Kuroda [1] criterion. For more on the elementary theory (such as the chain rule) see Kato [12]. The motion of H-smoothness is due to Kato [11]. The discussion in 14.12–14.15 is based on Kato [11], Lavine [2,3], and Reed and Simon [3]. Two-space scattering theory was developed by Kato [14]. For more on the S-matrix see the nice exposition of Kato [19]. For the Lax-Phillips theory see Lax and Phillips [2,3] (and also [4–7]) and Foiaş [4]. The remarks in 14.22 are taken from Goldstein and Sandefur [2]. Exercises 14.23.3, 5 are due, respectively, to Howland [2] and Lavine [2].

A very beautiful new method of scattering theory was developed in the seventies by V. Enss. This is treated in articles by Enss, Davies, Mourre, and Simon and in the books of Perry [1] and Reed and Simon [3].

Contributors to scattering theory (involving semigroups and unitary groups) include, among many others, Agmon, Birman, Davies, Enss, Hagedorn, Howland, Ikebe, Iorio, Kato, Lavine, O'Carroll, Pearson, Perry, Schechter, Simon, Weder, and Yajima.

Section 15.   The Cayley transform was introduced into functional analysis in the late twenties by von Neumann in his study of unbounded self-adjoint operators. Proposition 15.3 and Theorem 15.4 are due to Phillips [8]. For the classification of *m*-dissipative extensions of dissipative operators see Phillips [11] and Crandall and Phillips [1]. The analogous problem of self-adjoint extensions of symmetric operators was solved in the thirties by Stone and von Neumann. The techniques of M. Kreĭn [1] influenced Phillips's work. For a nice specific instance of 15.5 see Powers and Radin [1]. The elementary observation of 15.6 is related to the work of Feller in the 1950s and also to what is now commonly referred to as *Kato's inequality*; see, among other papers, Arendt [1], Kato [21], and Simon [12]. For random evolutions see Griego and Hersh ([2] as well as other articles), Keepler, Kertz, Kurtz, Papanicolaou, Pinsky, Quiring, and Varadhan. Theorem 15.8 and the argument of 15.9 are due to Griego and Hersh [1,2]; they were largely influenced by Kac [2]. Theorem 15.11 is due to Kurtz [6]. To interpret the final result of 15.12 one should recall the unified discussion of Nelson [5] on the Feynman path formula and Kac's version of it (involving Wiener measure) for the heat equation with a potential; see also Goldstein [22] and Simon [15]. Sections 15.13–15.17 give only the barest hints of the use of semigroups in control theory. For more, one may consult the following authors, among many others: Balas, Balakrishnan, Curtain, Datko, Fattorini, Lagnese, Lasiecka, Leigh, Levan, Lions, O'Brien, Rigby, Russell, Salamon, Slemrod, Triggiani, Vinter, and Wexler. Functional differential equations form a popular field. A small sampling of contributors to it consists of Hale, Kappel, Schappacher, Travis, and Webb. More discussion on nonlinear functional equations will be given in the next volume. Our discussion of the Navier-Stokes equations is based on the work of Fujita and Kato [1] and Kato and Fujita [1]. This is also discussed in the book of Henry [1]. Related matters are discussed in Pazy's book [5]. There are numerous related articles. We established the existence of a solution of (15.2) for $(t,x) \in [0,\tau) \times \Omega$. This is a *local* existence (and uniqueness) result since τ may be quite small. The questions of global existence [Can one take $\tau = \infty$? Is $\tau_{max} < \infty$ for certain f?] are fundamental open problems. Exercise 15.26.9 is related to the classical Poisson integral formula. (Take $A = d/dx$.) For Exercise 15.26.11 and related matters see Donaldson [1].

This section and Section 10 of Chapter I are merely brief guides to the literature and to the Reference section.

For more on applications to quantum theory see the articles by, among others, Albeverio, Bratelli, Carmona, Chernoff, Davies, Devinatz, Dollard, Faris, C. Friedman, Frigerio, Gill, Gorini, Høegh-Krohn, Howland, Kato, Lapidus,

Najman, Narnhofer, Radin, Robinson, Semenov, Simon, Spohn, Sudarshan, van Winter, and Verri.

For more on applications to classical physics see articles by, among others, Beale, Gilliam, Greenlee, Kaper, Kato, Majda, Rauch, Schulenberger, Slemrod, M. Taylor, and Wilcox.

For more on applications to probability theory see articles by, among others, Bass, Berg, M. Berger, Bharucha-Reid, Butzer, Chung, Ethier, Faraut, Feller, Feyel, Forst, Fukushima, Griego, Hawkes, Hersh, Heyer, Hirsch, Hunt, Kurtz, Liggett, Norman, Orey, Papanicolaou, Pfeiffer, Piech, Pinsky, D. Ray, Reuter, Revuz, Rosenkrantz, Roth, K. Sato, Sawyer, Silverstein, Sloan, Stroock, Varadhan, S. Watanabe, Wentzell, and Williams. In particular, Stroock and Varadhan use probability theory to show that certain dissipative second-order partial differential operators have m-dissipative closures.

The references contain articles on the applications of semigroups to areas not mentioned in the text, including harmonic analysis, integral equations, Banach space theory, and more.

References

For the references to the early work on semigroups see the bibliographies in Hille and Phillips [1] and Dunford and Schwartz [1]. For a much more complete list of references see Goldstein [24], which can be obtained from the Mathematics Research Library, Tulane University, New Orleans, LA 70118.

Adams, R. A.
[1] *Sobolev Spaces*, Academic Press, New York, 1975.
Agmon, S.
[1] *Elliptic Boundary Value Problems*, Van Nostrand, Princeton, N. J., 1965.
[2] Spectral properties of Schrödinger operators and scattering theory, *Ann. Sc. Norm. Sup. Pisa* 2 (1975), 151–218.
[3] *Lectures on Exponential Decay of Solutions of Second-Order Elliptic Equations: Bounds on Eigenfunctions of N-Body Schrödinger Operators*, Princeton U. Press, Princeton, N.J., 1982.
Agmon, S., A. Douglis, and L. Nirenberg
[1] Estimates near the boundary for solutions of elliptic partial differential equations, *Comm. Pure Appl. Math.* 12 (1959), 623–727.
Agmon, S. and L. Nirenberg
[1] Properties of solutions of ordinary differential equations in Banach space, *Comm. Pure Appl. Math.* 16 (1963), 121–239.
Akcoglu, M. A. and J. Cunsolo
[1] An ergodic theorem for semigroups, *Proc. Amer. Math. Soc.* 24 (1970), 161–170.
[2] An identification of ratio ergodic limits for semi-groups, *Z. Warsch. verw. Geb.* 15 (1970), 215–229.
Akhiezer, N. I. and M. I. Glazman
[1] *Theory of Linear Operators in Hilbert Space*, Vols. 1, 2, F. Ungar, New York, 1961 and 1963.
Albeverio, S. A. and R. Høegh-Krohn
[1] *Mathematical Theory of Feynman Path Integrals*, Lecture Notes in Math. No. 523, Springer, Berlin, 1976.
[2] Dirichlet forms and Markov semigroups on C*-algebras, *Comm. Math. Phys.* 56 (1977), 173–187.
[3] Hunt processes and analytic potential theory on rigged Hilbert spaces, *Ann. Inst. H. Poincaré* 13B (1977), 269–291.
[4] Dirichlet forms and diffusion processes on rigged Hilbert spaces, *Z. Warsch. verw. Geb.* 40 (1977), 1–57.
Alsholm, P.
[1] Wave operators for long-range scattering, *J. Math. Anal. Appl.* 59 (1977), 550–572.
Altomare, F.
[1] Operatori di Lion generalizzati e famiglie risolvanti, *Boll. Un. Mat. Ital.* 15 (1978), 60–79.
Amann, H.
[1] Dual semigroups and second order linear elliptic boundary value problems, *Israel J. Math.* 45 (1983), 225–254.
Amrein, W. O.
[1] Some questions in nonrelativistic quantum scattering theory. In *Scattering Theory in Mathematical Physics* (eds. J. La Vita and J. -P. Marchand), Reidel, Dordrecht, Holland (1974), 97–140.
[2] *Nonrelativistic Quantum Dynamics*, Reidel, Dordrecht, Holland, 1981.
Amrein, W. O. and V. Georgescu
[1] On the characterization of bound states and scattering states in quantum mechanics, *Helv. Phys. Acta* 46 (1973), 635–659.
Amrein, W. O., J. M. Jauch, and K. B. Sinha
[1] *Scattering Theory in Quantum Mechanics*, Benjamin, Reading, Mass., 1977.
Angelescu, N., G. Nenciu, and M. Bundaru
[1] On the perturbation of Gibbs semigroups, *Comm. Math. Phys.* 42 (1975), 29–30.

181

Arakawa, T. and J. Takeuchi
[1] On the potential operators associated with Bessel processes, *Z. Warsch. verw. Geb.* 40 (1977), 83–90.

Arendt, W.
[1] Kato's inequality and spectral decomposition for positive C_0-groups, *Manus. Math.* 40 (1982), 277–298.

Arendt, W., P. R. Chernoff, and T. Kato
[1] A generalization of dissipativity and positive semigroups, *J. Oper. Th.* 8 (1982), 167–180.

Arendt, W. and G. Greiner
[1] The spectral mapping theorem for one-parameter groups of positive operators on $C_0(X)$. In *Semesterbericht Funktionalanalysis Tübingen* (1982–83), 11–59.

Arosio, A.
[1] Equations différentielles opérationelles linéaires du duxième ordre: problème de Cauchy et comportement asymptotique lorsque $t \to \infty$, *C. R. Acad. Sci. Paris* 295 (1982), 83–86.

Athreya, K. B. and T. G. Kurtz
[1] A generalization of Dynkin's identity and some applications, *Ann. Prob.* 1 (1973), 570–579.

Avila, G. S. S.
[1] Spectral resolution of differential operators associated with symmetric hyperbolic systems, *Appl. Anal.* 1 (1972), 283–299.

Avrin, J.
[1] Singular first order perturbations of the heat equation, *Proc. Roy. Soc. Edinburgh* 96A (1984), 317–321.

Aziz, A., J. Wingate, and M. Balas, eds.
[1] *Control Theory of Systems Governed by Partial Differential Equations*, Academic Press, New York, 1977.

Babalola, V.
[1] Semigroups of operators on locally convex spaces, *Trans. Amer. Math. Soc.* 199 (1974), 163–179.

Babbitt, D. G.
[1] Wiener integral representations for certain semigroups which have infinitesimal generators with matrix coefficients, *J. Math. Mech.* 19 (1969/70), 1051–1067.

Baillon, J.-B.
[1] Caractère borné de certains générateurs de semigroupes linéaires dans les espaces de Banach, *C. R. Acad. Sci. Paris* 290 (1980), 757–760.

Balakrishnan, A. V.
[1] An operational calculus for infinitesimal generators of semigroups, *Trans. Amer. Math. Soc.* 91 (1959), 330–353.
[2] Fractional powers of closed operators and semigroups generated by them. *Pac. J. Math.* 10 (1960), 419–437.
[3] *Applied Functional Analysis*, Springer, New York, 1976.

Balas, M. (see also Aziz)
[1] Modal control of certain flexible dynamic systems, *SIAM J. Cont. Opt.* 16 (1978), 450–462.
[2] The Galerkin method and feedback control of linear distributed parameter systems, *J. Math. Anal. Appl.* 91 (1983), 527–546.
[3] Feedback control of dissipative hyperbolic distributed parameter systems with finite dimensional controllers, *J. Math. Anal. Appl.* 98 (1984), 1–24.

Ball, J. M.
[1] Strongly continuous semigroups, weak solutions, and the variation of constants formula, *Proc. Amer. Math. Soc.* 63 (1977), 370–373.

Ballotti, M. E.
[1] *Modern Versions of the Theorems of Kneser and Wiener*, Ph.D. Thesis, Tulane University, 1983.

Ballotti, M. E. and J. A. Goldstein
[1] Wiener's theorem and semigroups of operators, in *Lecture Notes in Math.* No. 1076, Springer, Berlin (1984), 16–22.

Banks, H. T., J. A. Burns, and E. M. Cliff.
[1] Parameter estimation and identification for systems with delays, *SIAM J. Control Optim.* 19 (1981), 791–828.

Baras, P. and J. A. Goldstein
[1] The heat equation with a singular potential, *Trans. Amer. Math. Soc.* 284 (1984) 121–139.

Baras, P., J. -C. Hassan, and L. Veron
[1] Compacité de l'opérateur définissant la solution d'une équation d'évolution non homogène, *C. R. Acad. Sci. Paris* 284 (1977), 799–802.

Barbu, V.
[1] Differentiable distribution semigroups, *Anal. Scuola Norm. Sup. Pisa* 23 (1969), 413–419.
[2] On the regularity of weak solutions of abstract differential equations, *Osaka J. Math.* 6 (1969), 49–56.
Bardos, C.
[1] Problèmes aux limites pour les équations aux derivées partielles du premier ordre a coefficients réels; théorèmes d'approximation; application a l'équation de transport, *Ann. Scient. Ec. Norm. Sup.* 3 (1970), 185–233.
[2] A regularity theorem for parabolic equations, *J. Functional Anal.* 7 (1971), 311–322.
Bardos, C. and L. Tartar
[1] Sur l'unicité retrograde des equations paraboliques et quelques questions voisines, *Arch. Rat. Mech. Anal.* 50 (1973), 10–25.
Bart, H. and S. Goldberg
[1] Characterization of almost periodic strongly continuous groups and semigroups, *Math. Ann.* 236 (1978), 105–116.
Bass, R.
[1] Adding and subtracting jumps from Markov processes, *Trans. Amer. Math. Soc.* 255 (1979), 363–376.
[2] Markov processes with Lipschitz semigroups, *Trans. Amer. Math. Soc.* 267 (1981), 307–320.
Batty, C. J. K.
[1] Dissipative mappings with approximately invariant subspaces, *J. Functional Anal.* 32 (1979), 336–341.
[2] Ground states of uniformly continuous dynamical systems, *Quart. J. Math. Oxford* 31 (1980), 37–47.
[3] A characterization of relatively bounded perturbations, *J. London Math. Soc.* 21 (1980), 355–364.
Batty, C. J. K. and E. B. Davies
[1] Positive semigroups and resolvents, *J. Oper. Th.* 10 (1983), 357–363.
Batty, C. J. K. and D. W. Robinson
[1] Positive one-parameter semigroups on ordered Banach spaces, *Proc. Roy. Soc. Edinburgh* 96A (1984), 221–296.
Beale, J. T.
[1] Spectral properties of an acoustic boundary condition, *Indiana U. Math. J.* 25 (1976), 895–917.
[2] Acoustic scattering from locally reacting surfaces, *Indiana U. Math. J.* 26 (1977), 199–222.
Beale, J. T. and S. I. Rosencrans
[1] Acoustic boundary conditions, *Bull. Amer. Math. Soc.* 80 (1974), 1276–1278.
Beals, R.
[1] On the abstract Cauchy problem, *J. Functional Anal.* 10 (1972), 281–299.
[2] Semigroups and abstract Gevrey spaces, *J. Functional Anal.* 10 (1972), 300–308.
[3] Laplace transform methods for evolution equations. In *Boundary Value Problems for Linear Evolution and Partial Differential Equations* (ed. H. G. Garnir), Reidel, Dordrecht, Holland (1977), 1–26.
Becker, M.
[1] *Über den Satz von Trotter mit Anwendungen auf die Approximationstheorie*, Forschungsbericht des Landes Nordrhein-Westfalen No. 2577, Westdeuscher Verlag, Germany, 1976.
[2] Linear approximation processes in locally convex spaces. I. Semigroups of operators and saturation, *Aeq. Math.* 14 (1976), 73–81.
Bellini-Morante, A.
[1] *Applied Semigroups and Evolution Equations*, Oxford U. Press, Oxford, 1979.
Beltrami, E. J. and F. Buianouckas
[1] A note on passive evolution systems, *J. Math. Anal. Appl.* 37 (1972), 327–330.
Belyi, A. G. and Yu. A. Semenov
[1] Kato's inequality and semigroup product-formulas, *Functional Anal. Appl.* 9 (1975), 320–321.
[2] One criterion of semigroup product convergence, *Lett. Math. Phys.* 1 (1976), 201–208.
Benchimol, C. D.
[1] A note on weak stabilizability of contraction semigroups, *SIAM J. Control Opt.* 16 (1978), 373–379.
Benzinger, H. E.
[1] Perturbation of the heat equation, *J. Diff. Eqns.* 32 (1979), 398–419.
[2] Rayleigh-Schrödinger perturbation of semigroups, *SIAM J. Math. Anal.* 13 (1982), 515–531.
[3] Strong resolvent convergence of diffusion operators. To appear.

Benzinger, H., E. Berkson, and T. A. Gillespie
  [1] Spectral families of projections, semigroups, and differential operators, *Trans. Amer. Math. Soc.*
      275 (1983), 431–475.
Berens, H. (see also Butzer)
Berens, H., P. L. Butzer and U. Westphal
  [1] Representation of fractional powers of infinitesimal generators of semigroups, *Bull. Amer.
      Math. Soc.* 74 (1968), 191–196.
Berens, H. and and U. Westphal
  [1] A Cauchy problem for a generalized wave equation, *Acta Sci. Math.* 29 (1968), 93–106.
Berg, C.
  [1] On the potential operators associated with a semigroup, *Studia Math.* 51 (1974), 111–113.
  [2] Sur les semigroups de convolution. In *Lecture Notes in Math.* No. 404, Springer, Berlin (1975),
      pp. 1–26.
Berg, C. and G. Forst
  [1] *Potential Theory on Locally Compact Abelian Groups*, Springer, Berlin, 1975.
Berger, C. A. and L. A. Coburn
  [1] One parameter semigroups of isometries, *Bull. Amer. Math. Soc.* 76 (1970), 1125–1129.
Berger, M. A. and A. Sloan
  [1] A Method of Generalized Characteristics, *Mem. Amer. Math. Soc.* No. 266 (1982).
  [2] Product formulas for solutions of initial value partial differential equations, I, II, *J. Diff. Eqns.*
      In press.
Berkson, E. (see also Benzinger)
  [1] One-parameter semigroups of isometries into H^p, *Pac. J. Math.* 86 (1980), 403–413.
  [2] On spectral families of projections in Hardy Spaces, *Proc. Japan Acad.* 58 (1982), 436–439.
Berkson, E., R. J. Fleming, J. A. Goldstein, and J. E. Jamison
  [1] One-parameter groups of isometries on C_p, *Rev. Roum. Math. Pures Appl.* 6 (1979), 863–868.
Berkson, E., R. J. Fleming, and J. E. Jamison
  [1] Group of isometries on certain ideals of Hilbert space operators, *Math. Ann.* 220 (1976), 151–
      156.
Berkson, E., R. Kaufman, and H. Porta
  [1] Möbius transformations of the disc and one-parameter groups of isometries of H^p, *Trans.
      Amer. Math. Soc.* 199 (1974), 223–239.
E. Berkson and H. Porta
  [1] One-parameter groups of isometries on the Hardy spaces of the torus, *Trans. Amer. Math. Soc.*
      220 (1976), 373–391.
  [2] Semigroups of analytic functions and composition operators, *Mich. Math. J.* 25 (1978), 101–
      115.
  [3] The group of isometries of H^p, *Ann. di Mat. Pure Appl.* 119 (1979), 231–238.
Berkson, E. and A. Sourour
  [1] The hermitian operators on some Banach spaces, *Studia Math.* 52 (1974), 33–41.
Bernier, C. and A. Manitius
  [1] On semigroups in $R^n \times L^p$ corresponding to differential equations with delays, *Can. J. Math.* 30
      (1978), 897–914.
Bers, L., F. John, and M. Schechter
  [1] *Partial Differential Equations*, Interscience, New York, 1964.
Berthier, A. M.
  [1] *Spectral Theory and Wave Operators for the Schrödinger Equation*, Pitman, London, 1982.
Beurling, A.
  [1] On analytic extension of semigroups of operators, *J. Functional Anal.* 6 (1970), 387–400.
Bharucha-Reid, A. T.
  [1] Markov branching processes and semigroups of operators, *J. Math. Anal. Appl.* 12 (1965), 513–
      536.
Bharucha-Reid, A. T. and H. Rubin
  [1] Generating functions and the semigroup theory of branching Markov processes, *Proc. Nat.
      Acad. Sci. USA* 44 (1958), 1057–1060.
Birman, M. Š.
  [1] Existence conditions for wave operators, *Amer. Math. Soc. Transl.* (2) 54 (1966), 91–118.
  [2] Scattering problems for differential operators with perturbation of the space, *Math. USSR
      Isvestija* 5 (1971), 459–474.
Bivar-Weinholtz, A. De and R. Piraux
  [1] Formula de Trotter pour l'opérateur de Schrodinger avec un potentiel singulier complexe, *C. R.
      Acad. Sc. Paris* 288 (1979), 539–542.

Blumenthal, R. M. and R. K. Getoor
[1] *Markov Processes and Potential Theory*, Academic Press, New York, 1968.
Bobisud, L. E.
[1] Large-time behavior of solutions of abstract wave equations, *J. Math. Anal. Appl.* 63 (1978), 168–176.
Bobisud, L. E. and J. Calvert
[1] Singularly perturbed differential equations in a Hilbert space, *J. Math. Anal. Appl.* 30 (1970), 113–127.
[2] Energy bounds and virial theorems for abstract wave equations, *Pac. J. Math.* 47 (1973), 27–37.
Bobisud, L. E. and R. Hersh
[1] Perturbation and approximation theory for higher-order abstract Cauchy problems, *Rocky Mtn. J. Math.* 2 (1972), 57–73.
Bochner, S. and K. Chandrasekharan
[1] *Fourier Transforms*, Ann. of Math. Studies No. 19, Princeton U. Press, Princeton, N.J. 1949
Bondy, D. A.
[1] An application of functional operator models to dissipative scattering theory, *Trans. Amer. Math. Soc.* 223 (1976), 1–43.
Bony, J. -M., Ph. Courrège, and P. Priouret
[1] Sur la forme intégro-différentielle du générateur infinitésimal d'un semi-groupe de Feller sur une varieté différentiable, *C. R. Acad. Sci. Paris* 263 (1966), 207–210.
[2] Semi-groupes de Feller sur une variété à bord compacte et problèmes aux limites integro-differentiels du second ordre donnant lieu au principe du maximum, *Ann. Inst. Fourier* 18 (1968), 369–521.
Bouleau, N.
[1] Perturbation positife d'un semi-groupe droit dans le cas critique; application à la construction de processus de Harris. In *Lecture Notes in Math.* No. 906, Springer, Berlin (1982), 53–87.
Bragg, L. R.
[1] Related non-homogeneous partial differential equations, *Appl. Anal.* 4 (1974), 161–189.
[2] Linear evolution equations that involve products of commutative operators, *SIAM J. Math. Anal.* 5 (1974), 327–335.
[3] The Riemann-Liouville integral and parameter shifting in a class of abstract Cauchy problems, *SIAM J. Math. Anal.* 7 (1976), 1–11.
[4] Some abstract Cauchy problems in exceptional cases, *Proc. Amer. Math. Soc.* 65 (1977), 105–112.
[5] The ascent method for abstract wave problems, *J. Diff. Eqns.* 381 (1980), 413–421.
[6] Abstract Cauchy problems that involve a product of two Euler-Poisson-Darboux operators, *J. Diff. Eqns.* 41 (1981), 426–439.
[7] Exceptional Cauchy problems and regular singular points, *Appl. Anal.* 14 (1983), 203–211.
[8] The decomposition and structure of solution operators of evolution type equations that involve sums and products of commutative operators, preprint.
Bragg, L. R. and J. W. Dettman
[1] Related problems in partial differential equations, *Bull. Amer. Math. Soc.* 74 (1968), 375–378.
[2] Related partial differential equations and their applications, *SIAM J. Appl. Math.* 16 (1968), 459–467.
[3] An operator calculus for related partial differential equations, *J. Math. Anal. Appl.* 22 (1968), 261–271.
[4] Analogous function theories for the heat, wave, and Laplace equations, *Rocky Mtn. J. Math.* 13 (1983), 191–214.
Bratelli, O., T. Digernes, and D. W. Robinson
[1] Positive semigroups on ordered Banach spaces, *J. Oper. Th.* 9 (1983), 371–400.
Bratteli, O., R. H. Herman, and D. W. Robinson
[1] Perturbations of flows on Banach spaces and operator algebras, *Comm. Math. Phys.* 59 (1978), 167–196.
Bratelli, O. and A. Kishimoto
[1] Generation of semigroups and two-dimensional quantum lattice systems, *J. Functional Anal.* 35 (1980), 344–369.
Bratelli, O., A. Kishimoto, and D. W. Robinson
[1] Positivity and monotonicity properties of C_0-semigroups. I, *Comm. Math. Phys.* 45 (1980), 67–84.

Bratelli, O. and D. W. Robinson
[1] *Operator Algebras and C* and W*-Algebras, Symmetry Groups, Decomposition of States*, Springer, New York, 1979.
[2] Positive C_0-semigroups on C^*-algebras, *Math. Scand.* 49 (1981), 259–274.

Brenner, P.
[1] The Cauchy problem for symmetric hyperbolic systems in L_p, *Math. Scand.* 19 (1966), 27–37.

Brenner, P. and V. Thomée
[1] On rational approximation of semigroups, *SIAM J. Numer. Anal.* 16 (1979), 683–694.
[2] On rational approximations of groups of operators, *SIAM J. Numer. Anal.* 17 (1980), 119–125.

Brezis, H. and T. Kato
[1] Remarks on the Schrödinger equation with singular complex potentials, *J. Math. Pures Appl.* 88 (1979), 137–151.

Brezis, H., W. Rosenkrantz, and B. Singer
[1] On a degenerate elliptic-parabolic equation occurring in the theory of probability, *Comm. Pure Appl. Math.* 24 (1971), 395–416.
[2] An extension of Khintchine's estimate for large deviations to a class of Markov chains converging to a singular diffusion, *Comm. Pure Appl. Math.* 24 (1971), 705–726.

Browder, F. E.
[1] On the spectral theory of elliptic differential operators, *Math. Ann.* 142 (1961), 20–130.

Buianouckas, F. (see Beltrami)

Bundaru, M. (see Angelescu)

Burak, T.
[1] On semigroups generated by restrictions of elliptic operators to invariant subspaces, *Israel J. Math.* 12 (1972), 79–93.

Burns, J. A. (see also Banks)

Burns, J. A. and T. L. Herdman
[1] Adjoint semigroup theory for a Volterra integro-differential system, *Proc. Amer. Math. Soc.* 81 (1975), 1099–1102.
[2] Adjoint semigroup theory for a class of functional differential equations, *SIAM J. Math. Anal.* 7 (1976), 729–745.

Butcher, G. H. and J. A. Donaldson
[1] Regular and singular perturbation problems for a singular abstract Cauchy problem, *Duke Math. J.* 42 (1975), 435–445.

Butzer, P. L. (see also Berens)

Butzer, P. L. and H. Berens
[1] *Semi-groups of Operators and Approximation*, Springer, New York, 1967.

Butzer, P. L. and W. Dickmeis
[1] Direct and inverse mean ergodic theorems with rates for semigroups of operators. In *Approximation and Function Spaces*, North Holland, Amsterdam (1981), pp. 191–206.

Butzer, P. L., W. Dickmeis, L. Hahn, and R. J. Nessel
[1] Lax-type theorems and a unified approach to some limit theorems in probability theory with rates, *Resultate der Math.* 2 (1979), 30–53.

Butzer, P. L. and L. Hahn
[1] A probabilistic approach to representation formulae for operators with rates of convergence, *Semigroup Forum*, 21 (1980), 257–272.

Butzer, P. L. and R. J. Nessel
[1] *Fourier-Analysis and Approximation I*, Birkhäuser, Basel, 1971.

Butzer, P. L. and S. Pawelke
[1] Semi-groups and resolvent operators, *Arch. Rat. Mech. Anal.* 30 (1968), 127–147.

Calvert, J. (see Bobisud)

Cannon, J. T.
[1] Convergence criteria for a sequence of semigroups, *Appl. Anal.* 5 (1975), 23–31.

Carmona, R.
[1] Regularity properties of Schrödinger and Dirichlet semigroups, *J. Functional Anal.* 17 (1974), 227–237.

Carmona, R. and A. Klein
[1] Exponential moments for hitting times of uniformly ergodic Markov processes, *Ann. Prob.* 11 (1983), 648–655.

Carmona, R. and B. Simon
[1] Pointwise bounds on eigenfunctions and wave packets in N-body quantum systems. V. Lower bounds and path integrals, *Comm. Math. Phys.* 80 (1981), 59–98.

Carr, J. and M. Z. M. Malhardeen
[1] Stability of nonconservative linear systems. In *Lecture Notes in Math.* No. 799, Springer, Berlin (1980), 45–68.
Carroll, R. W.
[1] *Abstract Methods in Partial Differential Equations*, Harper and Row, New York, 1969.
[2] *Transmutation and Operator-Differential Equations*, North Holland, Amsterdam, 1979.
Carroll, R. W. and R. E. Showalter
[1] *Singular and Degenerate Cauchy Problems*, Academic Acss, New York, 1976.
Certain, M.
[1] One-parameter semigroups holomorphic away from zero, *Trans. Amer. Math. Soc.* 187 (1974), 377–389.
Certain, M. and T. G. Kurtz
[1] Landau-Kolmogorov inequalities for semigroups and groups, *Proc. Amer. Math. Soc.* 63 (1977), 226–230.
Chandrasekharan, K. (see Bochner)
Chazarain, J.
[1] Problèmes de Cauchy au sens des distributions vectorielles et applications, *C. R. Acad. Sci. Paris* 266 (1968), 10–13.
[2] Problèmes de Cauchy abstraits et applications á quelques problèmes mixtes, *J. Functional Anal.* 7 (1971), 386–446.
Chebli, H.
[1] Operateurs de translation généralisée et semi-groupes de convolution. In *Lecture Notes in Math.* No. 404, Springer, Berlin, (1975), pp. 35–59.
Chen, G. and R. Grimmer
[1] Integral equations as evolution equations, *J. Diff. Eqns.* 45 (1982), 33–74.
Chernoff, P. R.(see also Arendt, Cirincione)
[1] Note on product formulas for operator semigroups, *J. Functional Anal.* 2 (1968), 238–242.
[2] Perturbations of dissipative operators of relative bound one, *Proc. Amer. Math. Soc.* 33 (1972), 72–74.
[3] Some remarks on quasi-analytic vectors, *Trans. Amer. Math. Soc.* 167 (1972), 105–113.
[4] Universally commutatable operators are scalars, *Mich. Math. J.* 20 (1973), 101–107.
[5] Essential self-adjointness of powers of generators of hyperbolic equations, *J. Functional Anal.* 12 (1973), 401–414.
[6] Product formulas, nonlinear semigroups and addition of unbounded operators. *Mem. Amer. Math. Soc.* 140 (1974).
[7] Quasi-analytic vectors and quasi-analytic functions, *Bull. Amer. Math. Soc.* 81 (1975), 637–646.
[8] Two counterexamples in semigroup theory on Hilbert space, Proc. Amer. Math. Soc. 56 (1976), 253–255.
[9] On the converse of the Trotter product formula, *Ill*, J. Math. 20 (1976), 348–353.
[10] Schrödinger and Dirac operators with singular potentials and hyperbolic equations, *Pac. J. Math.* 72 (1977), 361–382.
[11] The quantum n-body problem and a theorem of Littlewood, *Pac. J. Math.* 70 (1977), 117–123.
[12] Optimal Landau-Kolmogorov inequalities for dissipative operators in Hilbert and Banach spaces, *Adv. Math.* 34 (1979), 137–144.
Chernoff, P. R. and J. A. Goldstein
[1] Admissible subspaces and the denseness of the intersection of the domains of semigroup generators, *J. Functional Anal.* 9 (1972), 460–468.
Christensen, E.
[1] Generators of semigroups of completely positive maps, *Comm. Math. Phys.* 62 (1978), 167–171.
Christensen, E. and D. E. Evans
[1] Cohomology of operator algebras and quantum dynamical semigroups, *J. London Math. Soc.* 20 (1979), 358–368.
Chung, K. L.
[1] *Markov Chains with Stationary Transition Probabilities* (2nd ed.), Springer, New York, 1967.
[2] Lectures on Boundary Theory for Markov Chains, Princeton U. Press, Princeton, N. J., 1970.
Cioranescu, I.
[1] A characterization of distribution semigroups with finite growth order, *Rev. Roum. Math. Pures Appl.* 22 (1977), 1053–1058.
[2] On distribution semigroups of subnormal operators, *J. Oper. Th.* 5 (1981), 47–52.
[3] On periodic distribution groups. In *Operator Theory: Adv. Appl. Vol.* 2, Birkhauser, Basel-Boston (1981), 53–61.

Cioranescu, I. and L. Zsido
[1] Analytic generators for one-parameter groups, *Tôhoku Math. J.* 28 (1976), 327–362.
Cirincione, R. J. and P. R. Chernoff
[1] Dirac and Klein-Gordon equations: Convergence of solutions in the nonrelativistic limit, *Comm. Math. Phys.* 79 (1981), 33–46.
Clarkson, J. A.
[1] Uniformly convex spaces, *Trans. Amer. Math. Soc.* 40 (1936), 396–414.
Cliff, E. M. (see Banks)
Coburn, L. A. (see Berger, C. A.)
Coddington, E. and N. Levinson
[1] *Theory of Ordinary Differential Equations*, McGraw-Hill, New York, 1955.
Coffman, C. V., R. J. Duffin, and V. J. Mizel
[1] Positivity of weak solutions of non-uniformly elliptic equations, *Annali di Mat.* 104 (1975), 209–238.
Coleman, B. D. and V. J. Mizel
[1] Norms and semi-groups in the theory of fading memory, *Arch. Rat. Mech. Anal.* 23 (1966), 87–123.
[2] A general theory of dissipation in materials with memory, *Arch. Rat. Mech. Anal.* 27 (1968), 255–274.
[3] On the general theory of fading memory, *Arch. Rat. Mech. Anal.* 29 (1968), 18–31.
Cook, J. M.
[1] Convergence to the Møller wave-matrix, *J. Math. Phys.* 36 (1952), 82–87.
Cooper, J. L. B.
[1] One parameter semigroups of isometric operators, *Ann. Math.* 48 (1947), 827–842.
Cornea, A. and G. Licea
[1] *Order and Potential Resolvent Families of Kernels*, Lecture Notes in Math. No. 494, Springer, Berlin, 1975.
Costa, D. G.
[1] Equipartition of energy for a class of second order equations, *Proc. Amer. Math. Soc.* 64 (1977), 65–70.
Courrège, P. (see also Bony)
[1] Générateur infinitésimal d'un semi-groupe de convolution sure R^n, et formule de Lévy-Khinchine, *Bull, Sci. Math.* 88 (1964), 3–30.
Cowling, M. G.
[1] Harmonic analysis on semigroups, *Ann. Math.* 117 (1983), 267–283.
Crandall, M. G.
[1] Norm preserving extensions of linear transformations on Hilbert space, *Proc. Amer. Math. Soc.* 21 (1969), 335–340.
Crandall, M. G. and A. Pazy
[1] On the differentiability of weak solutions of a differential equation in Banach space, *J. Math. Mech.* 10 (1969), 1007–1016.
Crandall, M. G., A. Pazy, and L. Tartar
[1] Remarks on generators of analytic semigroups, *Israel J. Math.* 32 (1979), 363–374.
Crandall, M. G. and R. S. Phillips
[1] On the extension problem for dissipative operators, *J. Functional Anal.* 2 (1968), 147–176.
Cunsolo, J. (see Akcoglu)
Curtain, R. F.
[1] The spectrum determined growth assumption for perturbations of analytic semigroups, *Systems Control Lett.* 2 (1982/83), 106–109.
Cuthbert, J. R.
[1] On semigroups such that $T_t - I$ is compact for some $t > 0$, *Z. Warsch verw. Geb.* 18 (1971), 9–16.
[2] An inequality with relevance to the Markov group problem, *J. London Math. Soc.* 11 (1975), 104–106.
Daletskiĭ, Ju. L.
[1] Infinite dimensional elliptic operators and the corresponding parabolic equations, *Russ. Math. Surv.* 22 (1967), 1–53.
Daleckiĭ, Ju. L. and M. G. Kreĭn
[1] *Stability of Solutions of Differential Equations in Banach Spaces*, Translations of Math. Monographs, Vol. 43, Amer. Math. Soc., Providence, R. I., 1974.
Da Prato, G.
[1] Semigruppi di crescenza n, *Ann. Scuola Norm. Sup. Pisa* 20 (1966), 753–782.

[2] Semigruppi regolarizzabili, *Ricerche di Mat.* 15 (1966), 233–248.
[3] R-semigruppi analitici ed equazioni di evoluzione in L^p, *Ricerche di Mat.* 16 (1967), 233–249.
[4] Somme de générateurs infinitésimaux de classe C_0, *Rend. della classe Sci. fis. math. nat. (Acc. Naz. Lincei)* 35 (1968), 14–21.
[5] Somme di generatori infinitesimali di semigruppi analiticiti *Rend. Sem. Math. Univ. Padova* 40 (1968), 151–161.
[6] Somme di generatori infinitesimali di semi-gruppi di contrazioni di spazi di Banach riflessivi, *Boll. Un. Mat. Ital.* 1 (1968), 138–141.
[7] Quelques résultats d'existence unicité et regularité pour un probléme de la théorie du contrôl, *J. Math Pures Appl.* 52 (1973), 353–375.
[8] Sums of linear operators. In *Linear Operators and Approximation II*, ISNM Vol. 25, Birkhauser; Basel (1974), 461–472.
[9] *Applications Croissantes et Equations d'Évolution* dans les Espaces de Banach, Academic Press, New York, 1976.
[10] Maximal regularity results for abstract differential equations in Banach spaces. In *Evolution Equations and Their Applications* (eds. F. Kappel and W. Schappacher), Pitman, London (1982), pp. 52–62.
[11] Abstract differential equations and extrapolation spaces, in *Lecture Notes in Math.* No. 1076, Springer, Berlin (1984), 53–61.
[12] Some results on linear stochastic evolution equations in Hilbert spaces by the semi-group method. To appear.
Da Prato, G. and E. Giusti
[1] Une charatterizzazioni dei generatori di funzioni coseno astratte, *Boll. Un. Mat. Ital.* 22 (1967), 357–368.
Da Prato, G. and Grisvard, P.
[1] Sommes d'opérateurs linéaires et équations différentielles opérationnelles, *J. Math. Pures Appl.* 54 (1975), 305–387.
[2] On an abstract singular Cauchy problems, *Comm. P. D. E.* 3 (1978), 1077–1082.
[3] Maximal regularity for evolution equations by interpolation and extrapolation. To appear.
Da Prato, G. and M. Iannelli
[1] On a method for studying abstract evolution equations in the hyperbolic case, *Comm. P. D. E.* 1 (1976), 585–608.
[2] Linear abstract integro-differential equations of hyperbolic type in Hilbert spaces *Rend. Sem. Math. Univ. Padova* 62 (1980), 191–206.
[3] Linear integro-differential equations in Banach spaces, *Rend. Sem. Math. Univ. Padova* 62 (1980), 207–219.
[4] Distribution resolvents for Volterra equations in a Banach space, *J. Integral Eqns.* 6 (1984), 93–103.
Da Prato, G. and U. Mosco
[1] Semigruppi distribuzioni analitici, *Ann. Scuola Norm. Sup. Pisa* 19 (1965), 367–396.
[2] Regularizzazione dei semigruppi distribuzioni analitici, *Ann. Scuola Norm. Sup Pisa* 19 (1965), 563–576.
Da Prato, G. and E. Sinestrari
[1] Hölder regularity for non-autonomous abstract parabolic equations, *Israel J. Math.* 42 (1982), 1–19.
[2] Abstract evolution equations in spaces of continuous functions. To appear.
Darmois, G. F.
[1] *Evolution Equations in a Banach Space*, Ph. D. thesis, Univ. of California, Berkeley, 1975.
Dassios, G. and M. Grillakis
[1] Equipartition of energy in scattering theory, *SIAM J. Math. Anal.* 14 (1983), 915–924.
Datko, R.
[1] An extension of a theorem of A. M. Lyapunov to semi-groups of operators, *J. Math. Anal. Appl.* 24 (1968), 290–295.
[2] Extending a theorem of A. M. Liapunov to Hilbert space, *J. Math. Anal. Appl.* 32 (1970), 610–616.
[3] Uniform asymptotic stability of evolutionary processes in a Banach space, *SIAM J. Math. Anal.* 3 (1972), 428–445.
[4] The uniform asymptotic stability of certain linear differential difference equations, *Proc. Roy. Soc. Edinburgh* 74A (1975), 71–79.
[5] Stabilization of linear evolutionary processes. In *Dynamical Systems*, Vol. 2 (eds. L. Cesari, J. K. Hale, and J. P. LaSalle), Academic, Press, New York (1976), pp. 95–98.

[6] Linear autonomous neutral differential equations in a Banach space, *J. Diff. Eqns.* 25 (1977), 258–274.

Davies, E. B. (see also Batty)

[1] Some contraction semigroups in quantum probability, *Z. Warsch, verw. Geb.* 23 (1972), 261–273.
[2] *Quantum Theory of Open Systems*, Academic Press, New York, 1976.
[3] The classical limit for quantum dynamical semigroups, *Comm. Math. Phys.* 49 (1976), 113–129.
[4] Quantum dynamical semigroups and the neutron diffusion equation, *Comm. Math. Phys.* 11 (1977), 169–188.
[5] Asymptotic analysis of some abstract evolution equations, *J. Functional Anal.* 25 (1977), 81–101.
[6] Generators of dynamical semigroups, *J. Functional Anal.* 34 (1979), 421–432.
[7] On Enss' approach to scattering theory, *Duke Math. J.* 47 (1980), 171–185.
[8] *One Parameter Semigroups*, Academic Press, Londdon, 1980.
[9] Asymptotic modifications of dynamical semigroups on C^*-algebras, *J. London Math. Soc.* 24 (1981), 537–547.
[10] Metastable states of symmetric Markov semigroups. I, *Proc. London Math. Soc.* 451 (1982), 133–150.
[11] Metastable states of symmetric Markov semigroups. II, *J. London Math. Soc.* 26 (1982), 541–556.
[12] Dynamical stability of metastable states *J. Functional Anal.* 46 (1982), 373–386.
[13] Some norm bounds and quadratic form inequalities for Schrödinger operators, *J. Oper. Th.* 9 (1983), 147–162.
[14] The harmonic functions of mean ergodic Markov semigroups, *Math. Z.* 181 (1982), 543–552.
[15] Spectral properties of metastable Markov processes, *J. Functional Anal.* 52 (1983), 315–329.
[16] Hypercontractive and related bounds for double well Schrödinger Hamiltonians, *Quart. J. Math. Oxford* 34 (1983), 407–421.
[17] The rate of resolvent and semigroup convergence, *Quart. J. Math. (Oxford)* 35 (1984), 121–130.

Davies, E. B. and B. Simon

[1] Ultracontractive semigroups and some problems in analysis. In *Aspects of Mathematics and Its Applications*, North Holland. To appear.

Dawson, D. A.

[1] The critical measure diffusion process, *Z. Warsch. verw. Geb.* 40 (1977), 125–145.
[2] Stochastic measure diffusion processes, *Can. Math. Bull.* 22 (1979), 129–138.

De Graff J.

[1] A constructive approach to one-parameter semigroups of operators in Hilbert space, *Arch. Rat. Mech. Anal.* 43 (1971), 125–153.
[2] On a class of lateral boundary value problems associated with infinitesimal generators, *Arch. Rat. Mech. Anal.* 47 (1972), 330–342.

De Leeuw, K. (see Leeuw)

Delfour, M. C.

[1] The largest class of hereditary systems defining a C_0 semigroup on the product space, *Can. J. Math.* 32 (1980), 969–978.

Delfour, M. C. and A. Manitius

[1] The structural operator F and its role in the theory of retarded systems, I, *J. Math. Anal. Appl.* 73 (1980), 466–490.
[2] The structural operator F and its role in the theory of retarded systems, II, *J. Math. Anal. Appl.* 74 (1980), 359–381.

Dellacherie, C. and P. -A. Meyer

[1] *Probability and Potentials*, North Holland, New York, 1978.

Dembart, B.

[1] On the theory of semigroups of operators on locally convex spaces, *J. Functional Anal.* 16 (1974), 123–160.

Derndinger, R.

[1] Uber das Spektrum positiver Generatoran, *Math. Z.* 172 (1980), 281–293.
[2] Betragshalbgruppen normstetiger Operator-halbgruppen. In *Semesterbericht Funktional-analysis*, Tübingen (1983), 87–95.
[3] Positive Halbgruppen auf $C(X)$: Ein Gegenbeispiel. In *Semesterbericht Funktional-analysis*, Tübingen (1983), 97–100.

Derndinger, R. and R. Nagel

[1] Der Generator stark stetiger Verbandshalbgruppen auf $C(X)$ und dessen Spektrum, *Math. Ann.* 245 (1979), 159–177.

Desch, G. W. and W. Schappacher
[1] On relatively bounded perturbations of linear C_0-semigroups, *Ann. Scuol. Norm. Sup. Pisa*, to appear.
Dettman, J. W. (see also Bragg)
[1] The wave, Laplace and heat equations and related transforms, *Glasgow Math. J.* 11 (1970), 117–125.
[2] Perturbation techniques in related differential equations, *J. Diff. Eqns.* 14 (1973), 547–558.
[3] Related semi-groups and the abstract Hilbert transform. In *Func. Theor. Methods Diff. Eqns.* (1976), 94–108.
[4] Related singular problems and the generalized Hilbert transform, *Proc. Roy. Soc. Edinburgh* 79A (1977), 173–182.
[5] Saturation theorems connected with the abstract wave equations, *SIAM J. Math. Anal.* 9 (1978), 54–64.
Dettweiler, E.
[1] Branching semigroups on Banach lattices, *Math. Z.* 181 (1982), 411–434.
Devinatz, A.
[1] A note on semigroups of unbounded self-adjoint operators, *Proc. Amer. Math. Soc.* 5 (1954), 101–102.
[2] On an inequality of Tosio Kato for degenerate elliptic operators, *J. Func. Anal.* 32 (1979), 312–335.
Devinatz, A. and P. Malliavin
[1] On the possibility of defining the Chapman-Kolmogorov semigroup on L^∞, *Proc. Royal Soc. Edinburgh* 83A (1979), 327–331.
Di Blasio, G.
[1] Differentiability of solutions of some evolution equations in nonreflexive Banach space, *Ann. Mat. Pura Appl.* 121 (1979), 223–247.
Di Blasio, G., M. Iannelli, and E. Sinestrari
[1] An abstract partial differential equation with a boundary condition of renewal type. To appear.
Di Blasio, G., K. Kunisch, and E. Sinestrari
[1] L^2-regularity for parabolic partial integro-differential equations with delay in the highest order derivatives, *J. Math. Annal. Appl.* 102 (1984), 38–57.
Dickmeis, W. (see also Butzer)
[1] Lax-type equivalence theorems with orders for linear evolution operators, *Serdica* 7 (1982), 380–386.
Dickmeis, W. and R. J. Nessel
[1] Classical approximation processes in connection with Lax equivalence theorems with orders, *Acta Sci. Math.* 40 (1978), 33–48.
Digernes, T. (see Bratelli)
Ditzian, Z.
[1] Note on Hille's exponential formula, *Proc. Amer. Math. Soc.* 25 (1970), 351–352.
[2] Exponential formulas for semigroups of operators in terms of the resolvent, *Israel J. Math.* 9 (1971), 541–553.
[3] Some remarks on inequalities of Landau and Kolmogorov, *Aeq. Math.* 12 (1975), 145–151.
[4] Note on Hille's question, *Aeq. Math.* 15 (1977), 143–144.
[5] On inequalities of periodic functions and their derivatives, *Proc. Amer. Math. Soc.* 87 (1983), 463–466.
[6] Discrete and shift Kolmogorov type inequalities. To appear.
Dollard, J. D.
[1] Quantum mechanical scattering theory for short-range and Coulomb interactions, *Rocky Mtn. J. Math.* 1 (1971), 5–88.
Dollard, J. D. and C. N. Friedman
[1] Product integrals and the Schroedinger equation, *J. Math. Phys.* 18 (1977), 1598–1607.
[2] On strong product integration, *J. Functional Anal.* 28 (1978), 309–354.
[3] Product integrals, II: Contour integral, *J. Functional Anal.* 28 (1978), 355–368.
[4] *Product Integration*, Addison-Wesley, Reading, Mass., 1979.
[5] Product integration of measures and applications, *J. Diff. Eqns.* 31 (1979), 418–464.
Dolph, C. L.
[1] Positive real resolvents and linear passive Hilbert systems, *Ann. Acad. Sci. Fenn. Ser.* AI 336/9 (1963), 331–339.
Donaldson, J. A. (see also Butcher)
[1] A singular abstract Cauchy problem, *Proc. Nat. Acad. Sci.* 66 (1970), 269–274.

[2] New integral representations for solutions of Cauchy's problem for abstract parabolic equations, *Proc. Nat. Acad. Sci.* 68 (1971), 2025–2027.
[3] An operational calculus for a class of abstract operator equations, *J. Math. Anal. Appl.* 37 (1972), 167–184.
[4] The Cauchy problem for a first order system of abstract operator equations, *Bull. Amer. Math. Soc.* 81 (1975), 576–578.
[5] The abstract Cauchy problem. *J. Diff. Eqns.* 25 (1977), 400–409.

Donaldson, J. A. and R. Hersh
[1] A perturbation series for Cauchy's problem for higher-order abstract parabolic equations, *Proc. Nat. Acad. Sci.* 67 (1970), 41–42.

Doob, J. L.
[1] *Stochastic Processes*, Wiley, New York, 1953.
[2] Martingales and one-dimensional diffusion, *Trans. Amer. Math. Soc.* 78 (1955), 168–208.

Dorea, C. C. Y. (see also Rosenkrantz)
[1] Differentiability preserving properties of a class of semi-groups, *Z. Warsch. verw. Geb.* 36 (1976), 13–26.
[2] A semigroup characterization of the multi-parameter Wiener process, *Semigroup Forum* 26 (1983), 287–293.

Dorroh, J. R.
[1] Contraction semigroups in a function space, *Pac. J. Math.* 19 (1966), 35–38.
[2] Some properties of a partial differential operator, *Ill. J. Math.* 11 (1967), 177–188.
[3] A simplified proof of a theorem of Kato on linear evolution equations, *J. Math. Soc. Japan* 27 (1975), 474–478.

Douglas, R. G.
[1] On the C^*-algebra of a one-parameter semigroup of isometries, *Acta Math.* 128 (1972), 143–151.

Douglis, A. (see Agmon)

Drisch, T. and W. Hazod
[1] Analytische Vektoren von Faltungshalbgruppen, I, II, III, *Math. Z.* 172 (1980), 1–28; *Mh, Math.* 88 (1979), 107–122; *Mh. Math.* 89 (1980), 205–218.

Dubois, R. -M.
[1] Operateurs locaux localement dissipatifs à valeurs dans un espace de Hilbert. In *Lecture Notes in Math.* No. 906, Springer, Berlin (1982), 88–102.
[2] Problèmes mixtes abstraits et principe du maximum parabolique. In *Lecture Notes in Math.* No. 906, Springer, Berlin (1982), 103–113.
[3] Complexifications d'espaces de Banach reels et dissipativite, *Bull. Soc. Math. Belg.* 34B (1982), 121–135.

Duffin, R. J. (see also Coffman)
[1] Equipartition of energy in wave motion, *J. Math. Anal. Appl.* 32 (1970), 386–391.

Dunford, N. and J. T. Schwartz
[1] *Linear Operators, Part I: General Theory*, Interscience, New York, 1958.
[2] *Linear Operators, Part II: Spectral Theory*, Interscience, New York, 1963.
[3] *Linear Operators, Part III: Spectral Operators*, Wiley-Interscience, New York, 1971.

Dunford, N. and I. E. Segal
[1] Semigroups of operators and the Weierstrass theorem, *Bull. Amer. Math. Soc.* 52 (1946), 911–914.

Durrett, R.
[1] On Lie generators for one parameter semigroups, *Houston J. Math.* 3 (1977), 163–164.

Dynkin, E. B.
[1] Markov processes and semigroups of operators, *Theory Prob. Appl.* 1 (1956), 22–54.
[2] *Markov Processes*, Vols. 1 and 2, Springer, Berlin, 1965.
[3] General boundary conditions for denumerable Markov processes, *Theory Prob. Appl.* 12 (1967), 187–221.

Eastham, M. S. P. and H. Kalf
[1] *Schrödinger Type Operators with Continuous Spectra*, Pitman, London, 1982.

Edwards, D. A.
[1] A maximal ergodic theorem for Abel means of continuous parameter operator semigroups, *J. Functional Anal.* 7 (1971), 61–70.

Eilenberg, S.
[1] Sur les groupes compacts d'homéomorphies, *Fund. Math.* 28 (1937), 75–80.

Elliot, J.
[1] The equation of evolution in a Banach space, *Trans. Amer. Math. Soc.* 103 (1962), 470–483.
[2] Lateral conditions for semi-groups in partially ordered spaces, *J. Math. Mech.* 16 (1967), 1071–1093.
[3] Lateral conditions for semigroups involving mappings in L^p. I, *J. Math. Anal. Appl.* 25 (1969), 388–410.

Elliot, J. and M. L. Silverstein
[1] On boundary conditions for symmetric submarkovian resolvents, *Bull. Amer. Math. Soc.* 76 (1970), 752–757.

Ellis, R. S.
[1] Chapman-Enskog-Hilbert expansion for the Ornstein-Uhlenbeck process and the approximation of Brownian motion, *Trans. Amer. Math. Soc.* 199 (1974), 65–74.

Ellis, R. S. and M. A. Pinsky
[1] Limit theorems for model Boltzmann equations with several conserved quantities, *Indiana U. Math. J.* 23 (1973), 289–307.
[2] Asymptotic nonuniqueness of the Navier-Stokes equation in kinetic theory, *Bull. Amer. Math. Soc.* 80 (1974), 1160–1164.
[3] The first and second fluid approximations to the linearized Boltzmann equation, *J. Math. Pures Appl.* 54 (1975), 125–156.
[4] The projection of the Navier-Stokes equations upon the Euler equations, *J. Math. Pures Appl.* 54 (1975), 157–181.
[5] Asymptotic equivalence of the linear Navier-Stokes and heat equations in one dimension, *J. Diff. Eqns.* 17 (1975), 406–420.

Ellis, R. S. and W. A. Rosenkrantz
[1] Diffusion approximation for transport processes with boundary conditions, *Indiana U. Math. J.* 26 (1977), 1075–1096.

Emamirad, H. A.
[1] Semi-groupe ultra-contractif et schéma de Crank-Nicholson dans un espace de Banach, *C. R. Acad. Sci. Paris* 287 (1978), 343–345.

Embry, M. R. (see also Embry-Wardrop)

Embry, M. R. and A. Lambert
[1] Weighted translation semigroups, *Rocky Mtn. J. Math.* 7 (1977), 333–344.
[2] Subnormal weighted translation semigroups, *J. Functional Anal.* 24 (1977), 268–275.
[3] The structure of a special class of weighted translation semigroups, *Pac. J. Math.* 75 (1978), 359–371.
[4] Weighted translation semigroups on $L^2(O, \infty)$. In *Lecture Notes in Math.* No. 693, Springer, New York (1978), pp. 87–91.

Embry, M. R., A. L. Lambert, and L. J. Wallen
[1] A simplified treatment of the structure of semigroups of partial isometries, *Mich. Math. J.* 22 (1975), 175–179.

Embry-Wardrop, M. (see also Embry)
[1] Semigroups of quasi-normal operators, *Pac. J. Math.*, 101 (1982), 103–113. .
[2] Weighted translation semigroups with operator weights, *Acta Sci. Math.* 46 (1983), 249–260.
[3] The partially isometric factor of a semigroup, *Indiana U. Math. J.* 32 (1983), 893–901.
[4] The positive factor of a subnormal semigroup, *Houston J. Math.* 9 (1983), 387–395.
[5] Semi-groups of quasinormal operators, *Pac, J. Math.* 101 (1982), 103–113.

Engelbert, H. J.
[1] On the construction of semigroups from substochastic resolvents, *Math. Nachr.* 105 (1982), 171–192.

Enss, V.
[1] Asymptotic completeness for quantum mechanical potential scattering I. Short range potentials, *Comm. Math. Phys.* 61, (1979) 285–291.
[2] Asymptotic completeness for quantum mechanical potential scattering II. Singular and long range potentials, *Ann. Phys.* 119 (1979), 117–132.
[3] Two cluster scattering of N changed particles, *Comm. Math. Phys.* 61 (1978), 285–291.
[4] A new method for asymptotic completeness. In *Lecture Notes in Physics* No. 116, Springer, Berlin (1980), pp. 45–51.
[5] Geometric methods in spectral and scattering theory of Schrödinger operators. In *Rigorous Atomic and Molecular Physics* (eds. G. Velo and A. S. Wightman), Plenum, New York (1981).

[6] Propagation properties of quantum scattering states, *J. Functional Anal.* 52 (1983), 219–251.

Ethier, S. N.

[1] A class of degenerate diffusion processes occurring in population genetics, *Comm. Pure Appl. Math.* 29 (1976), 483–492.

[2] Differentiability preserving properties of Markov semigroups associated with one dimensional diffusion, *Z. Warsch. verw. Geb.* 45 (1978), 225–238.

[3] Limit theorem for absorption times of genetic models, *Ann. Prob.* 7 (1979), 622–638.

[4] A limit theorem for two-locus diffusion models in population genetics. *J. Appl. Prob.* 16 (1979), 402–408.

[5] A class of infinite dimensional diffusions occurring in population genetics, *Indiana U. Math. J.* 30 (1981), 925–935.

Ethier, S. N. and T. G. Kurtz

[1] The infinitely-many-neutral-alleles diffusion model, *Adv. Appl. Prob.* 13 (1981), 429–452.

Ethier, S. N. and M. F. Norman

[1] Error estimate for the diffusion approximation of the Wright-Fisher model, *Proc. Nat. Acad. Sci. USA* 74 (1977), 5096–5098.

Evans, D. E. (see also Christensen)

[1] Time dependent perturbations and scattering of strongly continuous groups on Banach spaces, *Math. Ann.* 221 (1976), 275–290.

[2] On the spectral type of one-parameter groups on operator algebras, *Proc. Amer. Math. Soc.* 61 (1976), 351–352.

[3] Irreducible quantum dynamical semigroups, *Comm. Math. Phys.* 54 (1977), 293–297.

[4] A review of semigroups of completely positive maps. In *Mathematical Problems in Theoretical Physics* (ed. K. Osterwalder), Springer, Berlin (1980), pp. 400–406.

[5] Dissipators for symmetric quasi-free dynamical semigroups on the CAR algebra, *J. Functional Anal.* 37 (1980), 318–330.

Evans, D. E. and H. Hanche-Olsen

[1] The generators of positive semigroups, *J. Functional Anal.* 32 (1979), 207–212.

Evans, D. E. and J. T. Lewis

[1] Dilations of dynamical semigroups, *Comm. Math. Phys.* 50 (1976), 219–227.

[2] Some semigroups of completely positive maps on the CCR algebra, *J. Functional Anal.* 26 (1977), 369–377.

Exner, P. and G. I. Kolerov

[1] Uniform product formulae with application to the Feyman-Nelson integral for open systems, *Lett. Math. Phys.* 6 (1982), 153–159.

Fan, K.

[1] Generalized Cayley transforms and strictly dissipative matrices, *Linear Alg. Appl.* 5 (1972), 155–172.

[2] Orbits of semi-groups of contractions and groups of isometries, *Abh. Math. Sem. U. Hamburg* 45 (1976), 245–250.

Fannes, M. and F. Rocca

[1] A class of dissipative evolutions with applications in thermodynamics of fermion systems, *J. Math. Phys.* 21 (1980), 221–226.

Faraut, J.

[1] Semi-groupes de mesures complexes et calcul symbolique sur les générateurs infinitésimaux de semi-groupes d'opérateurs, *Ann. Inst. Fourier* 20 (1972), 235–301.

Faraut, J. and K. Hazrallah

[1] Semi-groupes d'opérateurs invariants at opératuers dissipatifs invariants, *Ann. Inst. Fourier* 22 (1972), 147–164.

Faris, W. G.

[1] The product formula for semigroups defined by Friedrichs extensions, *Pac. J. Math.* 22 (1967), 47–70.

[2] Product formulas for perturbations of linear propagators, *J. Functional Anal.* 1 (1967), 93–108.

[3] *Self-adjoint Operators*, Lecture Notes in Math. Vol. 433, Springer, Berlin, 1975.

Fattorini, H. O.

[1] Ordinary differential equations in linear topological spaces I, *J. Diff. Eqns.* 5 (1969), 72–105.

[2] Ordinary differential equations in linear topological spaces II, *J. Diff. Eqns.* 6 (1969), 50–70.

[3] Extension and behavior at inifinity of solutions of certain linear operator differential equations, *Pac. J. Math.* 33 (1970), 583–615.

[4] A representation theorem for distribution semigroups, *J. Functional Anal.* 6 (1970), 83–96.

[5] Uniformly bounded cosine functions in Hilbert spaces, *Indiana U. Math. J.* (1970/71), 411–425.

[6] Un teorema de perturbacion para generadores de funciones coseno, *Rev. Un. Mat. Argentina* 25 (1971), 199–211.

[7] The abstract Goursat problem, *Pac J. Math.* 37 (1971), 51–83.

[8] The undetermined Cauchy problem in Banach spaces, *Math. Ann.* 200 (1973), 103–112.

[9] Two point boundary value problems for operational differential equations, *Ann. Scuola Norm. Sup. Pisa* 1 (1974), 63–79.

[10] Weak and strong extension of first-order differential operators in $\mathbb{R}^m$, *J. Diff. Eqns.* 34 (1979), 353–360.

[11] Some remarks on second order abstract Cauchy problems, *Funkcialaj Ekvacioj* 24 (1981), 331–344.

[12] A note on fractional derivatives of semigroups and cosine functions, *Pac. J. Math.* 109 (1983), 335–347.

[13] Singular perturbation and boundary layer for an abstract Cauchy problem, *J. Math. Anal. Appl.* 97 (1983), 529–571.

[14] *The Abstract Cauchy Problem*, Addison Wesley, Reading, Mass., 1983.

Fattorini, H. O. and A. Radnitz

[1] The Cauchy problem with incomplete initial data in Banach spaces, *Mich, Math. J.* 18 (1971), 291–320.

Feinsilver, P.

[1] *Special Functions, Probability Semigroups, and Hamiltonian Flows,* Lecture Notes in Math. No. 696, Springer, New York, 1978.

Feissner, G. F.

[1] Hypercontractive semi-groups and Sobolev's inequality, *Trans. Amer. Math. Soc.* 210 (1975), 51–62.

Feller, W.

[1] The parabolic differential equations and the associated semi-groups of operators, *Ann. Math.* 55 (1952), 468–519.

[2] Semi-groups of transformations in general weak topologies, *Ann. Math.* 57 (1953), 287–308.

[3] On the generation of unbounded semigroups of bounded linear operators, *Ann. Math* 58 (1953), 166–174.

[4] The general diffusion operator and positivity preserving semigroups in one dimension, *Ann. Math.* 10 (1954), 417–436.

[5] On second order differential operators, *Ann. Math.* 61 (1955), 90–105.

[6] Generalized second order differential operators and their lateral conditions, *Ill. J. Math.* 1 (1959), 459–504.

[7] On boundaries and lateral conditions for the Kolmogorov differential equations, *Ann. Math.* 65 (1957), 527–570.

[8] On the equation of the vibrating string, *J. Math. Mech.* 8 (1959), 339–348.

[9] On semi-Markov processes, *Proc. Nat. Acad. Sci.* 51 (1964) 653–659.

[10] *An Introduction to Probability Theory and Its Applications,* Vol. 2 (2nd ed.) Wiley, New York, 1971.

Feyel, D.

[1] Espaces complètement réticulés de pseudo-noyaux. Applications aux résolvantes et aux semi-groupes complexes. In *Lecture Notes in Math.* No. 681, Springer, Berlin (1978), pp. 54–80.

[2] Théorèmes de convergence presque-sûre existence de semi-groupes, *Adv. Math.* 34 (1979) 145–162.

[3] Propriétés de permanence du domaine d'un générateur infinitesimal. In *Lecture Notes in Math.,* No. 713, Springer, Berlin (1979), pp. 38–50.

[4] Compléments sur la convergence presque sûre des familles résolves. In *Lecture Notes in Math.* No. 713, Springer, Berlin (1979), 51–55.

Feynman, R. P.

[1] Space-time approach to non-relativistic quantum mechanics, *Rev. Mod. Phys.* 20 (1948), 367–387.

Fillmore, P. A. and J. P. Williams

[1] On operator ranges, *Adv. Math.* 7 (1971), 254–281.

Fleming, R. J. (see also Berkson)

Fleming, R. J., J. A. Goldstein, and J. E. Jamison

[1] One parameter groups of isometries on certain Banach spaces, *Pac. J. Math.* 64 (1976), 145–151.

Foguel, S.

[1] Dissipative Markov operators and the Dirichlet problem, *Indiana U. Math. J.* 29 (1980), 13–19.

Foiaş (see also Sz. -Nagy)
[1] On strongly continuous semigroups of spectral operators in Hilbert space, *Acta Sci. Math.* 19 (1958), 188–191.
[2] Remarques sur les semi-groupes distributions d'opérateurs normaux, *Portugal. Math.* 19 (1960), 227–242.
[3] Ergodic problems in functional spaces related to Navier-Stokes equations. In *Proc. Inter. Conf. Functional Anal. Rel. Top.* (Tokyo, 1969) 290–304.
[4] On the Lax-Phillips nonconservative scattering theory, *J. Functional Anal.* 19 (1975), 273–301.

Forst, G. (see also Berg)
[1] Sur le théorème de Hunt-Lion-Hirsch, *C. R. Acad. Sci. Paris* 289 (1979), 671–674.

Freedman, M. A.
[1] A faithful Hille-Yosida theorem for finite-dimensional evolutions. *Trans. Amer. Math. Soc.* 265 (1981), 563–573.
[2] Necessary and sufficient conditions for discontinuous evolutions with applications to Stieltjes integral equations, *J. Integral Eqns.* 5 (1983), 237–270.
[3] Operators of p-variation and the evolution representation problem, *Trans. Amer. Math. Soc.* 279 (1983), 95–112.
[4] Product integrals of continuous resolvents: Existence and nonexistence, *Israel J. Math.* 46 (1983), 145–160.

Freeman, J. M.
[1] The tensor product of semigroups and the operator equation $SX - XS = A$, *J. Math. Mech.* 19 (1970), 819–828.

Friedman, A.
[1] *Partial Differential Equations*, Holt, Rinehart and Winston, New York, 1969.
[2] Singular perturbations for partial differential equations, *Arch. Rat. Mech. Anal.* 4 (1968), 289–303.
[3] Probabilistic methods in partial differential equations, *Israel J. Math.* 13 (1972), 56–64.

Friedman, C. N. (see also Dollard)
[1] Perturbations of the Schrödinger equation by potentials with small support, *J. Functional Anal.* 10 (1972), 346–360.
[2] Semigroup product formulas, compressions, and continual observations in quantum mechanics, Indiana U. Math. J. 21 (1972), 1001–1011.

Friedrichs, K. O.
[1] Spectraltheorie halbbeschränkte Operatoren, I, II, III, *Math. Ann.* 109 (1934), 465–478 and 685–713; and 110 (1935), 777–779.
[2] Symmetric positive linear differential equations, *Comm. Pure Appl. Math.* 11 (1958), 333–418.
[3] On the differentiability of solutions of accretive linear differential equations. In *Contributions to Analysis* (ed. L. V. Ahlfors), Academic Press, New York (1974), pp. 147–150.

Frigerio, A.
[1] Quantum dynamical semigroups and approach to equilibrium, *Lett. Math. Phys.* 2 (1977), 79–87.
[2] Stationary states of quantum dynamical semigroups, *Comm. Math. Phys.* 63 (1978), 269–276.

Frigerio, A. and M. Verri
[1] Long-time asymptotic properties of dynamical semigroups on W*-algebras, *Math. Z.* 180 (1982), 275–286.

Fujie, Y. and H. Tanabe
[1] On some parabolic equations of evolution in Hilbert space, *Osaka J. Math.* 10 (1973), 115–130.

Fujita, H. (see also Kato)

Fujita, H. and T. Kato
[1] On the Navier-Stokes initial value problem. I, *Arch. Rat. Mech. Anal.* 16 (1964), 269–315.

Fujita, H. and H. Morimoto
[1] On fractional powers of the Stokes operator, *Proc. Japan Acad.* 46 (1971), 1141–1143.

Fujiwara, D.
[1] A characterization of exponential distribution semi-groups, *J. Math. Soc. Japan,* 18 (1966), 267–274.
[2] L^p-theory for characterizing the domain of the fractional powers of Δ in the half space, *J. Fac. Sci. U. Tokyo, Sect. I* 15 (1968), 167–177.

Fujiwara, D. and K. Uchiyama
[1] On some dissipative boundary value problems for the Laplacian, *J. Math. Soc. Japan.* 23 (1971), 625–635.

Fukushima, M.

[1] On boundary conditions for multi-dimensional Brownian motions with symmetric resolvent densities, *J. Math. Soc. Japan* 21 (1969), 58–93.

[2] Regular representations of Dirichlet spaces, *Trans. Amer. Math. Soc.* 155 (1971), 455–473.

[3] Dirichlet spaces and the strong Markov property, *Trans. Amer. Math. Soc.* 162 (1971), 185–224.

[4] On the generation of Markov processes by symmetric forms. In *Lecture Notes in Math.* No. 330, Springer, New York (1973), 46–79.

[5] On an L^p estimate of resolvents of Markov processes, *Publ. RIMS, Kyoto U.* 13 (1977), 277–284

[6] *Dirichlet Forms and Markov Processes*, North Holland, Amsterdam, 1980.

[7] (Ed.) *Functional Analysis in Markov Processes*, Lecture Notes in Math. No. 923, Springer, Berlin, 1982.

Gearhart, L.

[1] Spectral theory for contraction semigroups on Hilbert space, *Trans. Amer. Math. Soc.* 236 (1978), 385–394.

[2] The Weyl semigroup and left translation invariant subspaces, *J. Math. Anal. Appl.* 67 (1979), 75–91.

Gel'fand, I. M.

[1] On one-parameter groups of operators in normed spaces, *Dokl. Acad. Nauk SSSR* 25 (1939), 713–718.

Georgescu, V. (see Amrein)

Getoor, R. K. (see also Blumenthal)

[1] *Markov Processes: Ray Processes and Right Processes*, Lecture Notes in Math. No. 440, Springer, Berlin, 1975.

Gibson, A. G.

[1] A discrete Hille-Yosida-Phillips theorem, *J. Math. Anal. Appl.* 39 (1972), 761–770.

Giga, Y.

[1] The Stokes operator in L_p spaces, *Proc. Japan Acad.* 57 (1981), 85–89.

[2] Analyticity of the semigroup generated by the Stokes operator in L_r spaces, *Math. Z.* 178 (1981), 297–329.

Gikhman, I. I. and A. V. Skorokhod

[1] *Introduction to the Theory of Random Processes*, Saunders, Philadelphia, 1969.

Gilbarg, D. and N. S. Trudinger

[1] *Elliptic Partial Differential Equations of Second Order*, Springer, Berlin, 1977.

Gill, T. L.

[1] Time-ordered operators I: Foundations for an alternate view of reality, *Trans. Amer. Math. Soc.* 266 (1981), 161–181.

[2] New perspectives in time-ordered operators and divergences, I, II, *Hadronic J.* 3 (1980), 1575–1596 and 1597–1621.

[3] Time-ordered operators II, *Trans. Amer. Math. Soc.* 279 (1983), 617–634.

Gillespie, T. A. (see Benzinger)

Gilliam, D. S.

[1] Vector potentials for symmetric hyperbolic systems, *Rocky Mtn. J. Math.* 10 (1980), 575–583.

Gilliam, D. S. and J. R. Schulenberger

[1] A class of symmetric hyperbolic systems with special properties, *Comm. P. D. E.* 4 (1979), 509–536.

[2] Electromagnetic waves in a three-dimensional half space with a dissipative boundary, *J. Math. Anal. Appl.* 89 (1982), 129–185.

[3] *The Propagation of Electromagnetic Waves Through, Along and Over a Conducting Half Space.* To appear.

Gindler, H. A. and J. A. Goldstein

[1] Dissipative operator versions of some classical inequalities, *J. Analyse Math.* 28 (1975), 213–238.

[2] Dissipative operators and series inequalities, *Bull. Austral. Math. Soc.* 23 (1981), 429–442.

Giusti, E. (see also Da Prato)

[1] Funzione coseno periodiche, *Boll. Un. Mat. Ital.* 22 (1967), 478–485.

Glazman, I. M. (see Akhiezer)

Glimm, J. and A. Jaffe

[1] Singular perturbations of selfadjoint operators, *Comm. Pure Appl. Math.* 22 (1969), 401–414.

Globevnik, J.

[1] On a property of smooth operators, *Glasnik Math.* 7 (1972), 69–73.

Goldberg, S. (see also Bart)
Goldberg, S. and C. H. Smith
[1] Strong continuous semigroups of semi Fredholm operators, *J. Math. Anal. Appl.*, 64 (1978), 407–420.
Goldstein, J. A. (see also Ballotti, Baras, Berkson, Chernoff, Fleming, Gindler)
[1] Abstract evolution equations, *Trans. Amer. Math. Soc.* 141 (1969), 159–186.
[2] Semigroups and second order differential equations, *J. Functional Anal.* 4 (1969), 50–70.
[3] Time dependent hyperbolic equations, *J. Functional Anal.* 4 (1969), 31–49, and 6 (1970), 347.
[4] Some remarks on infinitesimal generators of analytic semigroups, *Proc. Amer. Math. Soc.* 22 (1969), 91–93.
[5] An asymptotic property of solutions of wave equations, *Proc. Amer. Math. Soc.* 23 (1969), 359–363.
[6] On a connection between first and second order differential equations in Banach spaces, *J. Math. Anal. Appl.* 30 (1970), 246–251.
[7] A Lie product formula for one parameter groups of isometries on Banach space, *Math Ann.* 186 (1970), 299–306.
[8] An asymptotic property of solutions of wave equations, II, *J. Math. Anal. Appl.* 32 (1970), 392–399.
[9] *Semigroups of Operators and Abstract Cauchy Problems*, Tulane University Lecture Notes, 1970.
[10] On the growth of solutions of inhomogeneous abstract wave equations, *J. Math. Anal. Appl.* 37 (1972), 650–654.
[11] Some counterexamples involving self-adjoint operators, *Rocky Mtn. J. Math.* 2 (1972), 142–149.
[12] A perturbation theorem for evolution equations and some applications, *Ill. J. Math.*, 18 (1974), 196–207.
[13] On the convergence and approximation of cosine functions. *Aeq. Math.*, 10 (1974), 201–205.
[14] Semigroup-theoretic proofs of the central limit theorem and other theorems of analysis, *Semigroup Forum* 12 (1976), 189–206, 388.
[15] Variable domain second order evolution equations, *Applicable Anal.* 5 (1976), 283–291.
[16] On the absence of necessary conditions for linear evolution operators, *Proc. Amer. Math. Soc.* 69 (1977), 77–80.
[17] A semigroup theoretic proof of the law of large numbers, *Semigroup Forum*, 15 (1977), 89–90.
[18] An operator semigroup version of the Edmunds-Moscatelli example, *J. London Math. Soc.* 17 (1978), 161–164.
[19] On the operator equation $AX + XB = Q$, *Proc. Amer. Math. Soc.* 78 (1978), 31–34.
[20] The universal addability problem for generators of cosine functions and operator groups, *Houston J. Math.* 6 (1980), 365–373.
[21] Some developments in semigroups of operators since Hille-Phillips, *Int. Eqns. Oper. Th.* 4 (1981), 350–365.
[22] Cosine functions and the Feymann-Kac Formula, *Quart. J. Math. Oxford*, 33 (1982), 303–307.
[23] Autocorrelation, equipartition of energy and random evolutions. In *Probability Measures on Groups* (ed. H. Heyer), Lecture Notes in Math. No. 928, Springer, Berlin (1982), 176–182.
[24] *A (More-or-Less) Complete Bibliography of Semigroups of Operators through 1984*, Tulane Univ., New Orleans, La., 1985.
Goldstein, J. A., M. K. Kwong, and A. Zettl
[1] Weighted Landau inequalities, *J. Math. Anal. Appl.* 95 (1983), 20–28.
Goldstein, J. A., C. Radin, and R. E. Showalter
[1] Convergence rates of ergodic limits for semigroups and cosine functions, *Semigroup Forum* 16 (1978), 89–95.
Goldstein, J. A. and S. I. Rosencrans
[1] Energy decay and partition for dissipative wave equations, *J. Diff. Eqns.* 36 (1980), 66–73 and 43 (1982), 156.
Goldstein, J. A. and J. T. Sandefur, Jr.
[1] Asymptotic equipartition of energy for differential equations in Hilbert space, *Trans. Amer. Math. Soc.* 219 (1976), 397–406.
[2] Abstract equipartition of energy theorems, *J. Math. Anal. Appl.* 67 (1979), 58–74.
[3] Equipartition of energy for symmetric hyperbolic systems. In *Constructive Methods in Analysis* (eds. C. V. Coffman and G. Fix), Academic Press, New York (1979), 395–411.
[4] Equipartition of energy for higher-order hyperbolic equations, *Proc. Nat. Acad. Sci. USA* 78 (1981), 698.
[5] Equipartition of energy for higher order abstract hyperbolic equations, *Comm. P. D. E.* 7 (1982), 1217–1251.

[6] Asymptotic properties of inhomogeneous wave equations. In *Trends in Theory and Practice of Nonlinear Differential Equations* (ed. V. Lakshmikantham), Marcel Dekker, New York (1984), pp. 185–196.

Gorini, V. (see also Verri)

Gorini, V., A. Kossakowski, and E. C. G. Sudarshan

[1] Completely positive dynamical semigroups of N-level systems, *J. Math. Phys.* 17 (1976), 821–825.

Görlich, E. and D. Pontzen

[1] An approximation theorem for semigroups of growth order ∞ and an extension of the Trotter-Lie formula, *J. Functional Anal.* 50 (1983), 414–425.

Graaf, J. De (see De Graaf)

Grady, M. D.

[1] Sufficient conditions for an operator-valued Feynman-Kac formula, *Trans. Amer. Math. Soc.* 223 (1976), 181–203.

Greenlee, W. M.

[1] Linearized hydrodynamic stability of a viscous liquid in an open container, *J. Functional Anal.* 22 (1976), 106–129.

[2] Linearized decay of disturbances of a viscous liquid in an open container, *Rocky Mtn. J. Math.* 10 (1980), 141–154.

Greiner, G. (see also Arendt)

[1] Zur Perron-Frobenius Theorie stark stetiges Halbgruppen, *Math. Z.* 177 (1981), 401–423.

[2] Spektrum and Asymptotic stark stetiger Halbgruppen positiver Operatoren, *Sitz. Heidelberg Akad. Wiss., Math.-Nat. Kl.* (1982), 3. Abh., 55–80.

[3] A spectral decomposition of strongly continuous groups of positive operators, *Quart. J. Math.* (*Oxford*) 35 (1984), 37–47.

[4] Spectral properties and asymptotic behavior of the linear transport equation, *Math. Z.* 185 (1984), 167–177.

[5] A typical Perron-Frobenius theorem with application to an age dependent population equation, in *Lecture Notes in Math.* No. 1076, Springer, Berlin (1984), 86–100.

Greiner, G. and R. Nagel

[1] La loi "zéro ou deux" et ses conséquences pour le comportement asymptotique des opérateurs positifs, *J. Math. Pures Appl.* 64 (1982), 261–273.

[2] On the stability of strongly continuous semigroups of positive operators on $L^2(\mu)$, *Ann. Scoul. Norm. Sup. Pisa* 10 (1983), 257–262.

Greiner, G., J. Voight, and M. Wolff

[1] On the spectral bound of the generator of semigroups of positive operators, *J. Oper. Th.* 5 (1981), 245–256.

Griego, R. J. and R. Hersh

[1] Random evolutions, Markov chains, and systems of partial differential equations, *Proc. Nat. Acad. Sci.* 62 (1969), 305–308.

[2] Theory of random evolutions with applications to partial differential equations, *Trans. Amer. Math. Soc.* 156 (1971), 405–418.

[3] Weyl's theorem for certain operator-valued potentials, *Indiana U. Math. J.* 27 (1978), 195–209.

Grillakis, M. (see Dassios)

Grimmer. R. (see also Chen)

[1] Resolvents for integral equations in abstract spaces. In *Evolution Equations and Their Applications* (eds. F. Kappel and W. Schappacher), Pitman, London (1982), pp. 101–120.

Grimmer, R. C. and F. Kappel

[1] Series expansions for resolvents of Volterra integrodifferential equations in Banach spaces, *SIAM J. Math. Anal.* 15 (1984), 595–604.

Grimmer, R. C. and W. Schappacher

[1] Weak solutions of integrodifferential equations and applications, *J. Int. Eqns.* 6 (1984), 205–229.

Grisvard, P. (see also Da Prato)

[1] Équations operationelles abstraites dans les éspaces de Banach et problémes aux limites dans des ouverts cylindrique, *Ann. Scuola Norm. Sup. Pisa* 21 (1967), 309–347.

[2] Équations différentialles abstraites, *Ann. Sci. Ecol. Norm. Sup.* 2 (1969), 311–395.

Groh, U. and F. Neubrander

[1] Stabilität starkstetiger, positiver Operator halbgruppen auf C^*-algebren, *Math. Ann.* 256 (1981), 509–516.

Guillement, J. P. and P. T. Lai
[1] Sur la charactérisation des semigroupes distributions, *J. Fac. Sci. Univ. Tokyo* 22 (1975), 299–318.
Gustafson, K.
[1] A perturbation lemma, *Bull. Amer. Math. Soc.* 72 (1966), 334–338.
[2] A note on left multiplication of semigroup generators, *Pac. J. Math.* 24 (1968), 463–465.
[3] Positive (noncommuting) operator products and semigroups, *Math. Zeit.* 105 (1968), 160–172.
[4] Doubling perturbation sizes and preservation of operator indices in normed linear spaces, *Proc. Camb. Phil. Soc.* 66 (1969), 281–294.
Gustafson, K. and G. Lumer
[1] Multiplicative perturbation of semigroup generators, *Pac. J. Math.* 41 (1972), 731–742.
Gustafson, K. and K. Sato
[1] Some perturbation theorems for nonnegative contraction semigroups, *J. Math. Soc. Japan* 21 (1969), 200–204.
Guzman, A.
[1] Growth properties of semigroups generated by fractional powers of certain linear operators, *J. Functional Anal.* 23 (1976), 331–352.
[2] Further study of growth of fractional-power semigroups, *J. Functional Anal.* 29 (1978), 133–141.
Gzyl, H.
[1] On generators of subordinate semigroups, *Ann. Prob.* 6 (1978), 975–983.
Hackman, M.
[1] The abstract time-dependent Cauchy problem, *Trans. Amer. Math. Soc.* 133 (1968), 1–50.
Hagood, J. W.
[1] The operator-valued Feynman-Kac formula with noncommutative operators, *J. Functional Anal.* 38 (1980), 99–117.
Hagedorn, G. A. and P. Perry
[1] Asymptotic completeness for certain three-body Schrödinger operators, *Comm. Pure Appl. Math.* 36 (1983), 213–232.
Hahn, L. (see Butzer)
Hale, J.
[1] *Theory of Functional Differential Equations*, Springer, New York, 1977.
Hanche-Olsen, H. (see Evans)
Hartman, P.
[1] *Ordinary Differential Equations*, Wiley, New York, 1964.
Hasegawa, M.
[1] On the convergence of resolvents of operators, *Pac. J. Math.* 21 (1967), 35–47.
[2] A note on the convergence of semi-groups of operators, *Proc. Japan Acad.* 40 (1964), 262–266.
[3] On contraction semi-groups and (di)-operators, *J. Math. Soc. Japan* 18 (1966), 303–330.
[4] On a property of the infinitesimal generators of semi-groups of linear operators, *Math. Ann.* 182 (1969), 121–126.
[5] Some remarks on the generation of semi-groups of linear operators, *J. Math. Anal. Appl.* 20 (1967), 454–463.
[6] On quasi-analytic vectors for dissipative operators, *Proc. Amer. Math. Soc.* 29 (1971), 81–84.
Hasegawa, M. and R. Sato
[1] A general ratio ergodic theorem for semigroups, *Pac. J. Math.* 62 (1976), 435–437.
Hasegawa, M., R. Sato, and S. Tsurumi
[1] Vector valued ergodic theorems for a one-parameter semigroup of linear operators, *Tôhoku Math. J.* 30 (1978), 95–106.
Hassan, J. -C. (see Baras)
Hazod, W. (see also Drisch)
[1] Eine Produktformel für Halbgruppen von Warscheinlichskeitmasse auf Lie-Gruppen, *Mh. Math.* 76 (1972), 295–299.
[2] *Stetige Faltungshalbgruppen von Warsheinlichs keitmassen und erzeugende Distributionen, Lecture Notes in Math.* Vol. 595, Springer, Berlin, 1977.
Hazrallah, K. (see Faraut)
Hejtmanek, J. (see Kaper)
Henry, D.
[1] *Geometric Theory of Semilinear Parabolic Equations*, Lecture Notes in Math. No. 840, Springer, Berlin, 1981.

Herbst, I.
[1] Contraction semigroups and the spectrum of $A_1 \otimes I + I \otimes A_2$, *J. Oper. Th.* 7 (1982), 61–78.
[2] The spectrum of Hilbert space semigroups, *J. Oper. Th.* 10 (1983), 87–94.

Herbst, I. W. and A. D. Sloan
[1] Perturbation of translation invariant positivity preserving semigroups on $L^2(R^N)$, *Trans. Amer. Math. Soc.* 236 (1978), 325–360.

Herdman, T. L. (see Burns)

Hering, H.
[1] Refined positivity theorem for semigroups generated by perturbed differential operators of second order with an application to Markov branching processes, *Math. Proc. Camb. Phil. Soc.* 83 (1978), 253–259.
[2] Uniform primitivity of semigroups generated by perturbed elliptic differential operators, *Math. Proc. Camb. Phil. Soc.* 83 (1979), 261–268.

Herman, R. H. (see Bratelli)

Herod, J. V.
[1] A product integral representation for an evolution system, *Proc. Amer. Math. Soc.* 29 (1971), 549–556.
[2] A pairing of a class of evolution systems with a class of generators, *Trans. Amer. Math. Soc.* 157 (1971), 247–260.

Herod, J. V. and R. W. McKelvey
[1] A Hille-Yosida theory for evolutions, *Israel J. Math.* 36 (1980), 13–40.

Hersh, R. (see also Bobisud, Donaldson, Griego, Papanicolaou)
[1] Direct solution of a general one-dimensional linear parabolic equation via an abstract Plancherel formula, *Proc. Nat. Acad. Sci.* 63 (1969), 648–654.
[2] Explicit solution of a class of higher-order abstract Cauchy problems, *J. Diff. Eqns.* 8 (1970), 570–579.
[3] Random evolutions: a survey of results and problems, *Rocky Mtn. J. Math.* 4 (1974), 443–477.
[4] The method of transmutations. In *Lecture Notes in Math.* No. 446, Springer, Berlin (1975), pp. 264–282.
[5] Stochastic solutions of hyperbolic equations. In *Lecture Notes in Math.* No. 446, Springer, Berlin (1975), pp. 283–300.

Hersh, R. and T. Kato
[1] High-accuracy stable difference schemes for well-posed initial-value problems, *SIAM J. Numer. Anal.* 16 (1979), 670–682.

Hersh, R. and G. Papanicolaou
[1] Non-commuting random evolutions, and an operator-valued Feynman-Kac formula, *Comm. Pure Appl. Math.* 25 (1972), 337–367.

Hersh, R. and M. Pinsky
[1] Random evolutions are asymptotically Gaussian, *Comm. Pure Appl. Math.* 25 (1972), 33–44.

Hess, H., R. Schrader, and D. A. Uhlenbrock
[1] Domination of semigroups and generalization of Kato's inequality, *Duke Math. J.* 44 (1977), 893–904.

Heyer, H.
[1] Infinitely divisible probability measures on compact groups. In *Lecture Notes in Math.* Vol. 247, Springer, Berlin (1970), pp. 55–249.
[2] Stetige Hemigruppen von Wahrscheinlichkeitsmassen und additive Prozesse auf uner lokal-kompakten Gruppe, *Nieuw Archif voor Wisk.* 27 (1979), 287–340.
[3] Infinitely divisible probability measures on compact groups. In *Lecture Notes in Math.* No. 247, Springer, Berlin (1972), pp. 55–249.
[4] Transient Feller semigroups on certain Gelfand pairs, *Bull. Inst. Math. Acad. Sinica* 11 (1983), 227–256.

Heyn, E.
[1] Die Differentialgleichung $dT/dt = P(t)T$ für Operatorfunktionen, *Math. Nachr.* 24 (1962), 281–330.
[2] Invarianz von Eigenschaffen in Bezug auf die Operatoren linel Halbgruppe, *Monatsh. Deutsch. Akad. Wiss. Berlin* 9 (1967), 273–279.
[3] Lineare Evolutionsgleichungen, *Math. Nachr.* 97 (1980), 297–311.
[4] Kegelinvarianz bei linearer Evolutionsgleichungen, *Math. Nachr.* 110 (1983), 333–350.

Hille, E.
[1] *Functional Analysis and Semi-groups*, Amer. Math. Soc. Coll. Publ., Vol. 31, New York, 1948.

[2] On the differentiability of semi-groups of operators, *Acta Sci. Math.* (Szeged) 12B (1950), 19–24.

[3] On the generation of semi-groups and the theory of conjugate functions, *Kungl. Fysiog Sällsk. I Lund Förd.* 21 (1951), 130–142.

[4] Une généralisation du problème de Cauchy, *Ann. Inst. Fourier* 4 (1952), 31–48.

[5] A note on Cauchy's problem, *Ann. Soc. Polon. Math.* 25 (1952), 56–68.

[6] The abstract Cauchy problem and Cauchy's problem for parabolic differential equations, *J. Anal. Math.* 3 (1953), 81–196.

[7] Quelques remarques sur l'équation de la chaleur, *Rend. Mat.* 15 (1956), 102–118.

[8] On the Landau–Kallman–Rota inequality, *J. Approx. Th.* 6 (1972), 117–122.

Hille, E. and R. S. Phillips

[1] *Functional Analysis and Semi-groups*, Amer. Math. Soc. Coll. Publ. Vol. 31, Providence, R. I., 1957.

Hirsch, F.

[1] Familles résolvantes, générateurs, cogénérateurs, potentiels, *Ann. Inst. Fourier (Grenoble)* 22 (1972), 89–210.

[2] Intégrales de résolvantes et calcul symbolique, *Ann. Inst. Fourier* 22 (1972), 239–264.

[3] Domaines d'opérateurs représentés comme intégrales de resolvantes, *J. Functional Anal.* 23 (1976), 199–217.

Hirsch, F. and J. -P. Roth

[1] Opérateurs dissipatifs et codissipatifs invariant sur un espace homogène. In *Lecture Notes in Math.* No. 404, Springer, Berlin (1975), pp. 229–245.

Hochberg, K. J.

[1] A signed measure on path space related to Wiener measure, *Ann. Prob.* 6 (1978), 433–458.

Høegh-Krohn. R. (see Albeverio, Simon)

Holbrook, J. A. R.

[1] A Kallman-Rota inequality for nearly Euclidean spaces, *Adv. Math.* 14 1(974), 335–345.

Holley, R. and D. W. Stroock

[1] Central limit phenomena of various interacting systems, *Ann. Math.* 110 (1979), 333–393.

[2] Diffusions on an infinite dimensional torus, *J. Functional Anal.* 42 (1981), 29–63.

Hooten, J. G.

[1] Dirichlet forms associated with hyper contractive semigroups, *Trans. Amer. Math. Soc.* 253 (1979), 237–256.

[2] Dirichlet semigroups on bounded domains, *Rocky Mtn. J. Math.* 12 (1982), 283–297.

Howland, J. S.

[1] Banach space techniques in the perturbation theory of self-adjoint operators with continuous spectra, *J. Math. Anal. Appl.* 20 (1967), 22–47.

[2] Nonexistence of asymptotic observables, *Proc. Amer. Math. Soc.* 35 (1972), 175–176.

[3] On a theorem of Gearhart, *Int. Eqns. Oper. Th.* 7 (1984), 138–142.

Hudson, R. L., P. D. F. Ion, and K. R. Parthasarathy

[1] Time orthogonal unitary dilations and noncommutative Feynman-Kac formulae, *Comm. Math. Phys.* 83 (1982), 261–280.

Hughes, R. J. (see also Kantorovitz)

[1] On the convergence of unbounded sequences of semigroups, *J. London Math. Soc.* 16 (1977), 517–528.

[2] Semigroups of unbounded linear operators in Banach spaces, *Trans. Amer. Math. Soc.* 230 (1977), 113–145.

[5] Singular perturbations in the interaction representation, II, *J. Functional Anal.* 49 (1982), 293–314.

[6] Some consequences of Duhamel's formula for weak solutions of time dependent Schrödinger equations, *Int. Eqns. Oper. Th.* 7 (1984), 206–217.

[7] A version of the Trotter product formula for quadratic-form perturbations. To appear.

Hughes, R. J. and S. Kantorovitz

[1] Boundary values of holomorphic semigroups of unbounded operators and similarity of certain perturbations, *J. Functional Anal.* 29 (1978), 253–273.

Hughes, R. J. and I. E. Segal

[1] Singular perturbations in the interaction representation, *J. Functional Anal.* 38 (1980), 71–98.

Hunt, G. A.

[1] Semi-groups of measures on Lie Groups, *Trans. Amer. Math. Soc.* 81 (1956), 264–293.

[2] Markov processes and potentials, I, II, *Ill. J. Math.* 1 (1957), 44–93 and 316–369.

[3] *Martingales et Processees de Markov*, Dunod, Paris, 1966.

Iannelli, M. (see also Da Prato, Di Blasio)
[1] On the Green function for abstract evolution equation, *Boll. Un. Mat. Ital.* 6 (1972), 154–174.
Ichinose, T. (see also Komaya)
[1] A product formula and its application to the Schrödinger equation, *Publ. RIMS, Kyoto Univ.* 16 (1980), 585–600.
Ikebe, T. and T. Kato
[1] Uniqueness of the self-adjoint extension of singular elliptic differential operators, *Arch. Rat. Mech. Anal.* 9 (1972), 77–92.
Ikeda, N., M. Nagasawa, and S. Watanabe
[1] Branching Markov Processes I, II, III, *J. Math. Kyoto Univ.* 8 (1968), 233–278, 8 (1968), 365–410, and 9 (1969), 95–160.
Ion, P. D. F. (see Hudson)
Iorio, Jr., R. J. and M. O'Carroll
[1] Asymptotic completeness for multi-particle Schrödinger Hamiltonians with weak potentials, *Comm. Math. Phys.* 27 (1972), 137–145.
Ishii, S.
[1] An approach to linear hyperbolic evolution equations by the Yosida approximation method, *Proc. Japan Acad.* 54 (1978), 17–20.
[2] Linear evolution equations $du/dt + A(t)u = 0$: a case where $A(t)$ is strongly uniform-measurable, *J. Math. Soc.* Japan 34 (1982), 413–424.
Itô, K.
[1] *Lectures on Stochastic Processes*, Tata Institute, Bombay, 1961.
[2] *Stochastic Processes*, Aarhus Univ. Lecture Notes, Aarhus, 1969.
Itô, K. and H. P. McKean
[1] *Diffusion Processes and Their Sample Paths*, Springer, Berlin, 1965.
Iwasaki, N.
[1] Local decay of solutions for symmetric hyperbolic systems with dissipative and coercive boundary conditions in exterior domains, *R. I. M. S.* Kyoto Univ. 5 (1969), 193–218.
Jaffe, A. (see Glimm)
Jamison, J. E. (see Berkson, Fleming)
Janenko, N. N.
[1] *Die Zweischenschrittmethode zur Lösung Mehrdimensionaler Probleme der Mathematischen Physik*, Lecture Notes in Math. Vol. 91, Springer, Berlin, 1969.
Janssen, A.
[1] Charakterisierung Stetiger Faltungshalbgruppen durch das Lévy-Mass, *Math. Ann.* 246 (1980), 233–240.
Janssen, A. and E. Siebert
[1] Convolution semigroups and generalized telegraph equations, *Math. Z.* 177 (1981), 519–532.
Jauch, J. M. (see also Amrein)
[1] Theory of the scattering operator, *Helv. Phys. Acta* 31 (1958), 127–158, 661–684.
Jensen, A. and T. Kato
[1] Spectral properties of Schrödinger operators and time-decay of the wave functions, *Duke Math. J.* 46 (1979), 583–611.
Jensen, A., E. Mourre, and P. Perry
[1] Multiple commutator estimates and resolvent smoothness in quantum scattering theory, *Ann. Inst. H. Poincaré* 41 (1984), 207–225.
Jiřina, M.
[1] On Feller's branching diffusion process, *Čas. Pěst. Mat.* 94 (1969), 84–90.
John, F. (see also Bers)
[1] *Partial Differential Equations*, 3rd ed., Springer, New York, 1978.
Jörgens, K. and J. Weidmann
[1] *Spectral Properties of Hamiltonian Operators*, Lecture Notes in Math., Vol. 313, Springer, Berlin, 1973.
Jørgensen, P. E. T.
[1] Trace states and KMS states for approximately inner dynamical one-parameter groups of *-automorphisms, *Comm. Math. Phys.* 53 (1977), 135–142.
[2] On one-parameter groups of automorphisms, and extensions of symmetric operators associated with unbounded derivations in operator algebras, *Tôhoku Math. J.* 30 (1978), 277–305.
[3] Ergodic properties of one-parameter automorphism groups of operator algebras, *J. Math. Anal. Appl.* 87 (1982), 354–372.

[4] Spectral theory for infinitesimal generators of one-parameter groups of isometries: the min-max principle and compact perturbations, *J. Math. Anal. Appl.* 90 (1982), 343–370.
[5] Compact symmetry groups and generators for sub-Markovian semigroups, *Z. Warsch. verw. Geb.* 63 (1983), 17–27.
[6] Monotone convergence of operator semigroups and the dynamics of infinite particle systems, *Adv. Math.* To appear.
[7] Semigroups of completely positive maps and existence of dynamics of infinite quantum systems. To appear.

Jørgensen, P. E. T. and R. T. Moore
[1] *Operator Commutation Relations*, Reidel, Dordrecht, Holland, 1984.

Kac, M.
[1] On some connections between probability theory and differential and integral equations, *Proc. Second Berkeley Symp. Math. Stat. Prob.*, U. Calif. Press (1951), 189–215.
[2] A stochastic model related to the telegrapher's equation, *Magnolia Petroleum Co. Colloq. Lectures* (1956); reprinted in *Rocky Mtn. J. Math.* 4 (1974), 497–510.
[3] *Probability and Related Topics in the Physical Sciences*, Interscience, New York, 1969.
[4] Probabilistic methods in some problems of scattering theory, *Rocky Mtn. J. Math.* 4 (1974), 511–537.

Kakutani, S. (see Yosida)

Kalf, H. (see Eastham)

Kallianpur, G. and V. Mandrekar
[1] Semi-groups of isometries and the representation and multiplicity of weakly stationary stochastic processes, *Ark. Mat.* 6 (1966), 319–335.

Kallman, R. R. and G. -C. Rota
[1] On the inequality $\|f'\|^2 \le 4\|f\| \| f''\|$. In *Inequalities* II (ed. O. Shisha), Academic Press, New York (1970) 187–192.

Kanda, S.
[1] Cosine families and weak solutions of second order differential equations, *Proc. Japan Acad.* 54 (1978), 119–123.

Kantorovitz, S. (see also Hughes)

Kantorovitz, S. and R. J. Hughes
[1] Spectral representation of local semigroups, *Math. Ann.* 259 (1982), 455–470.

Kaper, H.
[1] Long-time behavior of a nuclear reactor. In *Spectral Theory of Differential Operators* (eds. I. W. Knowles and R. T. Lewis), North Holland, New York (1981), pp. 247–251.

Kaper, H. G., C. G. Lekkerkerker, and J. Hejtmanek
[1] Editors of *Spectral Methods in Linear Transport Theory*, Birkhäuser, Boston, 1982.

Kappel, F. (see also Grimmer)
[1] An approximation scheme for delay equations. In *Nonlinear Phenomena in Abstract Spaces* (ed. V. Lakshmikantham), Academic Press, New York (1982), pp. 585–595.
[2] Finite dimensional approximation to systems with infinite dimensional state space. To appear.

Kato, T. (see also Arendt, Brezis, Fujita, Hersh, Ikebe, Jensen)
[1] Fundamental properties of Hamiltonian operators of Schrödinger type, *Trans. Amer. Math. Soc.* 70 (1951), 195–211.
[2] Integration of the equation of evolution in a Banach space, *J. Math. Soc. Japan* 5 (1953), 208–234.
[3] On the semi-groups generated by Kolmogoroff's differential equations, *J. Math. Soc. Japan* 6 (1954), 1–15.
[4] On linear differential equations in Banach spaces, *Comm. Pure Appl. Math.* 9 (1956), 479–486.
[5] Remarks on pseudo-resolvents and infinitesimal generators of semigroups, *Proc. Japan Acad.* 35 (1959), 467–468.
[6] Note on fractional powers of linear operators, *Proc. Japan Acad.* 36 (1960), 94–96.
[7] Abstract evolution equations of parabolic type in Banach and Hilbert spaces, *Nagoya Math. J.* 19 (1961), 93–125.
[8] A generalization of the Heinz inequality, *Proc. Japan Acad.* 37 (1961), 305–308.
[9] Fractional powers of dissipative operators, *J. Math. Soc. Japan* 13 (1961), 246–274.
[10] Fractional powers of dissipative operators, II, *J. Math. Soc. Japan* 14 (1962), 242–248.
[11] Wave operators and similarity for some non-selfadjoint operators, *Math. Ann.* 162 (1966), 258–279.
[12] *Perturbation Theory for Linear Operators*, Springer, New York, 1966.

[13] Some mathematical problems in quantum mechanics, *Progr. Theor. Phys. (Supplement)* 40 (1967), 3–19.
[14] Scattering theory with two Hilbert spaces, *J. Functional Anal.* 1 (1967), 342–369.
[15] Smooth operators and commutators, *Studia Math.* 31 (1968), 535–546.
[16] Linear evolution equations of "hyperbolic" type, *J. Fac. Sci. Univ. Tokyo* Sect. I, 27 (1970), 241–258.
[17] A characterization of holomorphic semigroups, *Proc. Amer. Math. Soc.* 25 (1970), 495–498.
[18] On an inequality of Hardy, Littlewood and Polyà, *Adv. Math.* 7 (1971), 217–218.
[19] Scattering theory. In *Studies in Applied Math.*, Vol. 7 (ed. A. H. Taub), Math. Assoc. of Amer. (1971), pp. 90–115.
[20] Scattering theory and perturbation of continuous spectra. In *Actes, Congrèss Intern. Math.*, 1970, Vol. 1, Gauthier-Villars, Paris (1971), pp. 135–140.
[21] Schrödinger operators with singular potentials, *Israel J. Math.* 13 (1972), 135–148.
[22] Linear evolution equations of "hyperbolic" type, II, *J. Math. Soc. Japan* 25 (1973), 648–666.
[23] A remark to the preceding paper by Chernoff, *J. Functional Anal.* 12 (1973), 415–417.
[24] On the Trotter-Lie product formula, *Proc. Japan Acad.* 50 (1974), 694–698.
[25] Singular perturbation and semigroup theory. In *Lecture Notes in Math.* No. 565, Springer, New York (1976), pp. 104–112.
[26] Remarks on Schrödinger operators with vector potentials, *Integral Eqns. Op. Th.* 1 (1978), 103–113.
[27] On some Schrödinger operators with a singular complex potential, *Ann. Scuol. Norm. Sup. Pisa* 5 (1978), 105–114.
[28] On the Cook-Kuroda criterion in scattering theory, *Comm. Math. Phys.* 67 (1979), 85–90.
[29] Trotter's product formula for an arbitrary pair of self-adjoint contraction semigroups. In *Advance in Math. Suppl. Studies*, Vol. 3, Academic Press, New York (1978), pp. 185–195.
[30] Two space scattering theory, with applications to many-body problems, *J. Fac. Sci. Univ. Tokyo* IA, 24 (1977), 503–514.
[31] Remarks on the selfadjointness and related problems for differential operators. In *Spectral Theory of Differential Operators*, (eds. I. W. Knowles and R. T. Lewis), North Holland, New York (1981), pp. 253–266.
[32] Holomorphic families of Dirac operators, *Math. Z.* 183 (1983), 399–406.
[33] Nonselfadjoint Schrödinger operators with singular first-order coefficients, *Proc. Roy. Soc. Edinburgh*, 96A (1984), 323–329.

Kato, T. and H. Fujita
[1] On the nonstationary Navier-Stokes system, *Rend. Sem. Mat. Univ. Padova* 32 (1962), 243–260.
Kato, T. and S. T. Kuroda
[1] Theory of simple scattering and eigenfunction expansions. In *Functional Analysis and Related Fields* (ed. F. E. Browder), Springer, Berlin (1970), pp. 99–131.
[2] The abstract theory of scattering, *Rocky, Mtn. J. Math.* 1 (1971) 127–171.
Kato T. and H. Tanabe
[1] On the abstract evolution equation, *Osaka Math. J.* 14 (1962), 107–133.
[2] On the analyticity of solution of evolution equations, *Osaka J. Math.* 4 (1967).
Kauffman, R. M.
[1] Unitary groups and differential operators, *Proc. Amer. Math. Soc.* 30 (1971), 102–106.
[2] Unitary groups and commutators, *Proc. Amer. Math. Soc.* 33 (1972), 95–100.
[3] A spectral decomposition theorem and its application to higher-order non-selfadjoint differential operators in $L_2[a,\infty)$, *Proc. London Math. Soc.* 40 (1980), 476–506.
[4] A differentiability property and its application to the spectral theory of differential operators, *Proc. Roy. Soc. Edinburgh* 88A (1981), 49–74.
Kaufman, R. (see Berkson)
Keepler, M.
[1] Backward and forward random evolutions, *Indiana U. Math. J.* 24 (1975), 937–947.
[2] Perturbation theory for backward and forward random evolutions, *J. Math. Kyoto Univ.* 16 (1976), 395–411.
Kendall, D. G.
[1] Unitary dilations of one-parameter semigroups of Markov transition operators, and the corresponding integral representations for Markov processes with a countable infinity of states, *Proc. London Math. Soc.* 9 (1959), 417–431.
[2] Some recent developments in the theory of denumberable Markov processes, *Trans. Fourth Prague Conf. Information Theory*, Academia, Prague (1967), pp. 11–27.

Kerscher, W. and R. Nagel
[1] Asymptotic behavior of one-parameter semigroups of positive operators, *Acta Appl. Math,* 2 (1984), 323–329.

Kertz, R. P.
[1] Perturbed semigroup limit theorems with applications to discontinuous random evolutions, *Trans. Amer. Math. Soc.* 199 (1974), 29–53.
[2] Limit theorems for discontinuous random evolutions with applications to initial value problems and to Markov process on *N*-lines, *Ann. Prob.* 2 (1974), 1054–1064.
[3] Limit theorems for semigroups with perturbed generators, with applications to multi-scaled random evolutions, *Trans. Amer. Math. Soc.* 27 (1978), 215–233.
[4] Random evolutions with underlying semi-Markov processes, *Publ. R. I. M. S. S. Kyoto Univ.* 14 (1978), 589–614.
[5] Semigroup perturbation theorems with an application to a singular perturbation problem in nonlinear ordinary differential equations, *J. Diff. Eqns.* 31 (1979), 1–15.

Kielhöfer, H.
[1] Global solutions of semilinear evolution equations satisfying an energy inequality, *J. Diff. Eqns.* 36 (1980), 188–222.

Kiffe, T. and M. Stecher
[1] Existence and uniqueness of solutions to abstract Volterra integral equations, *Proc. Amer. Math. Soc.* 68 (1978), 169–175.
[2] Properties and applications of the resolvent operator to a Volterra integral equation in Hilbert space, *SIAM J. Math. Anal.* 11 (1980), 82–91.

Kipnis, C.
[1] Majoration des semi-groupes de contractions de L^1 et applications, *Ann: Inst. H. Poincaré* 10 (1974), 369–384.

Kishimoto, A. (see also Bratelli)
[1] Universally weakly inner one-parameter automorphism groups of simple C^*-algebras, *Yokohama Math. J.* 29 (1981), 89–100.

Kishimoto, A. and D. W. Robinson
[1] Subordinate semigroups and order properties, *J. Austral. Math. Soc. Ser.* A. 31 (1981), 59–76.
[2] Positivity and monotonicity properties of C_0-semigroups, II, *Comm. Math. Phys.* 75 (1980), 85–101.

Kisyński, J.
[1] Sur les équations hyperboliques avec petit parametre, *Colloq. Math.* 10 (1963), 331–343.
[2] Sur les operateurs de Green des problèmes de Cauchy abstraits, *Studia Math.* 23 (1964), 285–328.
[3] A proof of the Trotter-Kato theorem on approximation of semigroups, *Colloq. Math.* 18 (1967), 181–184.
[4] On second order Cauchy's problem in a Banach space, *Bull. Acad. Polon. Sci.* 18 (1970), 371–374.
[5] On operator-valued solutions of d'Alembert's functional equation. I., *Studia Math.* 42 (1971), 43–66.
[6] On operator-valued solutions of d'Alembert's functional equation. II, *Colloq. Math.* 23 (1971), 107–114.
[7] On cosine operator functions and one-parameter groups of operators, *Studia Math.* 44 (1972), 93–105.
[8] On M. Kac's probabilistic formula for the solution of the telegraphist's equation, *Ann. Polon. Math.* 29 (1974), 259–272.
[9] Semi-groups of operators and some of their applications to partial differential equations. In *Control Theory and Topics in Functional Analysis,* Internat'l Atomic Energy Agency, Vienna (1976), pp. 305–405.
[10] Holomorphicity of semigroups of operators generated by sublaplacians on Lie groups. In *Lecture Notes in Math.* No. 561, Springer, Berlin (1976), pp. 283–297.
[11] On semigroups generated by differential operators on Lie groups, J. Functional Anal. 31 (1979), 234–244.

Klein, A. (see also Carmona)
[1] The semigroup characterization of Osterwalder-Schrader path spaces and the construction of Euclidean fields, *J. Functional Anal.* 27 (1978), 277–291.

Kluvánek, I.
[1] Réprésentations intégrales d'évolutions perturbées. *C. R. Acad. Sci. Paris* 288 (1979), 1065–1067.

[2] Operator valued measures and perturbations of semi-groups, *Arch. Rat. Mech. Anal.* 81 (1983), 161–180.

Knight, F. B.
[1] A path space for positive semigroups, *Ill. J. Math.* 13 (1969), 542–563.

Knowles, I.
[1] Dissipative Sturm-Liouville operators, *Proc. Roy. Soc. Edinburgh* 88A (1981), 329–343.

Kobayasi, K.
[1] On a theorem of linear evolution equations of hyperbolic type, *J. Math. Soc. Japan* 31 (1979), 647–654.

Kolerov, G. I. (see Exner)

Koller, H., M. Schechter, and R. A. Weder
[1] Schrödinger operators in the uniform norm, *Ann. Inst. H. Poincaré* 26 (1977), 303–311.

Komatsu, H.
[1] Semi-groups of operators in locally convex spaces, *J. Math. Soc. Japan* 16 (1964), 230–262.
[2] Fractional powers of operators, *Pacific J. Math.* 19 (1966), 285–346.
[3] Fractional powers of operators, II. Interpolation spaces, *Pac. J. Math.* 21 (1967), 89–111.
[4] Fractional powers of operators, III. Negative powers, *J. Math. Soc. Japan* 21 (1969), 205–220.
[5] Fractional powers of operators, IV. Potential operators, *J. Math. Soc. Japan* 21 (1969), 221–228.
[6] Fractional powers of operators, V. Dual operators, *J. Fac. Sci. Univ. Tokyo* Sect. I, 17 (1970), 373–396.

Komaya, T. and T. Ichinose
[1] On the Trotter product formula, *Proc. Japan Acad.* 57 (1981),95–100.

Kōmura, T.
[1] Semigroups of operators in locally convex spaces, *J. Functional Anal.* 2 (1968), 258–296.

Kōmura, Y.
[1] On linear evolution operators in reflexive Banach spaces, *J. Fac. Sci. Univ. Tokyo*, Sect. IA, 17 (1970), 529–542.

Konishi, Y.
[1] Cosine functions of operators in locally convex spaces, *J. Fac. Sci. Univ. Tokyo*, Sect. IA, 18 (1972), 443–463.

Kossakowski, A. (see also Gorini)
[1] On necessary and sufficient conditions for a generator of a quantum dynamical semigroup, *Bull. Acad. Polon. Sci. Ser. Math. Astr. Phys.* 20 (1972), 1021–1025.

Kraljevič, H. and S. Kurepa
[1] Semigroups on Banach spaces, *Glasnik Mat.* 5 (1970), 109–117.

Krasnoselskiĭ, M. A., S. G. Krein, and P. E. Sobolevskiĭ
[1] On differential equations with operators in Hilbert space, *Dokl. Acad. Nauk SSSR* 112 (1957), 990–993.

Krasnoselskiĭ, M. A. and P. E. Sobolevskiĭ
[1] Fractional powers of operators defined on Banach spaces, *Dokl. Acad. Nauk SSSR* 129 (1959), 499–502.

Kreĭn, M. G. (see also Daleckiĭ)
[1] The theory of self-adjoint extensions of semi-bounded hermitian transformations and its applications, *Mat. Sbornik* 20 (1947), 431–495 and 21 (1947), 365–404.

Kreĭn, S. G. (see also Krasnoselskiĭ)
[1] *Linear Differential Equations in Banach Spaces*, Translations of Math. Monographs, Amer. Math. Soc., Providence, R. I., 1971.
[2] *Linear Equations in Banach Spaces*, Birkhäuser, Boston, 1982.

Kreiss, H-O.
[1] Über Matrizen die beschränkte Halbgruppenerzeugen, *Math. Scand.* 7 (1959), 71–80.

Kriete, T. L.
[1] Canonical models and the self-adjoint part of a dissipative operator, *J. Funct. Anal.* 23 (1976), 39–94.

Kubokawa, Y.
[1] A local ergodic theorem for semi-groups in L_p, *Tôhoku Math. J.* 26 (1974), 411–422.
[2] Ergodic theorems for contraction semigroups, *J. Math. Soc. Japan* 27 (1975), 184–194.

Kunisch, K. (see also Di Blasio)

Kunisch, K. and W. Schappacher
[1] Necessary conditions for partial differential equations with delay to generate C_0-semigroups, *J. Diff. Eqns.* 50 (1983), 49–79.

Kunita, H.
[1] General boundary conditions for multi-dimensional diffusion processes, *J. Math. Kyoto Univ.* 10 (1970), 273–335.

Kuo, H.-H.
[1] Differential and stochastic equations in abstract Wiener space, *J. Functional Anal.* 12 (1973), 246–256.
[2] Potential theory associated with the Uhlenbeck-Ornstein process, *J. Functional Anal.* 21 (1976), 63–75.

Kuo, H. -H. and M. A. Piech
[1] Stochastic integrals and parabolic equations in abstract Wiener space, *Bull. Amer. Math. Soc.* 79 (1973), 478–482.

Kurepa, S. (see also Kraljevič)
[1] Uniformly bounded cosine functions in a Banach space, *Math. Balkanika* 2 (1972), 109–115.
[2] A weakly measurable selfadjoint cosine function, *Glasnik Mat.* 8 (1973), 73–79.
[3] Decomposition of weakly measurable semigroups and cosine operator functions, *Glasnik Mat.* 11 (1976), 91–95.
[4] Semigroups and cosine functions. In *Lecture Notes in Math.* No. 948, Springer, Berlin (1982), pp. 47–72.

Kuroda, S. T. (see also Kato)
[1] On the existence and the unitary property of the scattering operator, *Nuovo Cimento* 12 (1959), 431–454.

Kurtz, T. G. (see also Athreya, Certain, Ethier)
[1] Extensions of Trotter's operator semigroup approximation theorems, J. Functional Anal. 3 (1969), 354–375.
[2] A general theorem on the convergence of operator semigroups, *Trans. Amer. Math. Soc.* 148 (1970), 23–32.
[3] Solutions of ordinary differential equations as limits of pure jump processes, *J. Appl. Prob.* 7 (1970), 49–58.
[4] Limit theorems for sequences of jump Markov processes approximating ordinary differential equations, *J. Appl. Prob.* 8 (1971), 344–356.
[5] Comparison of semi-Markov and Markov processes, *Ann. Math. Stat.* 42 (1971), 991–1002.
[6] A random Trotter product formula, *Proc. Amer. Math. Soc.* 35 (1972), 147–154.
[7] A limit theorem for perturbed operator semigroups with applications to random evolutions, *J. Functional Anal.* 12 (1973), 55–67.
[8] Semigroups of conditioned shifts and approximation of Markov processes, *Ann. Prob.* 3 (1975), 618–642.
[9] An abstract averaging theorem, *J. Functional Anal.* 23 (1976), 135–144.
[10] Applications of an abstract perturbation theorem to ordinary differential equations, *Houston J. Math.* 3 (1977), 67–82.
[11] Strong approximation theorems for density dependent Markov chains, *Stoch. Proc. Appl.* 6 (1978), 223–240.
[12] A variational formula for the growth rate of a positive operator semigroup, *SIAM J. Math. Anal.* 10 (1979), 112–117.
[13] Diffusion approximations for branching processes. In *Branching Processes, Advances in Probability*, Vol. 5 (eds, A. Joffe and B. Ney), Marcel-Dekker, New York (1979), 262–292.
[14] Representations of Markov processes as multiparameter time changes, *Ann. Prob.* 8 (1980), 682–715.
[15] *Approximation of Population Processes*, SIAM, Philadelphia, 1981.

Kurtz, T. G. and M. Pierre
[1] A counterexample for the Trotter product formula, *J. Diff. Eqns.* 52 (1984), 407–414.

Kwong, M. K. (see also Goldstein)
Kwong, M. K. and A. Zettl
[1] Norm inequalities for dissipative operators on inner product spaces, *Houston J. Math.* 5 (1979), 543–557.
[2] Remarks on the best constants for norm inequalities among powers of an operator, *J. Approx. Theory* 26 (1979), 249–258.
[3] Ramifications of Landau's inequality, *Proc. Roy. Soc. Edinburgh* 86A (1980), 175–212.
[4] Norm inequalities of product form in weighted L_p spaces, *Proc. Roy. Soc. Edinburgh* 89A (1981), 293–307.
[5] Weighted norm inequalities of sum form involving derivatives, *Proc. Roy. Soc. Edinburgh* 88A (1981), 121–134.

[6] Norm inequalities involving derivatives. In *Lecture Notes in Math*. No. 846, Springer, New York (1981), pp. 227–243.

[7] Landau's inequality, *Rocky Mountain Math. J*. To appear.

Labrousse, J. -P.

[1] Une charactérisation topologique des générateurs infinitésimaux de semi-groupes analytiques et de contractions sur un espace de Hilbert, *Atti. Accad. Naz. Lincei, VIII. Ser. Rend. Cl. Sci. fis. mat. Natur.* 52 (1972), 631–636.

Ladas, G. E. and V. Lakshmikantham

[1] *Differential Equations in Abstract Spaces*, Academic Press, New York, 1972.

Lagnese, J. E.

[1] Note on a theorem of C. T. Taam concerning bounded solutions of nonlinear differential equations. *Proc. Amer. Math. Soc.* 20 (1969), 351–356.

[2] On equations of evolution and parabolic equations of higher order in *t*, *J. Math. Anal. Appl.* 32 (1970), 15–37.

[3] General boundary value problems for differential equations of Sobolev type, *SIAM J. Math. Anal.* 3 (1972), 105–119.

[4] Exponential stability of solutions of differential equations of Sobolev type. *SIAM J. Math. Anal.* 3 (1972), 625–636.

[5] Elliptic and parabolic boundary value problems of nonlocal type, *J. Math. Anal. Appl.* 40 (1972), 183–201.

[6] Approximation of solutions of differential equations in Hilbert space, *J. Math. Soc. Japan* 25 (1973), 132–143.

[7] Singular differential equations in Hilbert space, *SIAM J. Math. Anal.* 4 (1973), 623–637.

[8] Existence, uniqueness and limiting behavior of solutions of a class of differential equations in Banach space, *Pac. J. Math.* 53 (1974), 473–485.

[9] Rate of convergence in singular perturbations of hyperbolic problems, *Indiana U. Math. Soc.* 24 (1974), 417–432.

[10] The final value problem for Sobolev equations, *Proc. Amer. Math. Soc.* 56 (1976), 247–252.

[11] Boundary value control of a class of hyperbolic equations in a general region, *SIAM J. Cont. Opt.* 15 (1977), 973–983.

[12] Exact boundary controllability of a class of hyperbolic equations, *SIAM J. Cont. Opt.* 16 (1978), 1000–1017.

[13] On the support of solutions of the wave equation with applications to exact boundary value controllability, *J. Math. Pures Appl.* 58 (1979), 121–135.

[14] Boundary patch control of the wave equation in some non-star complemented regions, *J. Math. Anal. Appl.* 77 (1980), 364–380.

[15] Boundary stabilization of linear elastodynamic systems, *SIAM J. Cont. Opt.* 21 (1983), 968–984.

Lai, P. T. (see Guillement)

Lakshmikantham, V. (see Ladas)

Lambert, A. (see Embry)

Langer, H.

[1] A class of infinitesimal generators of one-dimensional Markov processes, *J. Math. Soc. Japan* 28 (1976), 242–249.

[2] Absolutstetigkeit der Übergangsfunktion einer Klasse eindimensional Fellerprozesse, *Math. Nachr.* 75 (1976), 101–112.

Langer, H., L. Partzsch, and D. Schütze

[1] Über verallgemeinerte gewöhnliche Differentialoperatoren mit nichtlokalen Randbedingungen und die von ihner erzeugten Markov-Prozesse, *Publ. R. I. M. S., Kyoto Univ.* 7 (1972), 659–702.

Langer, H. and W. Schenk

[1] A class of infinitesimal generators of one-dimensional Markov processes, II. Invariant measures, *J. Math. Soc. Japan* 31 (1980), 1–18.

Lapidus, M.

[1] Perturbation d'un semi-groupe par une groupe unitaire, *C. R. Acad. Sc. Paris* 291 (1980), 535–538.

[2] Generalization of the Lie-Trotter formula, *Int. Eqns. Oper. Th.* 4 (1981), 366–415.

[3] The problem of the Trotter-Lie formula for unitary groups of operators, *Séminaire Choquet, Publ. Math. Univ. P. et M. Curie* 46 (1982), 1701–1745.

[4] Perturbation and nonperturbation theorems in semigroup theory. Applications to the Schrödinger operator, *Bull. Sci. Math.* To appear.

[5] Perturbation theory and a dominated convergence theorem for Feynman integrals. To appear.

Lasiecka, I.
[1] State constrained control problems for parabolic systems: Regularity of optimal solutions, *Appl. Math. Optim.* 6 (1980), 1–29.
[2] Boundary control of parabolic systems: Finite-element approximation, *Appl. Math. Optim.* 6 (1980), 31–62.
[3] Unified theory for abstract parabolic boundary problems—a semigroup approach, *Appl. Math. Optim.* 6 (1980), 287–333.

Lasiecka, I. and R. Triggiani
[1] A cosine operator approach to modeling $L_2(0, T; L_2(\Gamma))$-Boundary input hyperbolic equations, *Appl. Math. Optim.* 7 (1981), 35–93.
[2] Dirichlet boundary stabilization of the wave equation with damping feedback of finite range, *J. Math. Anal. Appl.* 97 (1983), 112–130.

Lavine, R. B.
[1] Scattering theory for long range potentials, *J. Functional Anal.* 5 (1970), 368–382.
[2] Commutators and scattering theory, I. Repulsive interactions, *Comm. Math. Phys.* 20 (1971), 301–323.
[3] Commutators and scattering theory II. A class of one body problems, *Indiana U. Math. J.* 21 (1972), 643–656.
[4] Commutators and local decay. In *Scattering Theory in Mathematical Physics* (eds. J. A. LaVita and J. -P. Marchand), Reidel, Dordrecht, Holland (1974), 141–156.

La Vita, J. and J. -P. Marchand (eds.)
[1] *Scattering Theory in Mathematical Physics*, Reidel, Dordrecht, Holland, 1974.

Lax, P. D.
[1] Operator theoretic treatment of hyperbolic equations, *Bull. Amer. Math. Soc.* 58 (1952), 182.
[2] On Cauchy's problem for hyperbolic equations and the differentiability of the solutions of elliptic equations, *Comm. Pure Appl. Math.* 8 (1955), 615–653.

Lax, P. D. and A. N. Milgram
[1] Parabolic equation. In *Contributions to the Theory of Partial Differential Equations*, Ann. of Math. Studies No. 33, Princeton U. Press, Princeton, N. J. (1954), pp. 167–190.

Lax, P. D. and R. S. Phillips
[1] Local boundary conditions for dissipative symmetric linear differential operators, *Comm. Pure Appl. Math.* 13 (1960), 427–455.
[2] *Scattering Theory*, Academic Press, New York, 1967.
[3] Scattering theory, *Rocky Mtn. J. Math.* 1 (1971), 173–223.
[4] Scattering theory for the acoustic equation in an even number of space dimensions, *Indiana U. Math. J.* 22 (1972), 101–134.
[5] Scattering theory for dissipative hyperbolic systems, *J. Functional Anal.* 14 (1973), 172–235.
[6] Scattering theory for domains with non-smooth boundaries, *Arch. Rat. Mech. Anal.* 22 (1978), 83–98.
[7] The time delay operator and a related trace formula. In *Topics in Functional Analysis* (ed. I. Gohberg and M. Kac), Academic Press, New York (1978), pp. 197–215.

Lax, P. D. and R. D. Richtmeyer
[1] Survey of the stability of linear finite difference equations *Comm. Pure Appl. Math.* 9 (1956), 267–293.

Leeuw, K. De
[1] On the adjoint semi-groups and some problems in the theory of approximation, *Math. Z.* 73 (1966), 219–234.

Leigh, J. R.
[1] *Functional Analysis and Linear Control Theory*, Academic Press, London, 1980.

Le Jan, Y.
[1] Dual markovian semigroups and processes. In *Lecture Notes in Math.* No. 923, Springer, Berlin (1982), pp. 47–75.

Lekkerkerker, C. G. (see Kaper)

Lenard, A.
[1] Probabilistic version of Trotter's exponential product formula in Banach algebras, *Acta Sci. Math.* 32 (1971), 101–107.

Levan, N.
[1] On the unitary subspace of a dissipative perturbation of a contraction semigroup, *Indiana U. Math. J.* 28 (1979), 241–249.
[2] On the reduction of a contraction semigroup to a completely nonselfadjoint nonunitary one,

Numer. Funct. Anal. Opt. 1 (1979), 619–631.
[3] A note on uniformly bounded Hilbert space semigroups, *J. Math. Anal. Appl.* 77 (1980), 344–350.
[4] On a class of (C_0) semigroups in control theory. To appear.

Levan, N. and L. Rigby
[1] Strong stabilizability of linear contractive control systems on Hilbert space, *SIAM J. Cont. Opt.* 17 (1979), 23–35.
[2] On the reduction of a contraction semigroup to a completely nonunitary semigroup, *J. Math. Anal. Appl.* 67 (1979), 1–11.

Levine, H. A.
[1] On a theorem of Knops and Payne in dynamical linear thermoelasticity, *Arch. Rat. Mech. Anal.* 38 (1970), 290–307.
[2] Some uniqueness and growth theorems in the Cauchy problem for $Pu_{tt} + Mu_t + Nu = 0$ in Hilbert space, *Math. Z.* 126 (1972), 345–360.
[3] Uniqueness and growth of weak solutions to certain linear differential equations in Hilbert space, *J. Diff. Eqns.* 17 (1975), 73–81.
[4] An equipartition of energy theorem for weak solutions of evolutionary equations in Hilbert space: The Lagrange identity method, *J. Diff. Eqns.*, 24 (1977), 197–210.

Levinson, N. (see Coddington)
Lewis, J. T. (see Evans)
Licea, G. (see Cornea)
Lieb, E. and B. Simon
[1] Pointwise bounds on eigenfunctions and wave packets in N-body quantum systems.VI. Asymptotics in the two cluster region, *Adv. Appl. Math.* 1 (1980), 324–343.

Liggett, T. M.
[1] On convergent diffusions: The densities and the conditioned processes, *Indiana U. Math. J.* 20 (1970), 265–279.
[2] Existence theorems for infinite particle systems, *Trans. Amer. Math Soc.* 165 (1972), 471–482.
[3] A characterization of the invariant measures for an infinite particle system with interactions, *Trans. Amer. Math. Soc.* 179 (1973), 433–453.
[4] The stochastic evolution of infinite systems of interacting particles. In *Lecture Notes in Math.* No. 598, Springer, Berlin (1977), pp. 187–248.

Lin, C. S. -C. (see also Shaw)
[1] Wave operators and similarity for generators of semigroups in Banach spaces, *Trans. Amer. Math. Soc.* 139 (1969), 469–494.

Lin, C. S. -C. and S. Y. Shaw
[1] Ergodic theorems of semigroups and application, *Bull. Inst. Math. Acad. Sinica* 6 (1978), 181–188.

Lin, S. -C. (see Lin, C. S. -C.)
Lin, M.
[1] Semi-groups of Markov operators, *Boll. Un. Mat. Ital.* 6 (1972), 20–44.
[2] Weak mixing for semi-groups of Markov operators without finite invariant measures. In *Lecture Notes in Math.* No. 729, Springer, Berlin (1979), pp. 89–92.
[3] On local ergodic convergence of semi-groups and additive processes, *Israel J. Math.* 42 (1982), 300–308.

Lin, M., J. Montgomery, and R. Sine
[1] Change of velocity and ergodicity in flows and in Markov semi-groups, *Z. Warsch. verw. Geb.* 39 (1977), 197–211.

Lindblad, G.
[1] On the generators of quantum dynamical semigroups, *Comm. Math. Phys.* 48 (1976), 119–130.
[2] Dissipative operators and cohomology of operator algebras, *Lett. Math. Phys.* 1 (1976), 219–224.

Lion, G.
[1] Familles d'opérateurs et frontières en théorie du potentiel, *Ann. Inst. Fourier*, 16 (1966), 389–453.

Lions, J. -L.
[1] Une remarque sur des appplications du théorème de Hille-Yosida, *J. Math. Soc. Japan* 9 (1957), 62–70.
[2] Les semi-groupes distributions, *Portugal. Math.* 19 (1960), 141–164.
[3] *Equations Différentielles Opérationelles et Problèmes aux Limites*, Springer, Berlin, 1961.
[4] *Optimal Control of Systems Governed by Partial Differential Equations*, Springer-Verlag, New York, 1971.

[5] *Some Aspects of the Optimal Control of Distributed Parameter Systems*, Regional Conf. Series in Appl. Math., SIAM, Philadelphia, 1972.
[6] Some Aspects of the theory of linear evolution equations. In *Boundary Value Problems for Linear Evolution and Partial Differential Equations* (ed. H. G. Garnir), Reidel, Dordrecht, Holland (1977), 175–238.

Lions, J. -L. and E. Magenes
[1] *Nonhomogeneous Boundary Value Problems and Applications*, Vols. 1, 2, 3, Springer, Berlin, 1972.

Littman, W.
[1] The wave operator and L^p-norms, *J. Math. Mech.* 12 (1963), 55–68.

Ljubič, Ju. I.
[1] Conditions for the uniqueness of the solution to Cauchy's abstract problem *Soviet Math. Dokl.* 1 (1960), 110—113.
[2] Conditions for the denseness of the initial value manifold of an abstract Cauchy problem, *Soviet Math. Dokl.* 5 (1964), 384–387.

Loève, M.
[1] *Probability Theory* (3rd ed.), Van Nostrand, Princeton, N. J., 1963.

Loomis, I. H.
[1] *Isometries of Banach Spaces*, Ph.D. Thesis, Memphis State U., Memphis, Tenn. 1982.

Lumer, G. (see also Gustafson)
[1] Semi-inner product spaces, *Trans. Amer. Math. Soc.* 100 (1961), 29–43.
[2] Spectral operators, Hermitian operators, and bounded groups, *Acta Sci. Math.* (*Szeged*) 25 (1964), 75–85.
[3] Perturbations de générateurs infinitésimaux du type "changement de temps", *Ann. Inst. Fourier* 23 (1974), 271–279.
[4] Problème de Cauchy pour opérateurs locaux et "changement de temps", *Ann. Inst. Fourier* 25 (1975), 409–466.
[5] Équations de'évolution en norme uniforme (conditions necessaire et suffisantes de resolution et holomorphic), *Sem. Goulaouic-Schwartz*, exposé du 16 Novembre 1976.
[6] Problème de Cauchy avec valeurs an bord continues, comportement asymptotique, et applications. In *Lecture Notes in Math.* No. 563, Springer, Berlin (1976), 193–201.
[7] Principe du maximum et équations d'évolution dans L^2. In *Lecture Notes in Math.* No. 681, Springer, Berlin (1978), 143–156.
[8] Local operators, regular sets, and evolution equations of diffusion type. In ISNM 60, Birkhäuser, Basel (1981), 51–71.
[9] Local dissipativeness and closure of local operators. In *Operator Theory, Adv. Appl.*, Vol. 4, Birkhäuser, Basel (1982), 415–426.
[10] An exponential formula of Hille-Yosida type for propagators. To appear.

Lumer, G. and L. Paquet
[1] Semi-groupes holomorphes, product tensoriel de semi-groupes et equations d'évolution. In *Lecture Notes in Math.*, No. 713, Springer, Berlin (1979), pp. 156–177.

Lumer, G. and R. S. Phillips
[1] Dissipative operators in a Banach space, *Pac, J. Math.* 11 (1961), 679–698.

Lutz, D.
[1] Uber operatorwertige Lösungen der Funktionalgleichung des Cosinus, *Math. Z.* 171 (1980), 233–245.
[2] Periodische operatorwertige Cosinusfunktionen, *Resultate der Math.* 4 (1981), 75–83.
[3] Strongly continuous operator cosine functions, *Lecture Notes in Math.* No. 948, Springer, Berlin (1982), pp. 73–97.
[4] Some spectral properties of bounded operator cosine functions, *C. R. Math. Rep. Acad. Sci. Canada* 4 (1982), 81–85.
[5] An approximation theorem for cosine operator functions, *C. R. Math. Rep. Acad. Sci. Canada* 4 (1982), 359–362.
[6] Über die konvergenz Operatorwertiges Cosinusfunktionen mit gestörtem infinitesimalen Erzenger, *Periodica Math. Hung.* 14 (1983), 101–105.

Mac Camy, R. C., V. J. Mizel, and T. Seidman
[1] Approximate boundary controllability of the heat equation, I, II. *J. Math. Anal. Appl.* 23 (1968), 699–703 and 28 (1969), 482–492.

Mac Nerney, J. S.
[1] Integral equations and semigroups, *Ill. J. Math.* 7 (1963), 148–173.

Magnenes, E. (see Lions)

Majda, A.
[1] Outgoing solutions for perturbations of $-\Delta$ with applications to spectral and scattering theory, *J. Diff. Eqns.* 16 (1974), 515–547.
[2] The location of the spectrum of the dissipative acoustic operator, *Indiana U. Math. J.* 25 (1976), 973–987.
Majewski, W. A.
[1] Remarks on positive semigroups, *Lett. Math. Phys.* 6 (1982), 437–440.
Majewski, A. and D. W. Robinson
[1] Strictly positive and strongly positive semigroups, *J. Austral. Math. Soc.* 34A (1983), 36–48.
Malhardeen, M. Z. M. (see Carr)
Malliavin, P. (see Devinatz)
Mandl, P.
[1] *Analytical Treatment of One-Dimensional Markov Processes*, Springer, Berlin, 1968.
Mandrekar, V. (see Kallianpur)
Manitius, A. (see Bernier, Delfour)
Marchand, J. -P. (see La Vita)
Martin, JR., R. H.
[1] Product integral apprximations of solutions to linear operator equations, *Proc. Amer. Math. Soc.* 41 (1973), 506–512.
[2] Invariant sets for perturbed semigroups of linear operators, *Annali Mat. Pura Appl.* 105 (1975), 221–239.
Masani, P.
[1] Ergodic theorems for locally integrable semigroups of continuous linear operators on a Banach space, *Adv. Math.* 21 (1976), 202–228.
[2] Multiplicative product integration and the Trotter product formula *Adv. Math.* 40 (1981), 1–9.
Massey, F. J. (see also Rauch)
[1] Abstract evolution equations and the mixed problem for symmetric hyperbolic systems, *Trans. Amer. Math. Soc.* 168 (1972), 165–188.
Masson, D. and W. K. McClary
[1] Classes of C^∞ vectors and essential self-adjointness, *J. Functional Anal.* 10 (1972), 19–32.
Masuda, K.
[1] On the holomorphic evolution operators, *J. Math. Anal. Appl.* 39 (1972), 706–711.
McClary, W. K. (see Masson)
McGrath, S. A.
[1] A pointwise abelian ergodic theorem for L_p semigroups, $1 \leq p < \infty$, *J. Functional Anal.* 23 (1976), 195–198.
[2] On the local ergodic theorems of Krengel, Kubokawa, and Terrell, *Comment. Math Univ. Carolinae* 17 (1976), 49–59.
[3] Local ergodic theorems for noncommuting semigroups. *Proc. Amer. Math. Soc.* 79 (1980), 212–216.
McIntosh, A.
[1] On the comparability of $A^{1/2}$ and $(A^*)^{1/2}$, *Proc. Amer. Math. Soc.* 32 (1972), 430–434.
[2] On representing closed accretive sesquilinear forms as $(A^{1/2}u, A^{*1/2}v)$. In *Nonlinear Partial Differential Equations and Their Applications*, *Vol.* 3 (ed. H. Brezis and J. L. Lions), Pitman, London (1982), pp. 252–267.
McKean Jr., H. P. (see also Itô)
[1] *Stochastic Integrals*, Academic Press, New York, 1969.
[2] $-\Delta$ plus a bad potential, *J. Math. Phys.* 18 (1977), 1277–1279.
McKelvey, R. W. (see Herod)
Medeiros, L. A.
[1] An application of semigroups of class C_0, *Port. Math.* 26 (1967), 71–77.
Menzala, G. P. (see Perla Menzala)
Meyer, P. A. (see also Dellacherie)
[1] *Probabilities and Potentials*, Blaisdell, Waltham, Mass. 1966.
[2] Transformations of Markov processes, *Jber. Deutsch. Math. Verein.* 74 (1972), 86–92.
Michel, A. N. and R. K. Miller
[1] *Qualitative Analysis of Large Scale Dynamical Systems*, Academic Press, New York, 1977.
Milgram, A. N. (see Lax)
Miller, K.
[1] Logarithmic convexity results for holomorphic semigroups, *Pac. J. Math.* 58 (1975), 549–551.

Miller, R. K. (see also Michel)
[1] Linear Volterra integrodifferential equations as semigroups, *Funkcial. Ekvac.* 17 (1974), 39–55.
[2] Volterra integral equations in a Banach space, *Funkcial Ekvac.* 18 (1975), 163–193.
[3] Well posedness of abstract Voterra problems. In *Lecture Notes in Math.* No. 737, Springer, Berlin (1979), pp. 192–205.
Miyadera, I. (see also Shimizu)
[1] On one-parameter semi-groups of operators, *J. Math. Tokyo* 1 (1952), 23–26.
[2] Generation of a strongly continuous semi-groups of operators, *Tôhoku Math. J.* 4 (2) (1952), 109–114.
[3] Semi-groups of operators in Fréchet space and applications to partial differential equations, *Tôhoku Math. J.* 11 (1959), 162–183.
[4] A note on contraction semi-groups of operators, *Tôhoku Math. J.* 11 (1959), 98–105.
[5] On perturbation theory for semi-groups of operators, *Tôhoku Math. J.* 18 (1966), 299–310.
[6] On the generation of semigroups of linear operators, *Tôhoku Math. J.* 24 (1972), 251–261.
Miyadera, I., S. Ôharu, and N. Okazawa
[1] Generation theorems for semigroups of linear operators, *Publ. R. I. M. S. Kyoto Univ.* 8 (1972/73), 509–555.
Mizel, V. J. (see Coffman, Coleman, MacCamy)
Mizohota, S.
[1] Le problème de Cauchy pour les équations paraboliques, *J. Math. Soc. Japan* 8 (1956), 269–299.
Mizohata, S. and K. Mochizuki
[1] On the principle of limiting amplitude for dissipative wave equations, *J. Math. Kyoto Univ.* 6 (1966), 109–127.
Mochizuki, K. (see also Mizohata)
[1] Decay and asymptotics for wave equations with dissipative term. In *Lecture Notes in Physics*, No. 39, Springer, New York (1975), pp. 486–490.
[1] Scattering theory for wave equations with dissipative terms, *Publ. RIMS, Kyoto Univ.* 12 (1976), 383–390.
Mokobodzki, G.
[1] Sur l'algèbre continue dans le domaine éntendu d'un générateur infinitésimaux. In *Lecture Notes in Math.* No. 681, Springer, Berlin (1978), 168–187.
Montgomery, J. (see Lin)
Moore, R. T. (see also Jørgensen)
[1] Duality methods and perturbation of semigroups, *Bull. Amer. Math. Soc.* 73 (1967), 548–553.
[2] Measurable, Continuous and Smooth Vectors for Semigroups and Groups Representations, *Mem. Amer. Math. Soc.* No. 78, (1968), 1–80.
[3] Generation of equicontinuous semigroups by hermitian and sectorial operators, I and II, *Bull. Amer. Math. Soc.* 77 (1971), 224–229 and 368–373.
Morimoto, H. (see Fujita)
Morton, K. W. (see Richtmeyer)
Mosco, U. (see Da Prato)
Mourre, E. (see also Jensen)
[1] Link between the geometrical and the spectral transformation approaches in scattering theory, *Comm. Math. Phys.* 68 (1979), 91–94.
[2] Absence of singular continuous spectrum for certain self-adjoint operators, *Comm. Math. Phys.* 78 (1981), 391–408.
[3] Algebraic approach to some propagation properties of the Schrödinger equation. In *Mathematical Problems of Theoretical Physics* (eds. R. Schrader, E. Seiler, and P. A. Uhlenbrock), Lecture Notes in Physics No. 153, Springer, Berlin (1981).
[4] Operateurs conjugués et propriétés de propagation, *Comm. Math. Phys.* 91 (1983), 279–300.
Moyal, J. E.
[1] Mean ergodic theorems in quantum mechanics, *J. Math. Phys.* 10 (1969), 506–509.
[2] Particle populations and number operators in quantum theory, *Adv. Appl. Prob.* 4 (1972), 39–80.
Mueller, C. E. and F. B. Weissler
[1] Hypercontractivity for the heat semigroup for ultraspherical polynomials and on the n-sphere, *J. Functional Anal.* 48 (1982), 252–283.
Muhly, P. S.
[1] Toeplitz operators and semigroups, *J. Math. Anal. Appl.* 38 (1972), 312–319.
Nagasawa, M. (see Ikeda)
Nagel, R. (see also Derndinger, Greiner, Kerscher)
[1] Mittelergodische Halbgruppen Linearer Operatoren, *Ann. Inst. Fourier* 23 (1973), 75–87.

[2] Les semi-groupes d'operateurs positifs et leur generteurs. In *Initiation Seminar on Analysis*: G. Choquet-M. Rogalski-J. St. Raimond, Univ. Paris VI (1981), 10 pp.

[3] Zur charakterisierung stabiler Operatorhalbgruppen, *Semesterbericht Funktionalanalysis*, Univ. Tübingen (1981/82), 99–119.

[4] Sobolev spaces and semigroups. In *Semesterbericht Funktionalanalysis*, Tübingen (1983), pp. 1–19.

Nagel, R. and H. Uhlig

[1] An abstract Kato inequality for generators of positive operator semigroups on Banach lattices, *J. Oper. Th.* 6 (1981), 113–123.

Nagumo, M.

[1] Einige analytische Untersuchungen in linearen metrischen Ringen, *Jap. J. Math.* 13 (1936), 61–80.

[2] Perturbation and degeneration of evolutional equations in Banach spaces, *Osaka Math. J.* 15 (1963), 1–10.

Nagy, B.

[1] On the generators of cosine operator functions, *Publ. Math.* 21 (1974), 151–154.

[2] On cosine operator functions in Banach spaces, *Acta Sci. Math.* 36 (1974), 281–289.

[3] Spectral mapping theorems for semigroups of operators, *Acta Sci. Math.* 38 (1976), 343–351.

[4] Cosine operator functions and the abstract Cauchy problem, *Periodica Math. Hung.* 7 (1976), 15–18.

[5] Approximation theorems for cosine operator functions, *Acta Math. Acad. Sci. Hung.* 29 (1977), 69–76.

Nagy, B. Sz.-(see Sz.-Nagy)

Najman, B.

[1] Solution of a differential equation in a scale of spaces, *Glasnik Mat.* 34 (1979), 119–127.

[2] Trace class perturbations and scattering theory for equations of Klein-Gordon type, *Glasnik Mat.* 15 (1980), 79–86.

[3] Spectral properties of the operators of Klein-Gordon type, *Glasnik Mat.* 15 (1980), 97–112.

[4] Scattering theory for matrix operators, I, II, *Glasnik Mat.* 17 (1982), 97–110 and 285–302.

[5] Eigenvalues of the Klein-Gordon equation, *Proc. Roy. Soc. Edinburgh* 26 (1983), 181–190.

[6] The rate of convergence in singular perturbations of parabolic equations, in *Lecture Notes in Math.* No. 1076, Springer, Berlin (1984), 147–167.

Narnhofer, H.

[1] Self-adjoint operators, derivations and automorphisms on C^*-algebras, *J. Math. Phys.* 16 (1975), 2192–2196.

[2] Does there exist a scattering theory for time automorphism groups of C^*-algebras corresponding to two-body interactions?, *J. Math. Phys.* 20 (1979), 2502–2505.

Nathan, D. S.

[1] One-parameter groups of transformations in abstract vector spaces, *Duke Math. J.* 1 (1935), 518–526.

Navarro, C. B.

[1] Brief survey of semigroup theory and its applications to evolution problems, *Stochastics* 2 (1978), 1–34.

Nelson, E.

[1] The adjoint Markoff process, *Duke Math. J.* 25 (1958), 671–690.

[2] An existence theorem for second order parabolic equations, *Trans. Amer. Math Soc.* 88 (1958), 414–429.

[3] Representation of a Markovian semigroup and its infinitesimal generator, *J. Math. Mech.* 7 (1958), 977–987.

[4] Analytic vectors, *Ann Math.* 70 (1959), 572–615.

[5] Feynman integrals and the Schrödinger equation, *J. Math. Phys.* 5 (1964), 332–343.

[6] *Operator Differential Equations*, Lecture Notes (taken by John T. Cannon), Princeton University, 1965.

[7] *Dynamical Theories of Brownian Motion*, Princeton U. Press, Princeton, N. J., 1967.

[8] *Topics in Dynamics I: Flows*, Princeton U. Press, Princeton, N. J., 1969.

[9] Time-ordered operator products of sharp-time quadratic forms, *J. Functional Anal.* 11 (1972), 211–219.

Nelson, S. and R. Triggiani

[1] Analytic properties of cosine operators, *Proc. Amer. Math. Soc.* 74 (1979), 101–104.

Nenciu, G. (see Angelescu)

Nessel, R. J. (see Butzer, Dickmeis)

Neuberger, J. W.
[1] A quasi-analyticity condition in terms of finite differences, *Proc. London Math. Soc.* 14 (1964), 245–259.
[2] Analyticity and quasi-analyticity for one parameter semigroups, *Proc. Amer. Math Soc.* 25 (1970), 488–494.
[3] Lie generators for strongly continuous one parameter semigroups on a metric space, *Indiana U. Math. J.* 21 (1972), 961–971.
[4] Quasi-analyticity and semigroups, *Bull. Amer. Math. Soc.* 78 (1972), 909–922.
[5] Lie generators for one parameter semigroups of transformations, *J. Reine Angew. Math.* 258 (1973), 133–136.
[6] Quasi-analytic semigroups of bounded linear transformations, *J. London Math. Soc.* 7 (1973), 259–264.
Neubrander, F. (see also Groh)
[1] Laplace transform and asymptotic behavior of strongly continuous semigroups, *Semesterbericht Funktionalanalysis Tübingen* (1982/83), 139–161.
[2] Well-posedness of abstract Cauchy problems, *Semigroup Forum* 29 (1984), 74–85.
[3] *Well-posedness of Higher Order Abstract Cauchy Problems*, Ph.D. Thesis, Univ. Tübingen, 1984.
Von Neumann, J.
[1] Über die analytische Eigenschaften von Gruppen linearer Transformationen ihrer Darstellungen, *Math. Z.* 30 (1939), 3–42.
Neveu, J.
[1] Théorie des semi-groupes de Markov, *Univ. Calif. Publ. Statistics*, 2 (1958), 319–394.
Nirenberg, L. (see also Agmon)
[1] Remarks on strongly elliptic partial differential equations, *Comm. Pure Appl. Math.* 8 (1955), 648–674.
Norman, M. F. (see also Ethier)
[1] *Markov Processes and Learning Models*, Academic Press, New York, 1972.
[2] Approximation of stochastic processes by Gaussian diffusions, and applications to Wright-Fisher genetic models, *SIAM J. Appl. Math.* 29 (1975), 225–242.
[3] Diffusion approximation of non-Markovian processes, *Ann Prob.* 3 (1975), 358–364.
[4] A "psychological" proof that certain Markov semigroups preserve differentiability. In *Mathematical Psychology and Physiology* (ed. S. Crossberg), *Amer. Math. Soc.*, *Providence, R. I.* (1981), pp. 197–211.
Nur, H.
[1] Singular perturbations of differential equations in abstract spaces, *Pac. J. Math.* 36 (1971), 775–780.
Nussbaum, A. E.
[1] Integral representation of semigroups of unbounded self-adjoint operators, *Ann. Math.* 69 (1959), 133–141.
[2] Quasi-analytic vectors, *Ark. Mat.* 6 (1965), 179–191.
[3] A note on quasi-analytic vectors, *Studia Math.* 33 (1969), 305–309.
[4] Semigroups of subnormal operators, *J. London Math. Soc.* 14 (1976), 340–344.
[5] Multi-parameter local semi-groups of Hermitian operators. To appear.
O'Brien, Jr., R. E.
[1] Contraction semigroups, stabilization, and the mean ergodic theorem, *Proc. Amer. Math. Soc.* 71 (1978), 89–94.
[2] Weakly stable operators, *J. Math Anal. Appl.* 70 (1979), 170–179.
O'Carroll, M. (see also Iorio)
[1] On the inverse scattering problem for linear evolution equations. In *Contemporary Developments in Continuum Mechanics and Partial Differential Equations* (ed. G. M. de la Penha and L. A. Medeiros), North Holland, Amsterdam (1978), pp. 102–111.
Ôharu, S. (see also Miyadera)
[1] On the convergence of semigroups of operators, *Proc. Japan Acad.* 42 (1966), 880–884.
[2] Semigroups of linear operators in a Banach space, *Publ. R.I.M.S. Kyoto Univ.* 7 (1971), 205–260.
[3] The embedding problem for operator groups, *Proc. Japan Acad* 52 (1976), 106–108.
Ôharu, S. and H. Sunouchi
[1] On the convergence of semigroups of linear operators, *J. Functional Anal.* 6 (1970), 292–304.
Okazawa, N. (see also Miyadera, Takenaka)
[1] Operator semigroups of class (D_n), *Math. Japan.* 18 (1973), 33–51.
[2] Perturbations of linear *m*-accretive operators, *Proc. Amer. Math. Soc.* 37 (1973), 169–174.
[3] A generation theorem for semigroups of growth order α, *Tôhoku Math. J.* 26 (1974), 39–51.

[4] Remarks on linear *m*-accretive operators in a Hilbert space, *J. Math. Soc. Japan* 27 (1975), 160–165.

[5] Approximation of linear *m*-accretive operators in a Hilbert space, *Osaka J. Math.* 14 (1977), 85–94.

[6] Singular perturbations of *m*-accretive operators, *J. Math. Soc. Japan* 32 (1980), 19–44.

[7] On the perturbation of linear operators in Banach and Hilbert spaces, *J. Math. Soc. Japan* 34 (1982), 677–701.

Olubummo, A.

[1] A note on perturbation theory for semi-groups of operators, *Proc. Amer. Math. Soc.* 16 (1964), 818–822.

[2] Semigroups of multipliers associated with semigroups of operators, *Proc. Amer. Math. Soc.* 41 (1975), 161–168.

Olubummo, A. and R. S. Phillips

[1] Dissipative ordinary differential operators, *J. Math. Mech.* 14 (1965), 929–950.

Orey, S.

[1] *Probabilistic Methods in Partial Differential Equations*, U. of Minnesota Lecture Notes, 1978.

Ōuchi, S.

[1] Semigroups of operators in locally convex spaces, *J. Math. Soc. Japan* 25 (1973), 265–276.

Packel, E. W.

[1] A semi-group analogue of Foguel's counterexample, *Proc. Amer. Math. Soc.* 21 (1969), 240–244.

[2] A simplification of Gibson's theorem on discrete operator semigroups, *J. Math. Anal. Appl.* 39 (1972), 586–589.

Panchapagesan, T. V.

[1] Semigroups of scalar type operators in Banach spaces, *Pac. J. Math.* 30 (1969), 489–517.

Papanicolaou, G. C. (see also Hersh)

[1] Some problems and methods for the analysis of stochastic equations. In *Proc. Symp. Appl. Math.*, Vol. 6, Amer. Math. Soc. Providence, R. I. (1973), pp. 21–33.

[2] Stochastic equations and their applications, *Amer. Math. Monthly* 80 (1973), 526–544.

[3] Asymptotic analysis of transport processes, *Bull. Amer. Math. Soc.* 81 (1975), 330–392.

[4] Asymptotic analysis of stochastic equations. In *Studies in Probability Theory* (ed. M. Rosenblatt), Math. Assoc. Amer., Washington, D. C. (1978), pp. 111–179.

Papanicolaou, G. C. and R. Hersh

[1] Some limit theorems for stochastic equations and their applications, *Indiana U. Math. J.* 21 (1972), 815–840.

Papanicolaou, G. C. and S. R. S. Varadhan

[1] A limit theorem with strong mixing in Banach space and two applications to stochastic differential equations, *Comm. Pure Appl. Math.* 26 (1973), 497–523.

Paquet, L. (see also Lumer)

[1] Semi-groupes holomorphes en norme de sup. In *Lecture Notes in Math.*, No. 713, Springer, Berlin (1979), pp. 194–242.

[2] Semi-groupes généralisés et équations d'évolution. In *Lecture Notes in Math.*, No. 713, Springer, Berlin (1979), pp. 243–263.

[3] Opérateurs elliptiques sur les varietés non compactes, *J. Functional Anal.* 50 (1983), 267–284.

Parker, G. E.

[1] Semigroup structure underlying evolutions, *Internat. J. Math. Math. Sci.* 5 (1982), 31–40.

Parthasarathy, K. R. (see also Hudson)

Parthasarathy, K. R. and K. B. Sinha

[1] A random Trotter product formula. In *Statistics and Probability: Essays in Honor of C. R. Rao* (1982), pp. 553–565.

[2] Feynman path integrals of operator-valued maps, *J. Math. Phys.* 23 (1982), 1459–1462.

Partington, J. R.

[1] Hadamard-Landau inequalities in uniformly convex spaces, *Math. Proc. Camb. Phil. Soc.* 90 (1981), 259–264.

Partzsch, L. (see Langer)

Pawelke, S. (see Butzer)

Payne, L.

[1] *Improperly Posed Problems in Partial Differential Equations*, SIAM, Philadelphia, Pa., 1975.

Pazy, A. (see also Crandall)

[1] On the differentiability and compactness of semi-groups of linear operators, *J. Math. Mech.* 17 (1968), 1131–1142.

[2] Approximation of the identity operator by semi-groups of linear operators, *Proc. Amer. Math. Soc.* 30 (1971), 147–150.

[3] On the applicability of Lyapunov's theorem in Hilbert space, *SIAM. J. Math. Anal.* 3 (1972), 291–294.

[4] *Semi-groups of Linear Operators and Applications to Partial Differential Equations.* Univ. of Maryland Lecture Notes # 10, College Park, Md., 1974.

[5] *Semigroups of Linear Operators and Applications to Partial Differential Equations*, Springer, New York, 1983.

Peano, G.
[1] Intégration par séries des équations différentielles linéaires, *Math. Ann.* 32 (1888), 450–456. An earlier version appeared in *Atti Accad. Sci. Torino*, 20 February, 1887.

Pearson, D. B.
[1] Time-dependent scattering theory for highly singular potentials, *Helv. Phys. Acta* 47 (1974), 249–264.

[2] A generalization of the Birman trace theorem, *J. Functional Anal.* 28 (1978), 182–186.

[3] Singular continuous measures in scattering theory, *Comm. Math. Phys.* 60 (1978), 13–36.

Peetre, J.
[1] Sur la théorie des semi-groupes distributions, *Séminaire sure les Équations aux Derivées Partielles*, College de France, 1963–1964.

Perla Menzala, G.
[1] On inverse scattering for the Klein-Gordon equation with small potentials, *Funk. Ekvac.* 20 (1977), 61–70.

[2] On inverse scattering for the wave equation with a potential term via the Lax-Phillips theory: A simple proof, *J. Diff. Eqns.* 30 (1978), 41–48.

Perry, P. (see also Hagedorn, Jensen)
[1] *Scattering Theory by the Enss Method*, Harwood , New York, 1983.

Pfeifer, D.
[1] On a general probabilistic representation formula for semigroups of operators, *J. Math. Res. Exp.* 2 (1982), 93–98.

[2] On a probabilistic representation theorem of operator semigroups with bounded generator, *J. Math. Res. Exp.* 3 (1983).

[3] A semigroup-theoretic proof of Poisson's limit law, *Semigroup Forum* 26 (1983), 379–382.

[4] A note on probabilistic representations of operator semigroups, *Semigroup Forum* 28 (1984), 335–340.

Phillips, R. S. (see also Crandall, Hille, Lax, Lumer, Olubummo)
[1] Spectral theory for semigroups of linear operators, *Trans. Amer. Math. Soc.* 71 (1951), 393–415.

[2] On the generation of semi-groups of linear transformations, *Proc. Amer. Math. Soc.* 2 (1951), 234–237.

[3] On the generation of semi-groups of linear operators, *Pac. J. Math.* 2 (1952), 343–369.

[4] Perturbation theory for semi-groups of linear operators, *Trans. Amer. Math. Soc.* 74 (1953), 199–221.

[5] A note on the abstract Cauchy problem, *Proc. Nat. Acad. Sci. U.S.A.* 40 (1954), 244–248.

[6] Dissipative hyperbolic systems, *Trans. Amer. Math. Soc.* 86 (1957), 109–173.

[7] Dissipative operators and parabolic partial differential equations, *Comm. Pure Appl. Math.* 12 (1959), 249–276.

[8] Dissipative operators and hyperbolic systems of partial differential equations, *Trans. Amer. Math. Soc.* 90 (1959), 193–254.

[9] Semigroups of positive contraction operators, *Czech. Math. J.* 12 (1962), 294–313.

[10] Semi-groups of contraction operators. In *Equazioni Differenziale Astratte*, C.I.M.E., Rome, 1963.

[11] On dissipative operators. In *Lecture Series in Differential Equations*, Vol. 2 (ed' A' K. Aziz), Van Nostrand Rheinhold, New York (1969), pp. 65–119.

[12] Scattering theory for the wave equation with a short range perturbation, *Indiana U. Math. J.* 31 (1982), 609–639.

Piech, M. A. (see also Kuo)
[1] A fundamental solution of the parabolic equation on Hilbert space II: The semigroup property, *Trans. Amer. Math. Soc.* 150 (1970), 257–286.

[2] A product decomposition of the fundamental solution of a second order parabolic equation, *J. Diff. Eqns* 9 (1971), 443–452.

[3] Diffusion semigroups on abstract Wiener space, *Trans. Amer. Math. Soc.* 166 (1972), 411–430.

[4] Parabolic equations associated with the number operator, *Trans. Amer. Math. Soc.* 194 (1974), 213–222.

[5] The Ornstein-Uhlenbeck semigroup in an infinite dimensional L^2 setting, *J. Functional Anal.* 18 (1975), 271–285.

Pierre, M. (see Kurtz)

Pinsky, M. (see also Ellis, Hersh)

[1] Differential equations with a small parameter and the central limit theorem for functions defined on a finite Markov Chain, *Z. Wersch, verw. Geb.* 9 (1968), 101–111.

[2] Multiplicative operator functionals and their asymptotic properties. In *Advances in Probability*, Vol. 3, Marcel Dekker, New York (1974), pp. 1–100.

Piraux, R. (see Bivar-Wienholtz)

Pisier, G.

[1] Holomorphic semigroups and the geometry of Banach spaces, *Ann. Math.* 115 (1982), 375–392.

Pitt, L.

[1] Products of Markovian semi-groups of operators, *Z. Warsch. verw. Geb.* 12 (1969), 246–254.

Plamenevskiĭ, B. A.

[1] On the existence and asymptotics of solutions of differential equations with unbounded operator coefficients in a Banach space, *Math. USSR Isvestia* 6 (1972), 1327–1379.

Polichka, A. E. and P. E. Sobolevskiĭ

[1] A method for the approximate solution of Cauchy's problems for differential equations in Banach space with unbounded variable coefficients, *Diff. Eqns.* 12 (1977), 1191–1199.

Ponomarev, S. M.

[1] On convergence of semigroups, *Soviet Math. Dokl.* 13 (1972), 603–605.

Pontzen, D. (see Görlich)

Porta, H. (see Berkson)

Poulsen, E. T.

[1] Evolutionsgleichungen in Banach Räumen, *Math. Z.* 90 (1965), 286–309.

Powers, R. T. and C. Radin

[1] Average boundary conditions in Cauchy problems, *J. Functional Anal.* 23 (1976), 23–32.

Prato, G. Da (see Da Prato)

Priestly, W. M.

[1] C^2-preserving strongly continuous Markovian semigroups, *Trans. Amer. Math. Soc.* 180 (1973), 359–365.

Priouret, P. (see Bony)

Prugovečki, E.

[1] *Quantum Mechanics in Hilbert Space*, Academic Press, New York, 1971.

Prugovečki, E. and A. Tip

[1] Semi-groups of rank-preserving transformers on minimal norm ideals in $B(H)$, *Composito Math.* 30 (1975), 113–136.

Prüss, J.

[1] On semilinear evolution equations in Banach spaces, *J. Reine Angew. Math.* 303/304 (1978), 144–158.

Przebinda, T.

[1] Holomorphicity of a class of semigroups of measures operating on $L^p(G/H)$, *Proc. Amer. Math. Soc.* 87 (1983), 637–643.

Quiring, D.

[1] Random evolutions on diffusion processes, *Z. Warsch. verw. Geb.* 23 (1972), 230–244.

Radin, C. (see also Goldstein, Powers)

Radin, C. and B. Simon

[1] Invariant domains for the time-dependent Schrödinger equation, *J. Diff. Eqns.* 29 (1978), 289–296.

Radnitz, A. (see Fattorini)

Rankin III, S. M.

[1] A remark on cosine families, *Proc. Amer. Math. Soc.* 79 (1980), 376–378.

Rao, M.

[1] *Brownian Motion and Classical Potential Theory*, Lecture Note Series No. 42, Aarhus Univ., Aarhus, Denmark, 1977.

Rao, M. M.

[1] Inference in stochastic processes-V, Admissible means, *Sankhyà* 37 (1975), 538–549.

[2] *Stochastic Processes and Integration*, Sythoff and Noordhoff, Alphen aan den Rijn, The Netherlands, 1979.

Rauch, J.
[1] Energy and resolvent inequalities for hyperbolic mixed problems, *J. Diff. Eqns.* 11 (1972), 528–540.
[2] Qualitative behavior of dissipative wave equations in bounded domains, *Arch. Rat. Mech. Anal.* 62 (1976), 77–85.
Rauch, J. and F. J. Massey III
[1] Differentiability of solutions to hyperbolic initial-boundary value problems, *Trans. Amer. Math. Soc.* 189 (1974), 303–318.
Rauch, J. and M. Taylor
[1] Essential self-adjointness of powers of generators of hyperbolic mixed problems, *J. Functional Anal.* 12 (1973), 491–493.
[2] Exponential decay of solutions to hyperbolic equations in bounded domains, *Indiana U. Math. J.* 24 (1974), 79–86.
[3] Potential and scattering theory on wildly perturbed domains, *J. Functional Anal.* 18 (1975), 27–59.
[4] Decaying states of perturbed wave equations, *J. Math. Anal. Appl.* 54 (1976), 279–285.
Ray, D. B.
[1] On spectra of second-order differential operators, *Trans. Amer. Math. Soc.* 77 (1954), 299–321.
[2] Resolvents, transition functions and strongly Markovian processes, *Ann. of Math.* 70 (1959), 43–72.
Reed, M. and B. Simon
[1] *Methods of Modern Mathematical Physics, Vol. I: Functional Analysis*, Academic Press, New York, 1972.
[2] *Methods of Modern Mathematical Physics, Vol. II: Fourier Analysis and Self-adjointness*, Academic Press, New York, 1975.
[3] *Methods of Modern Mathematical Physics, Vol. III: Scattering Theory*, Academic Press, New York, 1979.
[4] *Methods of Mathematical Physics, Vol. IV; Analysis of Operators*, Academic Press, New York, 1978.
Reuter, G. E. H.
[1] A note on contraction semi-groups, *Math. Scand.* 3 (1955), 275–280.
[2] Denumerable Markov processes and the associated contraction semigroups on *l*, *Acta Math.* 97 (1957). 1–46.
[3] Denumerable Markov processes (II), *J. London Math. Soc.* 34 (1959), 81–91.
[4] Note on resolvents of denumerable submarkovian processes, *Z. Warach. verw. Geb.* 9 (1967), 16–19.
Revuz, D.
[1] Measures associees aux fonctionelles additives de Markov. I, *Trans. Amer. Math. Soc.* 148 (1970), 501–531.
[2] Sur la théorie du potentiel pour les processus de Markov récurrent, *Ann. Inst. Fourier* 21 (1972), 245–262.
[3] Sur les opérateurs potentiels au sens de Yosida, *Z. Warsch. verw. Geb.* 25 (1973). 199–203.
Richtmeyer, R. D. (see also Lax)
[1] *Principles of Advanced Mathematical Physics*, Vols. 1, 2, Springer, New York, 1978.
Richtmeyer, R. D. and K. W. Morton
[1] *Difference Methods for Initial Value Problems*, 2nd ed., Interscience, New York, 1967.
Riesz, F. and B. Sz.-Nagy
[1] *Functional Analysis* (English translation), F. Ungar, New York, 1955.
Rigby, L. (see Levan)
Robinson, D. W. (see also Batty, Bratelli, Kishimoto, Majewski)
[1] *The Thermodynamic Pressure in Quantum Statistical Mechanics*, Lecture Notes in Physics No. 9, Springer Berlin, 1971.
[2] Scattering theory with singular potentials. I. The two-body problem, *Ann. Inst. Henri Poincaré* 21 (1974), 185–215.
[3] The approximation of flows, *J. Functional Anal.* 24 (1977), 280–290.
[4] Propagation properties in scattering theory, *J. Austral Math. Soc.* 21B (1980), 474–485.
[5] Strongly positive semigroups and faithful invariant states, *Comm. Math. Phys.* 85 (1982), 129–142.
Rocca, F. (see Fannes)
Romanoff, N. P.
[1] On one parameter operator groups of linear transformations I, *Ann. Math.* 48 (1947), 216–233.

Rosencrans, S. I. (see also Beale, Goldstein)
[1] On Schwarzschild's criterion, *SIAM J. Appl. Math.* 17 (1969), 231–239.
[2] Diffusion transforms, *J. Diff. Eqns.* 13 (1973), 457–467.
[3] *Diffusion Processes*, Tulane Univ. Lecture Notes, New Orleans, La., 1978.
Rosenkrantz, W. A. (see also Brezis, Ellis)
[1] A convergent family of diffusion processes whose diffusion coefficients diverge, *Bull. Amer. Math. Soc.* 86 (1974), 973–976.
[2] An application of the Hille-Yosida theorem to the construction of martingales, *Indiana U. Math. J.* 24 (1974), 527–532.
[3] Limit theorems for solutions to a class of stochastic differential equations, *Indiana U. Math. J.* 24 (1975), 613–625.
[4] A strong continuity theorem for a class of semigroups of type Γ, with an application to martingales, *Indiana U. Math. J.* 25 (1976), 171–178.
Rosenkranz, W. A. and C. C. Y. Dorea
[1] Limit theorems for Markov processes via a variant of the Trotter-Kato theorem, *J. Appl. Prob.* 17 (1980), 704–715.
Rota, G. -C. (see also Kallman)
[1] An "alternierende verfahren" for general positive operators, *Bull. Amer. Math. Soc.* 68 (1962), 95–102.
[2] A limit theorem for the time-dependent evolution equation. In *Equazioni Differenziale Astratte*, C.I.M.E., Rome, 1963.
Roth, J. -P. (see also Hirsch)
[1] Approximation des opérateurs dissipatifs, *C. R. Acad. Sci. Paris* 276 (1973), 1285–1287.
[2] Opérateurs dissipatifs et semi-groupes dans les espaces de fonctions continuos, *Ann. Inst. Fourier* 26 (1976), 1–97.
[3] Les operateurs elliptiques comme générateurs infinitésimaux de semi-groups de Feller. In *Lecture Notes in Math.* No. 681, Springer, Berlin (1978), 234–251.
[4] Semi-groupes de mesures matricielles. In *Lecture Notes in Math.* No. 706, Springer, Berlin (1979), pp. 336–343.
[5] Recollement de semi-groupes de Feller locaux, *Ann. Inst. Fourier*, 30 (1980), 75–89.
Roth, W. J.
[1] Goursat problems for $u_{rs} = Lu$, *Indiana U. Math. J.* 22 (1973), 779–788.
Rubin, H. (see Bharucha-Reid)
Rudin, W.
[1] *Fourier Analysis on Groups*, Interscience, New York, 1962.
Ruelle, D.
[1] A remark on bound states in potential scattering theory, *Nuovo Cimento* 61A (1969), 655–662.
Russell, D. L.
[1] Controllability and stabilizability theory for linear partial differential equations: Recent progress and open questions, *SIAM Rev.* 20 (1978), 639–739.
Sakai, S.
[1] One one-parameter semigroups of *-automorphisms on operator algebras and their corresponding unbounded derivations, *Amer. J. Math.* 98 (1976), 427–440.
Salamon, D.
[1] *On Control and Observation of Neutral Systems*, Ph.D. Thesis, Univ. of Bremen, 1982.
Sandefur, Jr., J. T. (see also Goldstein)
[1] Higher order abstract Cauchy problems, *J. Math. Anal. Appl.* 60 (1977), 728–742.
[2] Asymptotic equipartition of energy for equations of elasticity, *Math. Meth Appl. Sci.* 5 (1983), 186–194.
Saneteka, N.
[1] A note on the abstract Cauchy problem in a Banach space, *Proc. Japan Acad.* 49 (1973), 510–513.
Sato, K. (see also Gustafson)
[1] On the generators of non-negative contraction semi-groups in Banach lattices, *J. Math. Soc. Japan* 20 (1968), 423–436.
[2] Lévy measures for a class of Markov semigroups in one dimension, *Trans. Amer. Math. Soc.* 148 (1970), 211–231.
[3] On dispersive operators in Banach lattices, *Pac. J. Math.* 33 (1970), 429–443.
[4] Positive pseudo-resolvents in Banach lattices, *J. Fac. Sci. Univ. Tokyo* 17 (1970), 305–313.
[5] A note on infinitesimal generators and potentials operators of contraction semigroups, *Proc. Japan Acad.* 48 (1972), 450–453.

[6] Potential operators for Markov processes, *Proc. Sixth Berkeley Symp. Math. Stat. Prob.* 3 (1972), 193–211.
[7] Core of potential operators for processes with stationary independent increments, *Nagoya Math. J.* 48 (1972), 129–145.
[8] Convergence to a diffusion of a multi-allelic model in population genetics, *Adv. Appl. Prob.* 10 (1978), 538–562.

Sato, K. and T. Ueno
[1] Multi-dimensional diffusion and the Markov process on the boundary, *J. Math. Kyoto Univ.* 4 (1965), 529–605.

Sato, R. (see also Hasegawa)
[1] Invariant measures for ergodic semigroups of operators, *Pac. J. Math.* 71 (1977), 173–192.
[2] Ergodic theorems for semigroups of positive operators, *J. Math. Soc. Japan* 29 (1977), 591–606.
[3] On local ergodic theorems for positive semigroups, *Studia Math.* 63 (1978), 45–55.
[4] Positive operators and the ergodic theorem, *Pac. J. Math.* 76 (1978), 215–219.
[5] Contraction semigroups in Lebesgue space, *Pac. J. Math.* 78 (1978), 251–259.
[6] On an individual ergodic theorem, *Math. J. Okayama Univ.* 24 (1982), 153–156.

Sauer, N.
[1] Linear evolution equations in two Banach spaces, *Proc. Roy. Soc. Edinburgh* 91A (1982), 287–303.

Sawyer, S. A.
[1] A formula for semigroups, with an application to branching diffusion processes, *Trans. Amer. Math. Soc.* 152 (1970), 1–38.

Schaefer, H. H.
[1] *Banach Lattices and Positive Operators*, Springer, Berlin, 1974.
[2] On the spectral bound of irreducible semi-groups. In *Semesterbericht Funktionalanalysis*, Tübingen (1983), pp. 21–28.

Schappacher, W. (see Desch, Grimmer, Kunisch)

Schechter, M. (see also Bers, Koller)
[1] *Spectra of Partial Differential Operators*, North Holland, Amsterdam, 1971.
[2] Scattering theory for elliptic operators of arbitrary order, *Comment. Math. Helv.* 49 (1974), 84–113.
[3] A unified approach to scattering, *J. Math. Pures Appl.* 53 (1974), 373–396.
[4] Scattering theory for second order elliptic operators, *Ann. Mat. Pura Appl.* 105 (1975), 313–331.
[5] A new criterion for scattering theory, *Duke Math. J.* 44 (1977), 863–872.
[6] Scattering in two Hilbert spaces, *J. London Math. Soc.* 19 (1979), 175–186.
[7] The invariance principle, *Comment. Math. Helv.* 54 (1979), 111–125.

Schechter, M. and R. A. Weder
[1] The Schrödinger operator with magnetic vector potential, *Comm. Partial Diff. Eqns.* 2 (1977), 549–561.

Schenk, W. (see Langer)

Schoene, A.
[1] Semigroups and a class of singular perturbation problems, *Indiana U. Math. J.* 20 (1970), 247–264.
[2] On the nonrelativistic limits of the Klein-Gordon and Dirac equations, *J. Math. Anal. Appl.* 71 (1979), 36–47.

Schonbek, T. P.
[1] Notes to a paper by C. N. Friedman, *J. Functional Anal.* 14 (1973), 281–294.

Schrader, R. (see Hess)

Schulenberger, J. R. (see also Gilliam)
[1] A local compactness theorem for wave propagation problems of mathematical physics, *Indiana U. Math. J.* 22 (1972), 429–433.
[2] On conservative boundary conditions for operators of constant deficit: The Maxwell operator, *J. Math. Anal. Appl.* 48 (1974), 223–249.

Schulenberger, J. R. and C. H. Wilcox
[1] Completeness of the wave operators for perturbations of uniformly propagative systems, *J. Functional Anal.* 7 (1971), 447–474.
[2] A coerciviness inequality for a class of nonelliptic operators of constant deficit, *Ann. Mat. Pura Appl.* 92 (1972), 77–84.

Schütze, D. (see Langer)

Schwartz, J. T. (see also Dunford)
[1] Some non-self-adjoint operators. *Comm. Pure Appl. Math.* 13 (1960), 609–639.

Schwartz, L.
[1] *Lecture on Mixed Problems in Partial Differential Equations and the Representation of Semi-groups*, Tata Institute Lecture Notes, Bombay, 1958.
[2] Processus de Markov et *désintégrations régulières*, *Ann. Inst. Fourier, Grenoble* 27 (1977), 211–277.
Segal, I. E. (see also Dunford, Hughes)
[1] Non-linear semi-groups, *Ann. of Math.* 78 (1963), 339–364.
[2] Singular perturbation of semigroup generators. In *Lineare Operatoren und Approximation* (ed. P. L. Butzer, J. -P. Kahane, and B. Sz-Nagy), Birkhauser, Basel (1972), 54–61.
Seidman, T. (see MacCamy)
Semenov, Yu. A. (see also Belyi)
[1] Schrödinger operators with L^p_{loc}-potentials, *Comm. Math. Phys.* 53 (1977), 277–284.
[2] On the Lie-Trotter theorems in L^p-spaces, *Lett. Math. Phys.* 1 (1977), 379–385.
[3] Wave operators for the Schrödinger equation with strongly singular short range potentials, *Lett. Math. Phys.* 1 (1977), 457–461.
[4] On the problem of convergence of a bounded-below sequence of symmetric forms for the Schrödinger operator, *Studia Math.* 64 (1979), 77–85.
Sentilles, F. D.
[1] Semigroups of operators in $C(S)$, *Can. J. Math.* 22 (1970), 47–54.
Shaw, S. -Y. (see also Lin)
[1] Ergodic projections of continuous and discrete semigroups, *Proc. Amer. Math. Soc.* 78 (1980), 69–76.
[2] Ergodic limits of tensor product semigroups, *J. Math. Anal. Appl.* 76 (1980), 432–439.
[3] Ergodic properties of operator semigroups in general weak topologies, *J. Functional Anal.* 49 (1982), 152–169.
Shaw, S. -Y. and C. S Lin
[1] A sufficient condition for convergence of $\lim_{t\to\pm\infty} t^{-2}\int_0^t \exp(-sT)A\exp(sS)ds$ for unbounded operators S, *Bull. Inst. Math. Acad. Sinica* 8 (1980), 401–406.
Shimizu, M. and I. Miyadera
[1] Perturbation theory for cosine families on Banach spaces, *Tokyo J. Math.* 1 (1978), 333–343.
Showalter, R. E. (see also Carroll, Goldstein)
[1] Partial differential equations of Sobolev-Galpern type, *Pac. J. Math.* 31 (1969), 787–793.
[2] Well-posed problems for a partial differential equation of order $2m + 1$, *SIAM J. Math. Anal.* 1 (1970), 214–232.
[3] Local regularity of solutions of Sobolev-Galpern partial differential equations, *Pac. J. Math.* 34 (1970), 781–787.
[4] Equations with operators forming a right angle, *Pac. J. Math.* 45 (1973), 357–362.
[5] The final value problem for evolution equations, *J. Math. Anal. Appl.* 47 (1974), 563–572.
[6] Degenerate evolution equations and applications, *Indiana U. Math. J.* 23 (1974), 655–677.
[7] Quasi-reversibility of first and second order parabolic evolution equations. In *Improperly Posed Boundary Value Problems* (ed. A. Carasso and A. P. Stone), Pitman London (1975), pp. 76–84. pp. 76–84.
[8] Regularization and approximation of second order evolution equations, *SIAM J. Math. Anal.* 7 (1976), 461–472.
[9] The Sobolev equation, I, II, *Applicable Anal.* 5 (1975), 15–22 and 81–89.
[10] Perturbation of maximal accretive operators with right angle, *Port. Math.* 36 (1977), 79–82.
[11] *Hilbert Space Methods for Partial Differential Equations*, Pitman, London, 1977.
Siebert, E. (see also Janssen)
[1] Supports of holomorphic convolution semigroups and densities of symmetric convolution semigroups on a locally compact group, *Arch. Math.* 36 (1981), 424–433.
[2] Fourier analysis and limit theorems for convolution semigroups on a locally compact group, *Adv. Math.* 39 (1981), 111–154.
Silverstein, M. L. (see also Elliot)
[1] Markov processes with creation of particles, *Z. Warsch. verw. Geb.* 9 (1968), 235–257.
[2] *Symmetric Markov Processes*, Lecture Notes in Math. No. 426, Springer, Berlin 1974.
[3] *Boundary Theory for Symmetric Markov Processes*, Lecture Notes in Math. No. 516, Springer, New York, 1976.
[4] Application of the sector condition to the classification of submarkovian semigroups, *Trans. Amer. Math. Soc.* 244 (1978), 103–146.
Simon, B. (see also Carmona, Davies, Lieb, Radin, Reed)
[1] The theory of semi-analytic vectors: A new proof of a theorem of Masson and McClary, *Indiana U. Math. J.* 12 (1971), 1145–1151.

[2] *Quantum Mechanics for Hamiltonians Defined as Quadratic Forms*, Princeton U. Press, Princeton, N. J. 1971.

[3] Topics in functional analysis. In *Mathematics of Contemporary Physics* (ed. R. F. Streater), Academic Press, London (1972), 17–76.

[4] Ergodic semigroups of positivity preserving self-adjoint operators, *J. Functional Anal.* 12 (1973), 335–339.

[5] Positivity of the Hamiltonian semigroup and the construction of Euclidean region fields, *Helv. Phys. Acta* 46 (1973), 686–696.

[6] Quantum dynamics: From automorphism to Hamiltonian. In *Studies in Mathematical Physics* (ed. E. H. Lieb, B. Simon, and A. S. Wightman), Princeton U. Press, Princeton, N. J. (1976), 327–349.

[7] Geometric methods in multiparticle quantum systems, *Comm. Math. Phys.* 55 (1977), 259–274.

[8] Scattering theory and quadratic forms: On a theorem of Schechter, *Comm. Math. Phys.* 53 (1977), 151–153.

[9] An abstract Kato's inequality for generators of positivity preserving semigroups, *Indiana U. Math. J.* 26 (1977), 1067–1073.

[10] Classical boundary conditions as a technical tool in modern mathematical physics, *Adv. Math.* 30 (1978), 268–281.

[11] A canonical decomposition for quadratic forms with applications to monotone convergence theorems, *J. Functional Anal.* 28 (1978), 377–385.

[12] Kato's inequality and the comparison of semigroups, *J. Functional Anal.* 32 (1979), 97–101.

[13] Maximal and minimal Schrödinger forms, *J. Oper. Th.* 1 (1979), 37–47.

[14] Brownian motion, L^p properties of Schrödinger operators and the localization of binding, *J. Functional Anal.* 35 (1980), 215–229.

[15] *Functional Integration and Quantum Physics*, Academic Press, New York, 1980.

[16] Phase space analysis of simple scattering systems: Extensions of some work of Enss, *Duke Math. J.* 46 (1980), 119–168.

[17] Large time behavior of the L^p norm of Schrödinger semigroups, *J. Functional Anal.* 40 (1981), 66–83.

[18] Spectral analysis of multiparticle Schrödinger operators. In *Spectral Theory of Differential Operators* (eds. I. W. Knowles and R. T. Lewis), North Holland, New York (1981), pp. 369–370.

[19] Schrödinger semigroups, *Bull. Amer. Math. Soc.* 7 (1982), 447–526.

Simon, B. and R. Høegh-Krohn

[1] Hypercontractive semigroups and two dimensional self-coupled Bose fields, *J. Functional Anal.* 9 (1972), 121–180.

Sinclair, A. M.

[1] *Continuous Semigroups in Banach Algebras*, Cambridge Univ. Press, Cambridge, England, 1981.

Sine, R. (see Lin)

Sinestrari, E. (see also Da Prato, Di Blasio)

[1] Accretive differential operators, *Boll. Un. Mat. Ital.* 13 (1976), 19–31.

[2] Hölder and little Hölder regularity results for evolution equations of parabolic type. In *Evolution Equations and Their Application* (eds. by F. Kappel and W. Schappacher), Pitman, London (1982), pp. 220–228.

[3] Continuous interpolation spaces and spatial regularity in nonlinear Volterra integrodifferential equations, *J. Int. Eqns.* 5 (1983), 287–308.

[4] On the abstract Cauchy problem of parabolic type in spaces of continuous functions, *J. Math. Anal. Appl.* To appear.

Singbal-Vedak, K.

[1] A note on semigroups of operators on a locally convex space, *Proc. Amer. Math. Soc.* 16 (1965), 696–702.

Singer, B. (see Brezis)

Sinha, K. B. (see Amrein, Parthasarathy)

Skorokhod, A. V. (see also Gikhman)

[1] Operator stochastic differential equations and stochastic semigroups, *Russian Math. Surveys* 37 (1982), 177–204.

Slemrod, M.

[1] The linear stabilization problem on Hilbert space, *J. Functional Anal.* 2 (1972), 334–345.

[2] A note on complete controllability stabilizatility for linear control systems in Hilbert space, *SIAM J. Control* 12 (1974), 500–508.

[3] Asymptotic behavior of C_0 semi-groups as determined by the spectrum of the generator. *Indiana U. Math. J.* 25 (1976), 783–792.

Sloan, A. D. (see also Berger, Herbst)
[1] A nonperturbative approach to nondegeneracy of ground states in quantum field theory, *Trans. Amer. Math. Soc.* 16 (1974), 161–191.
[2] Analytic domination by fractional powers with linear estimates, *Proc. Amer. Math. Soc.* 51 (1975), 94–96.
[3] An application of the nonstandard Trotter product formula, *J. Math. Phys.* 18 (1977), 2495–2496.

Smagin, V. V. and P. E. Sobolevskiĭ
[1] Comparison theorems for the norms of the solutions of linear homogeneous differential equations in Hilbert space, *Differencial'nye Uravnenija* 6 (1970), 2005–2010.

Smith, C. H. (see Goldberg)

Sobolev, S. L.
[1] *Some Applications of Functional Analysis in Mathematical Physics*, Transl. Math. Monographs, Vol. 7, Amer. Math. Soc., Providence, R. I., 1963.

Sobolevskiĭ, P. E. (see also Krasnoselskiĭ, Polichka, Smagin)
[1] On equations of parabolic type in a Banach space, *Trudy Moscow Mat. Obšč.* 10 (1961), 297–350 (Russian). Translated in *Amer. Math. Soc. Translations* Ser. 2, 49 (1966), 1–62.
[2] General boundary value problems for parabolic equations investigated by the method of differential equations in Banach space, *Soviet Math. Dokl.* 7 (1966), 1330–1334.
[3] Differential equations with unbounded operators generating nonanalytic subgroups, *Soviet Math. Dokl.* 9 (1968), 1386–1390.
[4] On semigroups of growth α, *Soviet Math. Dokl.* 12 (1971), 202–205.

Sohr, H.
[1] Störungskriterien im reflexiven Banachraum, *Math. Ann.* 233 (1978), 75–87.
[2] Über die Selbstadjungiertheit von Schrödinger-Operatoren, *Math. Z.* 160 (1978), 255–261.
[3] Über die Existenz von Wellenoperatoren für zeitabhängige Störungen, *Mh. Math.* 86 (1981), 63–81.
[4] Störungstheorische Regularitätsuntersuchungen, *Math. Z.* 179 (1982), 179–192.

Sourour, A. (see also Berkson)
[1] Semigroups of scalar type operators on Banach spaces, *Trans. Amer. Math. Soc.* 200 (1974), 207–232.
[2] On strong controllability of infinite-dimensional linear systems, *J. Math. Anal. Appl.* 87 (1982), 460–462.

Sova, M.
[1] Cosine operator functions, *Rozprawy Mat.* 49 (1966), 1–46.
[2] Problème de Cauchy pour equations hyperboliques opérationnelles a coefficients constants non-bornes, *Ann. Scuola Norm. Sup. Pisa* 22 (1968), 67–100.
[3] Problèmes de Cauchy paraboliques abstraites de classes supérieures et les semi-groupes distributions, *Richerche Mat.* 18 (1969), 215–238.
[4] Inhomogeneous linear differential equations in Banach spaces, *Čas. Pěst. Mat.* 103 (1978), 112–135.

Spellmann, J. W.
[1] Concerning the infinite differentiability of semigroup motions, *Pac. J. Math.* 30 (1969), 519–523.
[2] Concerning the domains of generators of linear semigroups, *Pac. J. Math.* 35 (1970), 503–509.

Spohn, H.
[1] The spectrum of the Liouville-von Neumann operator, *J. Math. Phys.* 17 (1976), 57–60.
[2] An algebraic condition for the approach to equilibrium of an open N-level system *Lett. Math. Phys.* 2 (1977), 33–38.
[3] Entropy production for quantum dynamical semigroups, *J. Math. Phys.* 19 (1978), 1227–1230.

Stafney, J. D.
[1] Integral representations of functional powers of infinitesimal generators, *Ill. J. Math.* 20 (1976), 124–133.
[2] A spectral mapping theorem for generating operators in second order linear differential equations, *J. Math. Anal. Appl.* 77 (1980), 433–450.

Stecher, M. (see Kiffe)

Stein, E. M. and G. Weiss
[1] *Introduction to Fourier Analysis on Euclidean Spaces*, Princeton U. Press, Princeton, N. J., 1971.

Stewart, H. B.
[1] Generation of analytic semigroups by strongly elliptic operators, *Trans Amer. Math. Soc.* 199 (1974), 141–162.
[2] Spectral theory of heterogeneous diffusion systems, *J. Math. Anal. Appl.* 54 (1976), 59–78.

[3] Generation of analytic semigroups by strongly elliptic under general boundary conditions, *Trans. Amer. Math. Soc.* 259 (1980), 299–310.

Stoica, L.

[1] *Local operators and Markov Processes,* Lecture Notes in Math. No. 816, Springer, Berlin, 1980.

Stone, M. H.

[1] Linear tranformations in Hilbert space I-III, *Proc. Acad. Sci. U.S.A.* 15 (1929), 198–200, 423–425, 16 (1930), 172–175.

[2] Linear Transformations in Hilbert Space, Amer. Math. Soc. Colloq. Publ. 15, Providence, R. I., 1932.

[3] On one-parameter unitary groups in Hilbert space, *Ann. Math.* (2) 33 (1932), 643–648.

Strang, G.

[1] Approximating semigroups and the consistency of difference schemes, *Proc. Amer. Math. Soc.* 20 (1969), 1–7.

[2] On numerical ranges and holomorphic semigroups, *J. Analyse Math.* 22 (1969), 299–318.

Stroock, D. W. (see also Holley)

[1] Diffusion processes associated with Lévy generators, *Z. Warsch. verw. Geb.* 32 (1975), 209–244.

[2] The Malliavin calculus and its application to second order parabolic partial differential equations: Parts I, II, *Math. Systems. Theory* 14 (1981), 25–65 and 141–171.

[3] On the spectrum of Markov semigroups and the existence of invariant measures. In *Lecture Notes in Math.* No. 923, Springer, Berlin (1982), pp. 286–307.

Stroock, D. W. and S. R. S. Varadhan

[1] Diffusion processes with continuous coefficients, I, II, *Comm. Pure Appl. Math.* 22 (1969), 345–400 and 479–530.

[2] Two limit theorems for random evolutions having nonergodic driving processes. In *Proc. Conf. Stochastic Differential Equations and Applications* (ed. J. D. Mason), Academic Press, New York (1977), pp. 241–253.

[3] *Multidimensional Diffusion Processes,* Springer, New York, 1979.

Sudarshan, E. C. G. (see Gorini)

Sunouchi, H. (see Ōharu)

Suryanarayana, P.

[1] The higher order differentiability of evolutions of abstract evolution equations, *Pac. J. Math.* 22 (1967), 543–561.

Svendsen, E. C.

[1] Unitary one-parameter groups with finite speed of propagation, *Proc. Amer. Math. Soc.* 84 (1982), 357–361.

[2] An integral formula for multidimensional evolution equations, *Proc. Amer. Math. Soc.* 92 (1984), 185–189.

Sz. -Nagy, B. (see also Foiaş, Riesz)

[1] On uniformly bounded linear transformations in Hilbert space, *Acta Sci. Math. Szeged* 11 (1947), 152–157.

[2] Suites faiblement convergentes de transformations normales de l'éspace Hilbertien, *Acta Math. Acad. Aci. Hung.* 8 (1957), 295–302.

Sz. -Nagy, B. and C. Foiaş

[1] *Harmonic Analysis of Operators on Hilbert Space,* Elsevier, New York, 1970.

Taira, K.

[1] Semigroups and boundary value problems, *Duke Math. J.* 49 (1982), 287–320.

Takenaka, T. and N. Okazawa

[1] A Phillips-Miyadera type perturbation theorem for cosine functions of operators *Tôhoku Math. J.* 30 (1978), 107–115.

Takeuchi, J. (see Arakawa)

Taksar, M. I.

[1] Enhancing of semigroups, *Z. Warsch. verw. Geb.* 63 (1983), 445–462.

Tanabe, H. (see also Fujie, Kato)

[1] A class of equations of evolution in a Banach space, *Osaka Math. J.* 11 (1959), 121–145.

[2] Remarks on the equations of evolution in a Banach space, *Osaka J. Math.* 12 (1960), 145–166.

[3] On the equations of evolution in a Banach space, *Osaka J. Math.* 12 (1960), 363–376.

[4] On regularity of solutions of abstract differential equations of parabolic type in Banach space, *J. Math Soc. Japan* 19 (1967), 521–542.

[5] *Equations of Evolution,* Pitman, London, 1979.

[6] Linear Volterra integral equations of parabolic type, *Hokkaido Math. J.* 12 (1983), 265–275.

Tanabe, H. and M. Watanabe
[1] Note on perturbation and degeneration of abstract differential equations in Banach space, *Funk. Ekvac.* 9 (1966), 163–170.
Tanaka, H. (see Yosida)
Tarasov, R. P. (see Bakaev)
Tartar, L. (see Bardos, Crandall)
Taylor, J. C.
[1] On the existence of sub-markovian resolvents, *Invent. Math.* 17 (1972), 85–93.
[2] A characterization of the kernel $\lim_{\lambda \downarrow 0} V_\lambda$ for submarkovian resolvents (V_λ), *Ann. Prob.* 3 (1975), 355–357.
Taylor, M. E. (see Rauch)
Terlinden, D. M.
[1] A spectral containment theorem analogous to the semigroup theory result $e^{t\sigma(A)} \subseteq \sigma(e^{tA})$, *Pac. J. Math.* 101 (1982), 493–501.
Terrel, T. R.
[1] The local ergodic theorem and semigroups of nonpositive operators, *J. Functional Anal.* 10 (1972), 424–429.
Thomée, V. (see Brenner)
Tip, A. (see Prugovečki)
Travis, C. C.
[1] Differentiability of weak solutions to an abstract inhomogeneous differential equation, *Proc. Amer. Math. Soc.* 82 (1981), 425–430.
Travis, C. C. and G. F. Webb
[1] Existence and stability for partial functional differential equations, *Trans. Amer. Math. Soc.* 200 (1974), 395–418.
[2] Compactness, regularity, and uniform continuity properties of strongly continuous cosine families, *Houston J. Math.* 3 (1977), 555–567.
[3] Second order differential equations in Banach spaces. In *Nonlinear Equations in Abstract Spaces* (ed. V. Lakshmikantham), Academic Press, New York (1978), 331–361.
[4] An abstract second order semilinear Volterra integrodifferential equation, *SIAM J. Math. Anal.* 10 (1979), 412–424.
[5] Perturbation of strongly continuous cosine family generators, *Colloq. Math.* 45 (1981), 227–285.
Triggiani, R. (see also Lasiecka, Nelson)
[1] On the stabilizability problem in Banach space, *J. Math. Anal. Appl.* 52 (1975), 383–403.
[2] On the relationship between first and second order controllable systems in Banach space, *SIAM J. Cont. Opt.* 16 (1978), 847–859.
Trotter, H. F.
[1] Approximation of Semi-groups of operators, *Pac. J. Math.* 8 (1958), 887–919.
[2] On the product of semi-groups of operators, *Proc. Amer. Math. Soc.* 10 (1959), 545–551.
[3] An elementary proof of the central limit thoerem, *Arch. Math.* 10 (1959), 226–234.
[4] Approximation and perturbation of semigroups. In *Linear Operators and Approximation Theory II* (eds. P. L. Butzer and B. Sz. -Nagy), Birkhäuser, Basel (1974), pp. 3–21.
Trudinger, N. S. (see Gilbarg)
Tsekanovskiĭ, E. R.
[1] Friedrichs and Krein extensions of positive operators and holomorphic contraction semigroups, *Functional Anal. Appl.* 15 (1981), 308–309.
Tsurumi, S. (see also Hasegawa)
[1] An ergodic theorem for a semigroup of linear contractions, *Proc. Japan Acad.* 49 (1973), 306–309.
Uchiyama, K. (see Fujiwara)
Ueno, T. (see also Sato)
[1] Wave equation with Wentzell's boundary condition and a related semigroup on the boundary I, II, *Proc. Japan Acad.* 49 (1973), 672–677 and 50 (1974), 281–286.
[2] Representation of the generator and the boundary condition for semigroups of operators of kernel type, *Proc. Japan Acad.* 59 (1983), 414–417.
[3] An integro-differential operator and the associated semigroup of operators, *Proc. Japan Acad.* 59 (1983), 465–468.
Uhlenbrock, D. A. (see also Hess)
[1] Perturbation of statistical semigroups in quantum statistical mechanics, *J. Math. Phys.* 12 (1971), 2503–2512.
Uhlig, H. (see Nagel)

Ushijima, T.
[1] Note on the integration of the wave equation, *Scientifics Papers of the College of General Education, Univ. Tokyo* 17 (1967), 155–159.
[2] On the strong continuity of distribution semi-groups, *J. Fac. Sci. Univ. Tokyo* 17 (1970), 363–372.
[3] On the abstract Cauchy problem and semi-groups of linear operators in locally convex spaces, *Scientific Papers of the College of General Eduction Univ. Tokyo* 21 (1971), 93–122.
[4] On the generation and smoothness of semi-groups of linear operators, *J. Fac. Sci. Univ. Tokyo* 19 (1972), 65–127 and 20 (1973), 187–189.
[5] Approximation theory for semigroups of linear operators and its application to approximation of wave equations, *Japan. J. Math.* 1 (1975/76), 185–224.
[6] A semi-group theoretical analysis of a finite element method for a linearized viscous shallow-water system, *Publ. RIMS, Kyoto U.* 19 (1983), 1305–1328.

Van Winter, C.
[1] Semigroups associated with analytic Schrödinger operators, *J. Math. Anal. Appl.* 94 (1983), 377–405.
[2] Groups generated by analytic Schrödinger operators, *J. Math. Anal. Appl.* 94 (1983), 406–434.

Varadhan, S. R. S. (see also Papanicolaou, Stroock)
[1] Asymptotic probabilities and differential equations, *Comm. Pure Appl. Math.* 19 (1966), 261–286.
[2] Diffusion processes in a small time interval, *Comm. Pure Appl. Math.* 20 (1967), 659–685.
[3] *Stochastic Processes*, Lecture Notes, New York U., 1968.

Velo, G. and A. S. Wightman (eds.)
[1] *Constructive Quantum Field Theory*, Springer, Berlin, 1973.

Venttsel, A. D. (see Wentzell)

Verri, M. (see also Frigerio, Gorini)

Verri, M. and V. Gorini
[1] Quantum dynamical semigroups and multipole relaxation of a spin in isotropic surroundings, *J. Math. Phys.* 19 (1978), 1803–1807.

Veron, L. (see Baras)

Veselić, K.
[1] On the nonrelativistic limit of the bound states of the Klein-Gordon equation, *J. Math. Anal. Appl.* 96 (1983), 63–84.

Vinter, R. B.
[1] Stabilizability and semigroups with discrete generators, *J. Inst. Math. Appl.* 20 (1977), 371–378.
[2] Semigroups on product spaces with applications to initial value problems with non-local boundary conditions. In *Control of Distributed Parameter Systems*, Pergamon, Oxford, England (1978), pp. 91–98.

Voight, J. (see also Greiner)
[1] On the perturbation theory for strongly continuous semigroups, *Math. Ann.* 229 (1977), 163–171.

Von Neumann, J. (see Neumann)

Von Wahl, W. (see Wahl)

Von Waldenfels, M. (see Waldenfels)

Waelbroeck, L.
[1] Les semi-groups différentiables, *Deuxième Colloq. Anal. Fonct.*, Centre Belg. Rech. Math. (1964), 97–103.
[2] Differentiability of Hölder-continuous semigroups, *Proc. Amer. Math. Soc.* 21 (1969), 451–454.

Von Wahl, W.
[1] The equation $u' + A(t)u = f$ in a Hilbert space and L^p-estimates for parabolic equations, *J. London Math. Soc.* 25 (1982), 483–497.

Von Waldenfels, M.
[1] Positive Halbgruppen auf einem n-dimensionaler Torus, *Arch. Math.* 15 (1964), 191–203.

Walker, J. A.
[1] *Dynamical Systems and Evolution Equations*, Plenum, New York, 1980.

Wallen, L. J. (see also Embry)
[1] Semigroups of partial isometries, *J. Math. Mech.* 19 (1970), 745–750.
[2] Decomposition of semigroups of partial isometries, *Indiana U. Math. J.* 20 (1970), 207–212.
[3] Translational representations of one-parameter semigroups, *Duke Math. J.* 42 (1975), 111–119.

Wang, F. J. S.
[1] The convergence of a branching Brownian motion used as a model describing the spread of an epidemic, *J. Appl. Prob.* 17 (1980), 301–312.

[2] Diffusion approximations of age-and-position dependent branching processes, *Stoch. Proc. and Their Appl.* 13 (1982), 59–74.
Wardrop, M. (see Embry, Embry-Wardrop)
Watanabe, J.
[1] Ergodic theorems for dynamical semi-groups on operator algebras, *Hokkaido Math. J.* 8 (1979), 176–190.
[2] Asymptotic behavior and eigenvalues of dynamical semigroups on operator algebras, *J. Math. Anal. Appl.* 86 (1982), 411–424.
Watanabe, M. (see also Tanabe)
[1] On the differentiability of semigroups of linear operators in locally convex spaces, *Sci. Rep. Niigata Univ.*, Ser. A, 9 (1972), 23–34.
[2] On the characterization of semigroups of linear operators, *Sci. Rep. Niigata Univ.* Ser. A. 10 (1973), 43–50.
[3] A perturbation theory for abstract evolution equations of second order, *Proc. Japan Acad.* 58 (1982), 143–146.
[4] A new proof of the generation theorem of cosine families in Banach spaces, *Houston J. Math.*, 10 (1984), 285–290.
Watanabe, S. (see also Ikeda)
[1] On stochastic differential equations for multi-dimensional diffusion processes with boundary conditions, I, II, *J. Math. Kyoto Univ.* 11 (1971), 169–180 and 545–551.
[2] Asymptotic behavior and eigenvalues of dynamical semigroups on operator algebras, *J. Math. Anal. Appl.* 86 (1982), 411–424.
Watanabe, T. (see also Yosida)
[1] Approximation of uniform transport process on a finite interval to Brownian motion, *Nagoya Math. J.* 32 (1968), 297–314.
[2] On the boundary condition of transport semigroup, *Nagoya Math. J.* 37 (1970), 219–241.
Webb, G. F. (see also Travis)
[1] Linear functional differential equations with L^2 initial functions, *Funk. Ekvac.* 19 (1976), 65–77.
[2] Volterra integral equations as functional differential equations on infinite intervals, *Hiroshima Math. J.* 7 (1977), 61–70.
[3] Regularity of solutions to an abstract inhomogeneous linear differential equation, *Proc. Amer. Math. Soc.* 62 (1977), 271–277.
[4] A representation formula for strongly continuous cosine functions, *Aeq. Math.* 21 (1980), 251–256.
Weder, R. A. (see also Koller, Schechter)
[1] Second order operators in the uniform norm, *Comm. Partial Diff. Eqs.*, 3 (1978), 381–406.
[2] The unified approach to spectral analysis, *Comm. Math. Phys.* 60 (1978), 291–299.
[3] The unified approach to spectral analysis. II, *Proc. Amer. Math. Soc.* 75 (1979), 81–84.
[4] Spectral analysis, scattering theory and eigenfunctions expansions for strongly propagative systems. To appear.
Weidmann, J. (see Jörgens)
Weilenmann, J.
[1] Continuity properties of fractional powers, of the logarithm, and of holomorphic semigroups, *J. Functional Anal.* 27 (1978), 1–20.
Weiss, B.
[1] Abstract vibrating systems, *J. Math. Mech.* 17 (1967), 241–255.
Weiss, G. (see Stein)
Weissler, F. B. (see also Mueller)
[1] Logarithmic Sobolev inequalities for the heat-diffusion semigroup, *Trans. Amer. Math. Soc.*, 237 (1978), 255–269.
[2] Two-point inequalities, the Hermite semigroup, and the Gauss-Weierstrass semigroup, *J. Functional Anal.*, 32 (1979), 102–121.
[3] Logarithmic Sobolev inequalities and hyper-contractive estimates on the circle, *J. Functional Anal.* 37 (1980), 218–234.
Wentzell, A. D. (= Venttsel, A. D.)
[1] Semigroups of operators associated with generalized second order differential operators, *Dokl. Akad. Nauk USSR* 111 (1956), 269–272.
[2] On boundary conditions for multi-dimensional diffusion processes, *Theory Prob. Appl.* 4 (1959), 164–177.
[3] *A Course in the Theory of Stochastic Process*, McGraw-Hill, New York, 1981.
Westphal, U. (see also Berens)
[1] Ein kalkül für gebrochene Potenzen infinitesimaler Erzeuger von Halbgruppen und Gruppen

von Operatoren. Teil I: Halbgruppenerzeuger, Teil II: Gruppenerzeuger, *Comp. Math.* 22 (1970), 67–103 and 104–136.

Wexler, D.
[1] Lyapunov functions for evolution equations in Hilbert spaces. In *Nonlinear Phenomena in Mathematical Sciences* (ed. V. Lakshmikantham), Academic Press, New York (1982), pp. 997–1002.

Wightman, A. S. (see Velo)

Wilcox, C. H. (see also Schulenberger)
[1] Initial-boundary value problems for linear hyperbolic partial differential equations of the second order, *Arch. Rat. Mech. Anal.* 10 (1962), 361–400.
[2] Wave operators and asymptotic solutions of wave propagation problems of classical physics, *Arch. Rat. Mech. Anal.* 22 (1966), 37–78.
[3] Scattering states and wave operators in the abstract theory of scattering, *J. Functional Anal.* 12 (1973), 257–274.
[4] *Scattering Theory for the d'Alembert Equation in Exterior Domains*, Lecture Notes in Math. No. 442, Springer, New York, 1975.
[5] Asymptotic wave functions and energy distributions in strongly propagative anisotropic media, *J. Math. Pures Appl.* 57 (1978), 275–321.
[6] Theory of Bloch waves, *J. D'Anal. Math.* 33 (1978), 146–167.

Williams, D.
[1] On operator semigroups and Markov groups, *Z. Warsch. verw. Geb.* 13 (1969) 280–285.
[2] Brownian motions and diffusions as Markov processes, *Bull. London Math. Soc.* 6 (1974), 257–303.

Williams, J. P. (see Fillmore)

Wingate, M. (see Aziz)

Wolff, M. (see also Greiner)
[1] On C_0-semigroups of lattice homomorphisms on a Banach lattice, *Math. Z.* 164 (1978), 69–80.
[2] A remark on the spectral bound of the generator of semigroups of positive operators with applications to stability theory. In *Functional Analysis and Approximation*, ISNM, 60, Birkhäuser, Basel (1981), 39–50.

Wüst, R.
[1] Generalisations of Rellich's theorem on the perturbation of (essentially) self-adjoint operators, *Math. Z.* 119 (1971), 276–280.

Yagi, A.
[1] On the abstract linear evolution equations in Banach spaces, *J. Math. Soc. Japan* 28 (1976), 290–303.
[2] On the abstract evolution equation of parabolic type, *Osaka J. Math.* 14 (1977), 557–568.
[3] Differentiability of families of the fractional powers of self-adjoint operators associated with sesquilinear forms, *Osaka J. Math.* 20 (1983), 265–284.

Yajima, K.
[1] The limiting absorption principle for uniformly propagative systems, *J. Fac. Sci. Univ. Tokyo Sec. IA* 21 (1974), 119–131.
[2] Nonrelativisitic limit of the Dirac theory, scattering theory, *J. Fac. Sci. Univ. Tokyo Sec. IA* 23 (1976), 517–523.
[3] The quasi-classical limit of quantum scattering theory, *Comm. Math. Phys.* 69 (1979), 101–129.

Yamada, A.
[1] On the correspondence between potential operators and semigroups associated with Markov processes, *Z. Warsch. verw. Geb.* 15 (1970), 230–238.

Yosida, K.
[1] On the group embedded in the metrical complete ring, *Japan. J. Math.* 13 (1936), 7–26.
[2] On the differentiability and the representation of one-parameter semi-groups of linear operators, *J. Math. Soc. Japan* 1 (1948), 15–21.
[3] An operator-theoretical treatment of temporally homogeneous Markoff processes, *J. Math. Soc. Japan* 1 (1949), 244–253.
[4] An operator-theoretical integration of the wave equation, *J. Math. Soc. Japan* 8 (1956), 79–92.
[5] An operator-theoretical integration of the temporally inhomogeneous wave equation, *J. Fac. Sci. Univ. Tokyo* 7 (1957), 463–466.
[6] On the differentiability of semi-groups of linear operators, *Proc. Japan Acad.* 34 (1958), 337–340.
[7] Fractional powers of infinitesimal generators and the analyticity of the semi-groups generated by them, *Proc. Japan Acad.* 36 (1960), 86–89.

[8] On the integration of the equation of evolution, *J. Fac. Sci. Univ. of Tokyo* 9 (1963), 397–402.

[9] Time dependent evolution equation in a locally convex space, *Math. Ann.* 162 (1965), 83–86.

[10] *Functional Analysis*, Springer, Berlin, 1965.

[11] A perturbation theorem for semigroups of linear operators, *Proc. Japan Acad.* 42 (1966), 645–647.

[12] Positive resolvents and potentials, *Z. Warsch. verw. Geb.* 8 (1967), 210–218.

[13] The existence of the potential operator associated with an equicontinuous semigroup of class (C_0), *Studia Math.* 31 (1968), 531–533.

[14] On the potential operators associated with Brownian motions, *J. D'Analyse Math.* 23 (1970), 461–465.

[15] Abstract potential operators on Hilbert space, *Publ. R. I. M. S., Kyoto Univ.* 8 (1972), 201–205.

[16] *Lecture on Semi-Group Theory and its Applications to Cauchy's Problem in Partial Differential Equations*, Tata Inst. of Fund. Research, Bombay, 1975.

[17] Some aspects of E. Hille's contribution to semigroup theory, *Int. Eqns. Oper. Th.* 4 (1981), 311–329.

Yosida, K. and S. Kakutani

[1] Operator-theoretical treatment of Markoff's process and mean ergodic theorem, *Ann. of Math.* 42 (1941), 188–228.

Yosida, K., T. Watanabe, and H. Tanaka

[1] On the pre-closedness of the potential operator, *J. Math. Soc. Japan* 20 (1968), 419–421.

Yoshikawa, A.

[1] An abstract formulation of Sobolev type imbedding theorems and its applications to elliptic boundary value problems, *J. Fac. Sci. Univ. Tokyo Sec. IA* 17 (1970), 543–558.

[2] An operator theoretical remark on the Hardy-Littlewood-Sobolev inequality, *J. Fac. Sci. Univ. Tokyo Sec. IA* 17 (1970), 559–566.

[3] On the perturbation of closed operators in a Banach space, *J. Fac. Sci. Hokkaido Univ.* 22 (1972), 50–61.

[4] Note on singular perturbation of linear operators, *Proc. Japan Acad.* 48 (1972), 595–598.

[5] On the logarithm of closed linear operators, *Proc. Japan Acad.* 49 (1973), 169–173.

Zabczyk, J.

[1] A note on C_0-semigroups, *Bull. Acad. Polon. Sci. Sér. Math. Astron. Phys.* 23 (1975), 895–898.

Zaidman, S.

[1] Sur la perturbation presque-périodique des groupes et semi-groupes de transformations d'un espace de Banach, *Rend. di Mat.* 16 (1957) 197–206.

[2] *Differential Equations in Hilbert Spaces*, Pitman, London, 1979.

[3] Remarks on the well-posed weak Cauchy problem, *Boll. Un. Mat. Ital.* 17 (1980), 1012–1022.

Zettl, A. (see Goldstein, Kwong)

Zsidó, L. (see Ciorănescu)

Zygmund, A.

[1] *Trigonometric Series*, Vols. 1, 2, (2nd ed.), Cambridge U. Press, London, 1968.

Index of Symbols

T_α	semigroup generated by a fractional power	62
x^α	monomial	92
$\mathscr{X}$	Banach space	13
$\mathscr{X}^*$	dual of $\mathscr{X}$	24
χ_J	characteristic (or indicator) function of the set J	24
$W_\pm$	wave operators	148
$W_\pm$	generalized wave operators	149
$W^{m,p}(\Omega)$	Sobolev space	134
$W_0^{m,p}(\Omega)$	Sobolev space	134
$W^{2m,p}(\Omega; B)$	Sobolev space	138
$\|\cdot\|$ or $\|\cdot\|_{\mathscr{X}}$	norm is $\mathscr{X}$	13
$\overline{}$	closure (bar)	13
$\langle \cdot, \cdot \rangle$	inner product on a Hilbert space	25
$\langle \cdot, \cdot \rangle$	pairing between $\mathscr{X}$ and $\mathscr{X}^*$	25
$\blacksquare$	end of proof	ix

Author Index

Subject Index

242